The Theory of Complex Angular Momenta

This book provides a unique and rigorous introduction to the theory of complex angular momenta, based on the methods of field theory. It comprises an English translation of the series of lectures given by V.N. Gribov in 1969, when the physics of high energy hadron interactions was being created. Besides their historical significance, these lectures contain material which is highly relevant to research today. The basic physical results and the approaches Gribov developed are now being rediscovered in a new context: in the microscopic theory of hadrons provided by quantum chromodynamics. The ideas and calculation techniques presented in this book are useful for analysing high energy hadron scattering phenomena, deep inelastic lepton–hadron scattering, the physics of heavy ion collisions, and kinetic phenomena in phase transitions, and will be instrumental in the analysis of electroweak processes at the next generation particle accelerators, such as LHC and TESLA.

VLADIMIR NAUMOVICH GRIBOV received his PhD in theoretical physics in 1957 from the Physico-Technical Institute in Leningrad where he had worked since 1954. From 1962 to 1980 he was the head of the Theory Division of the Particle Physics Department of that institute, which in 1971 became the Leningrad Institute for Nuclear Physics. In 1980 he moved to Moscow where he became head of the particle physics section of the Landau Institute for Theoretical Physics. From 1981 he regularly visited the Research Institute for Particle and Nuclear Physics in Budapest where he was a scientific adviser until his death in 1997. Vladimir Gribov was one of the leading theoretical physicists of his time, who made seminal contributions to many fields, including quantum electrodynamics, neutrino physics, non-Abelian field theory, and, in particular, the physics of hadron interactions at high energies.

T0254277

CAMBRIDGE MONOGRAPHS ON
MATHEMATICAL PHYSICS

General editors: P. V. Landshoff, D. R. Nelson, S. Weinberg

[†] Issued as a paperback

The Theory of Complex Angular Momenta

Gribov Lectures on Theoretical Physics

V. N. GRIBOV

UNIVERSITY PRESS

CAMBRIDGE UNIVERSITY PRESS
Cambridge, New York, Melbourne, Madrid, Cape Town, Singapore, São Paulo

Cambridge University Press
The Edinburgh Building, Cambridge CB2 8RU, UK

Published in the United States of America by Cambridge University Press, New York

www.cambridge.org
Information on this title: www.cambridge.org/9780521818346

First published 2003
This digitally printed version 2007

A catalogue record for this publication is available from the British Library

ISBN 978-0-521-81834-6 hardback
ISBN 978-0-521-03703-7 paperback

Contents

Foreword

This book presents the lecture course given in 1969 at the Ioffe Physico-technical Institute in Leningrad (now St Petersburg, Russia) by one of the key figures of twentieth century theoretical physics, Vladimir N. Gribov.

The original motivation, brought up by A. Aronov, was for Gribov to present a course on the theory of complex angular momenta that would be understandable for condensed matter physicists. (Those were the days when condensed matter and particle physicists worked hand in hand, shared thoughts, methods and seminars.) The audience consisted of young bright theorists (though some had little or no experience in this specific field). So, to appreciate the book fully and learn from it, the reader should be familiar with the basics of quantum field theory (relativistic invariance, quantum theory of scattering, Feynman diagrams).

The lecture notes were taken down by A. Anselm, Ya. Azimov, A. Bukhvostov, I. Dyatlov, G. Danilov, E. Levin, G. Frolov, V. Gorshkov, L. Lipatov, E. Malkov, V. Shekhter and I.M. Shmushkevich (then in charge of the Ioffe PTI theory department), and edited by L. Frankfurt. E. Kuraev initiated their publication in the form of two voluminous preprints (in Russian) issued by the Kharkov PTI. This informal publication, known as the 'Kharkov preprint', immediately became a rarity.

More than 25 years later the effort of preparing an English version of the course was undertaken by E. Kuraev, B. Shaikhatdenov and V. Bytev. The lectures were then edited by G. Korchemsky and L. Lipatov. The final English translation was carried out by V. von Schlippe; the authors of the following Introduction share responsibility for the final scientific editing.

The lectures were delivered in 1969 when modern high energy particle physics was being created. In spite of the fact that the lectures, obviously, cover only the first decade of the development of the field – the 1960s – Gribov's view of the problematics of high energy interactions presented in this book has not lost its relevance to this day. On the contrary, the basic physical results and the approaches he developed are being rediscovered nowadays in a new context: in the microscopic theory of hadrons provided by quantum chromodynamics (QCD). Gribov's ideas and calculation techniques presented in this book are instrumental in analysing high energy hadron scattering phenomena, electroweak processes at the next generation particle accelerators (LHC, TESLA), the small-x regime of deep inelastic lepton–hadron scattering (HERA), physics of heavy ion collisions (RHIC, LHC), kinetic phenomena in phase transitions, etc.

V. Gribov encouraged his audience to participate actively in creation of this course. So technical derivations of some results presented in the book were actually performed and written down by his 'pupils'.

There is another feature that makes this book unusual and should be mentioned. It has to do with the Gribov style, in what concerns his personal relation with mathematical symbols commonly used for separating left- and right-hand sides of equations. The $=$ sign in a Gribov chalk equation (which, by the way, never looked like one) might mean 'equal', as well as 'approximately equal', 'asymptotically equal', 'proportional', 'of the order of', all the way down to 'similar to' and even 'reminiscent of'. It is important to stress that this style did not derive from sloppiness. When it came to deriving new physical results, according to Alexei Anselm, for many years Gribov's collaborator,

> Working with BH [*Gribov's nickname*] you had a strange feeling that numbers were his personal friends: all those factors of 2 and π simply knew their place in Gribov's formulae.

However, Gribov-the-lecturer would skip, without regret, a numerical coefficient or even a functional factor if that were irrelevant for the discussion of the physical issue under focus.

This would have never constituted a problem in a live lecture: from the context of the discussion it was always easy to gauge the actual meaning of that specific wiggle in a given relation. Not so in a written text, where the magic of a printed symbol can deceive a reader. So the editors have attempted a certain diversification of the 'Gribov wiggle', by introducing $\simeq$, $\propto$, etc. where they saw it appropriate. However, this does not guarantee that, having met an $=$ sign, a reader won't discover, now and then, that it actually stands for 'approximately equal' or ... (see above).

Having said that, we proudly invite the reader into the laboratory of one of the creators of high energy particle physics.

Yuri Dokshitzer

Introduction

Yuri Dokshitzer and Leonid Frankfurt

In the late 1950s, when Gribov, then a young researcher at the Ioffe Physico-technical Institute, became interested in the physics of strong hadron interactions, there was no consistent picture of high energy scattering processes, not to mention *a theory*. Apart from the Pomeranchuk theorem – an asymptotic equality of particle and antiparticle cross sections [1] – not much was theoretically understood about processes at high energies.

Gribov's 1961 paper 'Asymptotic behaviour of the scattering amplitude at high energies' (submitted to *Nucl. Phys.* on June 28, 1960) in which he proved an inconsistency of the *black disk model* of diffractive hadron–hadron scattering may be considered a first building block of the modern theory of high energy particle interactions [2].

Gribov's use of the so-called double dispersion representation for the scattering amplitude, suggested by S. Mandelstam back in 1958 [3], demonstrated the combined power of the general principles of relativistic quantum theory – unitarity (conservation of probability), analyticity (causality) and the relativistic nature (crossing symmetry) – as applied to high energy interactions.

The then-standard black disk model viewed a hadron as an object with a finite interaction radius that did not depend on collision energy, and employed for the imaginary part of the scattering amplitude the factorized expression

$$A_1(s,t) = s\, f(t). \qquad (0.1)$$

By studying the analytic properties in the cross-channels, Gribov showed that the model (0.1) for diffraction in the physical region of s-channel scattering contradicts the unitarity relation for partial waves in the crossing t-channel. To solve the puzzle, he suggested the behaviour of the amplitude (for large s and finite t) in the general form

$$A_1(s,t) = s^{q(t)}\, B_t(\ln s)\,, \qquad (0.2)$$

where B_t is a slow (non-exponential) function of $\ln s$ (decreasing fast with t), and $q(0) = 1$ ensures the approximate constancy of the total cross section, $\sigma^{\text{tot}}(s) \simeq \text{const}$.

1

In this first paper Gribov analysed the constant exponent, $q(t) = 1$, and proved that the cross section in this case has to decrease at high energies, $B_t(\ln s) < 1/\ln s$, to be consistent with the t-channel unitarity. He remarked on the possibility $q(t) \neq$ const as 'extremely unlikely' since, considering the t-dependence of the scattering amplitude, this would correspond to a strange picture of the radius of a hadron infinitely increasing with energy. He decided to 'postpone the treatment of such rapidly chan ging functions until a more detailed investigation is carried out'.

He published the results of such an investigation the next year in the letter to *ZhETF* 'Possible asymptotic behaviour of elastic scattering'. In his letter Gribov discussed the asymptotic behaviour 'which in spite of having a few unusual features is theoretically feasible and does not contradict the experimental data' [4]. Gribov was already aware of the finding by T. Regge [5] that in non-relativistic quantum mechanics

$$A(s,t) \propto t^{\ell(s)}, \qquad (0.3)$$

in the unphysical region $|t| \gg s$ (corresponding to large imaginary scattering angles $\cos \Theta \to \infty$), where $\ell(s)$ is the position of the pole of the partial wave f_ℓ in the complex plane of the orbital momentum ℓ.

T. Regge found that the poles of the amplitude in the complex ℓ-plane were intimately related with bound states/resonances. It is this aspect of the Regge behaviour that initially attracted the most attention:

> S. Mandelstam has suggested and emphasized repeatedly since 1960 that the Regge behavior would permit a simple description of dynamical states (private discussions). Similar remarks have been made by R. Blankenbecker and M.L. Goldberger and by K. Wilson   (quoted from [6]).

Gribov learned about the Regge results from a paper by G. Chew and S. Frautschi [7] which still advocated the wrong black disk diffraction model (0.1) but contained a *footnote* describing the main Regge findings.

The structure of the Regge amplitude (0.3) motivated Gribov to return to the consideration of the case of the t-dependent exponent in his general high energy ansatz (0.2) that was dictated by t-channel unitarity.

By then M. Froissart had already proved his famous theorem that limits the asymptotic behaviour of the total cross sections [8],

$$\sigma^{\text{tot}} \propto s^{-1} |A_1(s,0)| < \text{const} \cdot \ln^2 s. \qquad (0.4)$$

Thus, having accepted $\ell(0) - 1$ for the rightmost pole in the ℓ-plane as the condition 'that the strongest possible interaction is realized', Gribov formulated 'the main properties of such an asymptotic scattering behaviour':

- the total interaction cross section is constant at high energies,

- the elastic cross section tends to zero as $1/\ln s$,

- the scattering amplitude is essentially imaginary,

- the significant region of momentum transfer in elastic scattering shrinks with increasing energy, $\sqrt{-t} \propto (\ln s)^{-1/2}$.

He also analysed the s-channel partial waves to show that for small impact parameters $\rho < R$ their amplitudes fall as $1/\ln s$, while the interaction radius R increases with energy as $\rho \propto \sqrt{\ln s}$. He concluded:

> this behaviour means that the particles become grey with respect to high energy interaction, but increase in size, so that the total cross section remains constant.

These were the key features of what has become known as the 'Regge theory' of strong interactions at high energies. On the opposite side of the Iron Curtain, the basic properties of the Regge pole picture of forward/backward scattering were formulated half a year later by G. Chew and S. Frautschi in [9]. In particular, they suggested 'the possibility that the recently discovered ρ meson is associated with a Regge pole whose internal quantum numbers are those of an $I = 1$ two-pion configuration', and conjectured the universal high energy behaviour of backward $\pi^+\pi^0$, K^+K^0 and pn scattering due to ρ–reggeon exchange. G. Chew and S. Frautschi also stressed that the hypothetical Regge pole with $\alpha(0) = 1$ responsible for forward scattering possesses quantum numbers of the *vacuum*.

Dominance of the vacuum pole automatically satisfies the Pomeranchuk theorem. The name 'pomeron' for this vacuum pole was coined by Murray Gell-Mann, who referred to Geoffrey Chew as an inventor.

Shrinkage of the diffractive peak was predicted, and was experimentally verified at particle accelerator experiments in Russia (IHEP, Serpukhov), Switzerland (CERN) and the US (FNAL, Chicago), as were the general relations between the cross sections of different processes that followed from the Gribov factorization theorem [10].

In non-relativistic quantum mechanics the interaction Hamiltonian allows for scattering partial waves to be considered as analytic functions of complex angular momentum ℓ (provided the interaction potential is analytic in r).

Gribov's paper 'Partial waves with complex orbital angular momenta and the asymptotic behaviour of the scattering amplitude' showed that the partial waves with complex angular momenta can be introduced in a relativistic theory as well, on the basis of the Mandelstam double dispersion representation. Here it is the *unitarity in the crossing channel* that replaces Hamiltonian dynamics and leads to analyticity of the partial waves in ℓ. The corresponding construction is known as the 'Gribov–Froissart projection' [11].

A few months later Gribov demonstrated that the simplest (two-particle) t-channel unitarity condition indeed generates the moving *pole* singularities in the complex ℓ-plane. This was the *proof* of the Regge hypothesis in relativistic theory [12].

The 'Regge trajectories' $\alpha(t)$ combine hadrons into families: $s_h = \alpha(m_h^2)$, where s_h and m_h are the spin and the mass of a hadron (hadronic resonance) with given quantum numbers (baryon number, isotopic spin, strangeness, etc.) [9]. Moreover, at negative values of t, that is in the physical region of the s-channel, the very same function $\alpha(t)$ determines the scattering amplitude, according to (0.2). It looks *as if* high energy scattering were due to t-channel exchange of a 'particle' with spin $\alpha(t)$ that varies with momentum transfer t – the 'reggeon'.

Thus, the high energy behaviour of the scattering process $a + b \rightarrow c + d$ is linked with the spectrum of excitations (resonances) of low-energy scattering in the dual channel, $a + \bar{c} \rightarrow \bar{b} + d$. This intriguing relation triggered many new ideas (bootstrap, the concept of duality). Backed by the mysterious *linearity* of Regge trajectories relating spins and squared masses of observed hadrons, the duality ideas, via the famous Veneziano amplitude, gave rise to the concept of hadronic strings and to development of string theories in general.

A number of theoretical efforts were devoted to understanding the approximately constant behaviour of the total cross sections at high energies.

To construct a full theory that would include the pomeron trajectory with the maximal 'intercept' that respects the Froissart bound, $\alpha_P(0) = 1$, and would be consistent with unitarity and analyticity proved to be very difficult. This is because multi-pomeron exchanges become essential, which generate branch points in the complex plane of angular momentum ℓ. The simplest branch point of this kind (the two-pomeron cut) was first discovered by Mandelstam in his seminal paper of 1963 [13]. The result was generalized, very elegantly, by V.N. Gribov, I.Ya. Pomeranchuk and K.A. Ter-Martirosian [14]. They showed that Mandelstam's t-channel unitarity analysis could be recast as demonstrating the presence of an ℓ-plane contribution from the four-particle state whose generalization would be the contribution of the N-pomeron cut from the $2N$-particle state.

The t-channel unitarity analysis assumed extensive multi-particle complex angular momentum theory, however. When attempts to develop the needed formalism floundered, Gribov decided that a diagrammatic approach might be more straightforward.

He then developed the general diagram technique known as the Gribov reggeon calculus by considering 'hybrid diagrams' with Regge pole amplitudes connected by the non-planar couplings of Mandelstam. By the end of the 1960s, he had thus formulated the rules for constructing the field theory of interacting pomerons – the Gribov reggeon field theory (RFT). In doing so, he had reduced the problem of high energy scattering to a non-relativistic quantum field theory of interacting particles in 2+1 dimensions. For a long time the Gribov RFT was regarded as '*t*-channel' in origin. Later, as the structure of the '*s*-channel' imaginary parts was understood it was realized that the pomeron interaction diagrams directly reflected particle production processes with large rapidity gaps.

With the advent of non-Abelian QFTs, and QCD in particular, Gribov's approaches and calculation techniques were applied in 1976 by his pupils who demonstrated that vector mesons (gluons; intermediate bosons W, Z) *reggeize* in perturbation theory (L. Lipatov; L. Frankfurt and V. Sherman), and so do fermions (quarks; V. Fadin and V. Sherman). The vacuum singularity has also been analysed in perturbative QCD, which analysis resulted in the scattering cross section of two small transverse-size objects *increasing* with energy in a power-like fashion in the restricted energy range (the so-called 'hard' or 'BFKL' pomeron [15]).

A lot of theoretical effort is being invested these days in the programme, formulated by Lipatov, of constructing and solving a (2+1)-dimensional effective QCD pomeron dynamics – a direct offspring of the Gribov RFT.

The last lectures are devoted to the discussion of the problems of the so-called *weak* and *strong* reggeon coupling scenarios.

The problem of high energy behaviour of soft interactions remained unsolved, although some viable options were suggested. In particular, in 'Properties of Pomeranchuk poles, diffraction scattering and asymptotic equality of total cross sections' [16] Gribov showed that a possible consistent solution of the RFT in the weak coupling regime calls for the formal asymptotic equality of *all* total cross sections of strongly interacting particles.

In 1968 V.N. Gribov and A.A. Migdal demonstrated, in a general field theory framework, that in the strong coupling regime the scaling behaviour of the Green functions emerged [17]. Their technique helped to build the quantitative theory of second order phase transitions and to analyse critical indices characterizing the long range fluctuations near the critical point.

In the context of interacting reggeons, the study of the strong coupling regime (pioneered by A.B. Kaidalov and K.A. Ter-Martirosian) led to the introduction of the 'bare' pomeron with $\alpha_P(0) > 1$. The RFT based on *t*-channel unitarity should enforce the *s*-channel unitarity as well. The combination of increasing interaction radius and the amplitudes in the impact parameter space which did not fall as $1/\ln s$ (as in the one-pomeron picture) led to logarithmically increasing asymptotic cross sections, resembling the Froissart regime (and respecting the Froissart bound (0.4)).

Nowadays, the popularity of the notion of the 'supercritical' bare pomeron with $\alpha_P(0) > 1$ is based on experiment (increasing total hadron cross sections). Psychologically, it is also supported by the BFKL finding.

Gribov diffusion in the impact parameter space giving rise to energy increase of the interaction radius and to the reggeon exchange amplitude, coexisting fluctuations as a source of branch cuts, duality between hadrons and partons, a common basis for hard and soft elastic, diffractive and inelastic process – these are some of the key features of high energy phenomena in quantum field theories, which are still too hard a nut for QCD to crack.

Added to the main text of the lectures are Gribov's three seminal works produced in the 1970s. They are as follows.

A. The translation of the Gribov lecture at the Leningrad Nuclear Physics Institute Winter School in 1973, in which the understanding of the space–time evolution of high energy hadron–hadron and lepton–hadron processes, in particular the nature of the reggeon exchange from the s-channel point of view, has been achieved. This lecture gives a perfect insight into Gribov's extraordinary way of approaching complicated physical problems of a general nature. He outlined here the general phenomena and typical features that were characteristic for high energy processes in any quantum field theory. The power of Gribov's approach lies in applying the universal picture of fluctuating hadrons to both soft and hard interactions.

B. The paper written in collaboration with V. Abramovsky and O. Kancheli in which the general quantitative relation between the shadowing phenomenon in hadron–hadron scattering, the cross section of diffractive processes and inelastic multi-particle production had been discovered. This is one of the best-known applications of the Gribov RFT known as the 'AGK cutting rules'.

C. Gribov's last work in the subject which was devoted to the intermediate energy range and dealt with interacting hadron fluctuations ('heavy pomeron').

References

[1] I.Ya. Pomeranchuk, *ZhETF* **34** (1958) 725
    [*Sov. Phys. JETP* **7** (1958) 499].

[2] V.N. Gribov, *Nucl. Phys.* **22** (1961) 249.

[3] S. Mandelstam, *Phys. Rev.* **112** (1958) 1344.

[4] V.N. Gribov, *ZhETF* **41** (1961) 667.
     [Sov. Phys. *JETP* **14** (1962) 478].

[5] T. Regge, *Nuovo Cim.* **14** (1959) 951, **18** (1960) 947.

[6] S.C. Frautschi, M. Gell-Mann and F. Zachariasen,
     Phys. Rev. **126** (1962) 2204.

[7] G.F. Chew and S.C. Frautschi, *Phys. Rev. Lett.* **5** (1960) 580.

[8] M. Froissart, *Phys. Rev.* **123** (1961) 1053.

[9] G.F. Chew and S.C. Frautschi, *Phys. Rev. Lett.* **7** (1961) 394.

[10] V.N. Gribov and I.Ya. Pomeranchuk, *Phys. Rev. Lett.* **8** (1962) 343;
     ZhETF **42** (1962) 1141 [*Sov. Phys. JETP* **15** (1962) 788L].

[11] V.N. Gribov, *ZhETF* **41** (1961) 1962
     [*Sov. Phys. JETP* **14** (1962) 1395];

     M. Froissart, Report to the La Jolla Conference on the Theory of Weak and
     Strong Interactions, La Jolla, 1961 (unpublished).

[12] V.N. Gribov, *ZhETF* **42** (1962) 1260 [*Sov. Phys. JETP* **15** (1962) 873].

[13] S. Mandelstam, *Nuovo Cim.* **30** (1963) 1148.

[14] V.N. Gribov, I.Ya. Pomeranchuk and K.A. Ter-Martirosian,
     Phys. Lett. **9** (1964) 269;   *Yad. Fiz.* **2** (1965) 361.

[15] V.S. Fadin, E.A. Kuraev and L.N. Lipatov, *Phys. Lett.* **B60** (1975) 50;
     ZhETF **72** (1977) 377 [*Sov. Phys. JETP* **45** (1977) 199].

[16] V.N. Gribov, *Yad. Fiz.* **17** (1973) 603.

[17] V.N. Gribov and A.A. Migdal, *ZhETF* **55** (1968) 4.

Yuri Dokshitzer
Leonid Frankfurt

1

High energy hadron scattering

In these lectures the theory of complex angular momenta is presented. It is assumed that readers are familiar with the methods of modern quantum field theory (QFT). Nevertheless we shall briefly recall its basic principles.

1.1 Basic principles

The main experimental fact underlying the theory is the existence of strong interactions between particles of non-zero masses. The theory is constructed for quantities which have a direct physical meaning.

1.1.1 Invariant scattering amplitude and cross section

Such quantities are the scattering amplitudes,

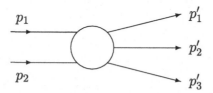

which are supposed to be functions of the kinematical invariants only: $A(p_1, \ldots, p_n) = A(p_i^2, p_i p_k)$. For simplicity, let us begin by considering the scattering of neutral, spinless particles as shown in Fig. 1.1. We use a normalization of the scattering amplitudes such that the kinematical factors arising from the wave functions of the external particles are factorized out. The cross section of any process can be defined in terms of

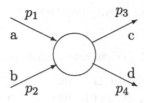

Fig. 1.1.   Two-particle scattering

the invariant amplitude A as follows:

$$\mathrm{d}\sigma_n = (2\pi)^4 \delta\left(p_1 + p_2 - \sum_i p_i'\right) |A|^2 \prod_{i=1}^{n} \frac{\mathrm{d}^3 p_i'}{2p_{i0}'(2\pi)^3} \frac{1}{I},$$

$$I = 4p_{10}p_{20}J = 4\sqrt{(p_1 p_2)^2 - m_1^2 m_2^2}. \tag{1.1}$$

Here the factor $(2\pi)^4 \delta()$ originates from energy–momentum conservation, $\mathrm{d}^3 p_i'/2p_{i0}'(2\pi)^3$ from the phase space volume; I is the Møller factor which combines the flux density J of the initial particles and $(2p_{10}\, 2p_{20})^{-1}$ coming from their wave functions.

1.1.2 Analyticity and causality

It is assumed that the scattering amplitude A is an analytic function of its arguments (for instance it cannot contain terms like $\Theta(p_{i0})$). This assumption is a manifestation of the causality principle. Without analyticity, the scattered waves could appear at their source before being emitted. Additionally, it is natural to conjecture at this point that the growth of the scattering amplitude, as one of the invariants tends to infinity for fixed values of the remaining invariants, is polynomially bounded,

$$|A(p_1, \dots, p_n)| < (p_i p_j)^N.$$

This assumption is closely related to causality and the locality of the interaction. One needs it in order to write the dispersion representation for the amplitudes (to be able to close the integration contour over an infinitely large circle).

1.1.3 Singularities

It is also assumed that all singularities of the amplitude on the physical sheet have the meaning of reaction thresholds, i.e. they are determined by physical masses of the intermediate state particles. In terms of Feynman diagrams they are the Landau singularities.

1.1.4 Crossing symmetry

We will clarify the meaning of crossing, taking as an example a four-particle amplitude. Since this amplitude depends on the kinematical invariants (and not on the sign of p_{i0}), the same analytic function describes the reaction

$$a(p_1) + b(p_2) \rightarrow c(p_3) + d(p_4) \qquad \text{for} \quad p_{10}, p_{20}, p_{30}, p_{40} > 0$$

as well as

$$a(p_1) + \bar{c}(-p_3) \rightarrow \bar{b}(-p_2) + d(p_4) \qquad \text{for} \quad p_{10}, p_{40} > 0, \ p_{20}, p_{30} < 0$$

and

$$a(p_1) + \bar{d}(-p_4) \rightarrow \bar{b}(-p_2) + c(p_3) \qquad \text{for} \quad p_{10}, p_{30} > 0, \ p_{20}, p_{40} < 0.$$

For an unstable particle, there is the additional reaction $a \rightarrow \bar{b} + c + d$ ($p_{10}, p_{30}, p_{40} > 0, p_{20} < 0$).

In fact, the crossing symmetry implies the CPT-theorem – invariance of the amplitude A with respect to the combination of charge conjugation C, space reflection P and time reversal T.

Crossing symmetry follows from the first three assumptions. It can be shown that the same assumptions allow us to prove the spin-statistics relation theorem (the Pauli theorem).

1.1.5 The unitarity condition for the scattering matrix

Unitarity has a simple physical meaning: the sum of probabilities of all processes which are possible at a given energy is equal to unity, $SS^+ = 1$. If $S = 1 + \mathrm{i}\,A$, then

$$\mathrm{i}\,(A - A^+) = -AA^+.$$

Representing the amplitude A as the sum of its real and imaginary parts, $A = \operatorname{Re} A + \mathrm{i} \operatorname{Im} A$, the unitarity condition takes the form

$$2 \operatorname{Im} A = AA^+. \tag{1.2}$$

1.2 Mandelstam variables for two-particle scattering

Let us show how all the above principles work in the case of the four-particle amplitude.

Although the amplitude of the $2 \to 2$ process depends evidently on two independent variables, that is the energy of the incoming particles and the scattering angle, it is more convenient to consider A as a function of three Mandelstam variables

$$s = (p_1 + p_2)^2, \quad t = (p_1 - p_3)^2, \quad u = (p_1 - p_4)^2.$$

They are related to each other by

$$s + t + u = \sum_{i=1}^{4} m_i^2$$

where the sum runs over the masses of all particles participating in the collision.

For the sake of simplicity, in what follows we restrict ourselves to the case of equal particle masses, $m_i = \mu$.

The Mandelstam variables have a simple physical meaning. For instance, in the centre-of-mass system (cms) of the reaction $a + b \to c + d$ (the so-called s-channel), s is the square of the total energy of the colliding particles and $t = -(\boldsymbol{p}_1 - \boldsymbol{p}_3)^2$ is the square of the momentum transfer from a to c. In the cms of the reaction $a + \bar{c} \to \bar{b} + d$ (t-channel), t plays the role of the total energy squared, and s is the square of momentum transfer. The variables u and t, respectively, play similar rôles in the u-channel reaction $a + \bar{d} \to \bar{b} + c$.

1.2.1 The Mandelstam plane

It is convenient, following Landau, to represent the kinematics of the three reactions graphically on the Mandelstam plane. We use here the well known geometrical fact that the sum of the distances from a point on the plane to the sides of an equilateral triangle does not depend on the position of the point. Therefore, taking into account the condition $s + t + u = 4\mu^2$, let us measure s, t and u as the distances to the sides of the triangle.

It is easy then to represent the physical region of any reaction on such a plane. For instance, the physical region of the reaction $a + \bar{c} \to \bar{b} + d$ corresponds to $t \geq 4\mu^2$, $s \leq 0$, $u \leq 0$ and it is shown on Fig. 1.2 as the upper shaded area. The physical regions of the other reactions can be identified in a similar manner.

In the case of the scattering of *identical* neutral particles the amplitude in each physical region is the same and it satisfies the unitarity condition separately in each region.

Examining the Mandelstam plane Fig. 1.2 we notice an interesting feature: as we move from positive to negative values of s (from the physical

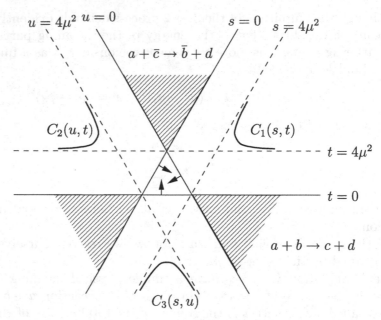

Fig. 1.2.   Crossing reactions on the Mandelstam plane

region of the s-channel to the u-channel), the energy dependence of the scattering amplitude turns into the angular dependence.

1.2.2 Threshold singularities on the Mandelstam plane

Let us discuss now singularities of the amplitude. As an illustration, we consider elastic scattering of neutral pions: $\pi^0 + \pi^0 \rightarrow \pi^0 + \pi^0$. We will assume that (in accordance with experiment) pions are the lightest stable hadrons and that there is no bound state of two neutral pions. Then, the amplitude has no singularities at $s < 4\mu^2$. The first threshold lies at $s = (2\mu)^2$. It corresponds to the two-particle intermediate state. The next, three-particle threshold could have appeared at $s = (3\mu)^2$. In reality, the second threshold in the pion scattering amplitude is situated at $s = (4\mu)^2$ – the four-particle state, since the transition of two pions into three is forbidden by G-parity conservation.

Similar singularities in *energy* are known to appear in quantum mechanics, for instance the threshold singularity at $s \rightarrow 4\mu^2$.

There is however a principal difference between relativistic and non-relativistic theories in the interpretation of the singularities in *momentum transfer*.

In quantum mechanics such singularities are determined by the potential. For instance, the Yukawa potential

$$V(r) \propto \frac{\exp(-\alpha r)}{r}$$

corresponds to a pole of the scattering amplitude in the plane of the squared momentum transfer k:

$$A(k^2) \propto \frac{1}{k^2 + \alpha^2}.$$

In the relativistic theory the rôle of the potential is played by energy singularities in the t-channel, thresholds at $t = 4\mu^2$, $16\mu^2$ and so on.

Let us illustrate this statement by considering the box diagram

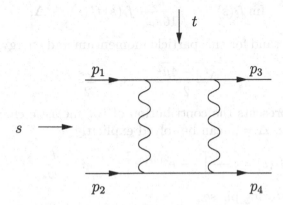

whose contribution we may interpret as defining the potential in the *next-to-Born* approximation. It is easy to see that the radius of this potential is $r = 1/2\mu$.

Thus, the assumption that all the singularities of the scattering amplitude are determined by the masses of real particles implies that there are no potentials with an infinite radius (since all hadrons have non-zero masses).

1.3 Partial wave expansion and unitarity

In order to obtain more concrete results, we must exploit analyticity and unitarity of the S-matrix.

Due to conservation of angular momentum, the unitarity condition for scattering amplitudes with given angular momentum ℓ becomes diagonal. It is convenient, therefore, to expand the s-channel amplitude into partial waves:

$$A(s,t) = \sum_{\ell=0}^{\infty} f_\ell(s)(2\ell+1)P_\ell(z), \qquad (1.3a)$$

where $P_\ell(z)$ is the Legendre polynomial and z is the cosine of the scattering angle:

$$z = \cos \Theta_s = 1 + \frac{2t}{s - 4\mu^2} = \frac{u - t}{u + t}. \qquad (1.3b)$$

From (1.3b) it becomes obvious that in the physical region of s-channel $(t, u \leq 0)$ we have $-1 \leq z \leq 1$, as expected.

Substituting the expansion (1.3a) into (1.2) and using well known orthogonality properties of Legendre polynomials, it is straightforward to derive the unitarity condition for partial amplitudes $f_\ell(s)$. It acquires a particularly simple form*

$$\mathrm{Im}\, f_\ell(s) \;=\; \frac{k_s}{16\pi\omega_s}\, f_\ell(s) f_\ell^*(s) \;+\; \Delta, \qquad (1.4a)$$

where p and ω stand for cms particle momentum and energy, respectively,

$$k_s = \frac{\sqrt{s - 4\mu^2}}{2}, \quad \omega_s = \frac{\sqrt{s}}{2}. \qquad (1.4b)$$

In (1.4a) Δ represents the contribution of the inelastic channels, $\Delta > 0$. The elastic case, $\Delta = 0$, can be solved explicitly:

$$f_\ell(s) = \mathrm{i}\, \frac{8\pi}{v} \left[1 - \mathrm{e}^{2\mathrm{i}\, \delta_\ell(s)} \right], \quad v = \frac{k_s}{\omega_s}, \qquad (1.5a)$$

with δ_l the scattering phase.

The solution of the elastic unitarity condition has the same form as in non-relativistic quantum mechanics except for the velocity factor $v = k/\omega$ which arises due to relativistic normalization of the amplitude A.

In the general case the solution of (1.4a) can be parametrized with the help of the 'elasticity parameter' $\eta_\ell(s) \leq 1$:

$$f_\ell(s) = \mathrm{i}\, \frac{8\pi}{v} \left[1 - \eta_\ell \cdot \mathrm{e}^{2\mathrm{i}\, \delta_\ell} \right], \quad \eta_\ell^2 = 1 - \frac{v}{4\pi}\Delta. \qquad (1.5b)$$

From (1.5) it follows that partial wave amplitudes are bounded from above:

$$\mathrm{Im}\, f_\ell \leq |f_\ell| \leq 16\pi\, v^{-1} \qquad (\eta_\ell = 1). \qquad (1.6a)$$

Maximal inelasticity of the scattering in a given partial wave corresponds to $\eta_\ell = 0$. In the high energy limit this leads to the restriction

$$\mathrm{Im}\, f_\ell \leq |f_\ell| \leq 8\pi \qquad (\eta_\ell = 0). \qquad (1.6b)$$

---

* Actual derivation of the unitarity condition for partial wave amplitudes uses the relation between the angles of initial, intermediate and final state particles and the known orthogonality properties of Legendre polynomials.

In this case the amplitude (1.5b) is purely imaginary, so that the elastic scattering is but a 'shadow' of inelastic channels. The model

$$f_\ell = \begin{cases} \mathrm{i}\dfrac{8\pi}{v}, & \eta_\ell = 0, & \text{for } \ell < \ell_0 = k_s R, \\[2mm] 0, & \eta_\ell = 1, \delta_\ell = 0, & \text{for } \ell > \ell_0, \end{cases}$$

is known as the 'black disk' model for diffractive scattering. At high energies $s \simeq 4k_s^2 \gg \mu^2$ ($v \simeq 1$) when $\ell_0 \gg 1$, it leads to the forward scattering amplitude (see (1.3a))

$$A(s,0) = \sum_\ell (2\ell+1)f_\ell \simeq \ell_0^2 \cdot 8\pi\mathrm{i} \simeq \mathrm{i}\,s \cdot 2\pi R^2,$$

which, according to the optical theorem, results in

$$\sigma_{\text{tot}} = \frac{\operatorname{Im} A(s,0)}{v\,s} \simeq 2\pi R^2 = \pi R^2\Big|_{\text{inelastic}} + \pi R^2\Big|_{\text{diffraction}}.$$

This is the pattern of diffraction off an absorbing disk of radius R.

1.3.1 Threshold behaviour of partial wave amplitudes

It is well known from quantum mechanics that for potentials of finite range, r_0, the partial waves behave like $(kr_0)^\ell$ as $k \to 0$. It can be easily seen that a similar result holds in the theory of the S matrix.

Indeed, the singularity in t of the amplitude $A(s,t)$, the closest to the physical region in the s-channel, is located at $t = 4\mu^2$. Therefore the series (1.3a) should be convergent for z up to $z_0 = 1 + 4\mu^2/2k_s^2$.

For $t > 0$ and $s \to 4\mu^2$, one gets $z \to \infty$ and $P_\ell(z)$ grows as $P_\ell(z) \sim z^\ell$. For the series (1.3a) to converge, one has to require that f_ℓ should fall with ℓ like $\left(2k_s^2/4\mu^2\right)^\ell$ but not faster since at $t = 4\mu^2$ the series has to be divergent.

1.3.2 Singularities of $\operatorname{Im} A$ on the Mandelstam plane (Karplus curve)

Repeating the same arguments for the *imaginary part* of the s-channel amplitude $\operatorname{Im} A$ we would get

$$\operatorname{Im} f_\ell(s) \propto k_s^{2\ell}, \qquad k_s \to 0.$$

This cannot be true, however, since it contradicts the unitarity condition: $\operatorname{Im} f_\ell \propto k_s^{4\ell+1}$ follows from (1.4a). Substituting this behaviour into (1.3a), we observe that the series for $\operatorname{Im}_s A(s,t)$ remains convergent at $t = \mu^2$. We conclude that singularities in t of the *imaginary part* of the amplitude are located *above* $t = 4\mu^2$, and their position depends on s.

Actually, using the unitarity condition one can find the exact form of the line of singularities of $\operatorname{Im}_s A(s,t)$ on the Mandelstam plane, known as Karplus (or Landau) curve.

Let us sketch its derivation in the region $4\mu^2 \le s \le 16\mu^2$, $t > 4\mu^2$ where the two-particle unitarity condition is valid ($\Delta = 0$ in (1.4a)).

For $t > 0$ we have $z > 1$ and the Legendre polynomials increase exponentially with ℓ:

$$P_\ell(\cosh\alpha) \overset{\ell\to\infty}{\simeq} \frac{e^{(\ell+\frac{1}{2})\alpha}}{\sqrt{2\pi\ell\sinh\alpha}}, \quad \cosh\alpha \equiv z = 1 + \frac{t}{2k_s^2} > 1. \qquad (1.7)$$

To ensure convergence of (1.3a) for $t < 4\mu^2$, partial waves have to fall as

$$f_\ell \sim e^{-\ell\alpha_0}, \quad \cosh\alpha_0 = 1 + \frac{4\mu^2}{2k_s^2}. \qquad (1.8)$$

Due to the unitarity condition (1.4a) the imaginary part falls even faster: $\operatorname{Im} f_\ell \sim \exp(-2\ell\alpha_0)$.

Consider now the series (1.3a) for $\operatorname{Im}_s A(s,t)$. With t increasing, the growing factor $\exp(\ell\alpha)$, originating from the Legendre polynomials, eventually overtakes the falling factor $\exp(-2\ell\alpha_0)$ due to $\operatorname{Im} f_\ell$. At this point the series becomes divergent, and $\operatorname{Im}_s A(s,t)$ develops a singularity.

Thus, the line of singularities of $\operatorname{Im}_s A(s,t)$ for $4\mu^2 \le s \le 16\mu^2$ is given by the equation $\alpha = 2\alpha_0$. In terms of the variables s and t this equation takes the form

$$\frac{t}{16\mu^2} = \frac{s}{s - 4\mu^2}, \quad 4\mu^2 \le s \le 16\mu^2.$$

In the complementary region $4\mu^2 \le t \le 16\mu^2$, $s \ge 4\mu^2$, the Karplus curve can be found using the symmetry of $A(s,t)$ under the permutation $s \leftrightarrow t$:

$$\frac{s}{16\mu^2} = \frac{t}{t - 4\mu^2}, \quad 4\mu^2 \le t \le 16\mu^2.$$

This example illustrates how the unitarity condition determines the analyticity domain of the scattering amplitude.

The lines of singularities C_i of the amplitude $A(s,t)$ are drawn on the Mandelstam plane in Fig. 1.2.

The fact that the Karplus curve $C_1(s,t)$ has finite asymptotes (in our example, the lines $s = 4\mu^2, t \to \infty$, and $t = 4\mu^2, s \to \infty$) is obvious, since otherwise the partial wave amplitudes would decrease with increasing ℓ faster than any exponential, which is in contradiction with the standard behaviour $f_\ell \sim \exp(-\alpha\ell)$ for $\ell \to \infty$.

In reality, the Karplus curves for $\pi\pi$ scattering are not symmetric with respect to s and t, which is a consequence of the pions being pseudoscalars (see the following lectures and the footnote on page 27).

1.4 The Froissart theorem

In 1958 Froissart showed that the analytic properties of the scattering amplitude together with the unitarity condition put certain restrictions on the asymptotic behaviour of $A(s,t)$ in the physical region. Let us show that asymptotically

$$\operatorname{Im} A(s,t)|_{t=0} \le \text{const} \cdot s \ln^2 \frac{s}{s_0}, \qquad s \to \infty.$$

First let us estimate f_ℓ at large s using the fact that the singularity of $\operatorname{Im}_s A(s,t)$ closest to the physical region of the s-channel is situated at $t = 4\mu^2$. As was shown above, at large ℓ the partial wave amplitude falls exponentially. Since for $k_s^2 \propto s \gg t$ (1.8) gives $\alpha \simeq \sqrt{t}/k_s$, we have

$$f_\ell(s) \simeq c(s,\ell) \exp\left(-\frac{\ell}{k_s}\sqrt{4\mu^2}\right), \qquad \ell, s \to \infty, \tag{1.9}$$

where $c(s,\ell)$ is slowly (non-exponentially) varying with ℓ.

Let us now assume that for t arbitrarily close to $4\mu^2$ the amplitude grows with s not faster than some power. Then the same is valid for $\operatorname{Im} c(s,\ell)$. Indeed, $\operatorname{Im} f_l$ is positive due to the unitarity condition, and so is $P_\ell(1 + t/2k_s^2)$ for $t \ge 0$. Therefore for each partial wave we have an estimate[†]

$$\left(\frac{s}{s_0}\right)^N > \operatorname{Im} A(s,t) = \sum_{\ell=0}^{\infty} \operatorname{Im} f_\ell(s)(2\ell+1)P_\ell\left(1 + \frac{t}{2k_s^2}\right)$$

$$> \operatorname{Im} c(s,\ell)\left(2\pi\ell\frac{\sqrt{t}}{k_s}\right)^{-1/2} \exp\left\{\frac{\ell}{k_s}\left(\sqrt{t} - \sqrt{4\mu^2}\right)\right\}. \tag{1.10}$$

Since (1.10) holds for arbitrary positive $t < 4\mu^2$, we conclude that

$$\operatorname{Im} c(s,\ell) < (s/s_0)^N,$$

and finally, modulo an irrelevant pre-exponential factor,

$$\operatorname{Im} f_\ell(s) \lesssim \left(\frac{s}{s_0}\right)^N \exp\left(-\frac{2\mu}{k_s}\ell\right). \tag{1.11}$$

(Using the unitarity condition one can derive a similar estimate for $\operatorname{Re} f_\ell$.)

[†] the series converges inside the so-called Lehman ellipse in the z plane

We are now in a position to estimate the imaginary part of the forward scattering amplitude:

$$\text{Im}\, A(s, t = 0) = \sum_{\ell=0}^{\infty} \text{Im}\, f_\ell(s)\, (2\ell + 1)$$

$$\leq 8\pi \sum_{\ell=0}^{L} (2\ell + 1) + \sum_{\ell=L+1}^{\infty} \text{Im}\, f_\ell(s)(2\ell + 1). \qquad (1.12)$$

Here we have extracted the finite sum $\ell < L$ in which partial waves are large, $\text{Im}\, f_\ell \simeq |f_\ell| = \mathcal{O}(1)$, and estimated its contribution from above by substituting for $\text{Im}\, f_\ell$ its maximal value allowed by unitarity, see (1.6b):

$$\sum_{\ell=0}^{L} (2\ell + 1) \simeq L^2.$$

The border value of the angular momentum L above which partial wave amplitudes become small, $\text{Im}\, f_{\ell>L} \ll 1$, and fall exponentially with ℓ according to (1.11) can be found by setting

$$\left(\frac{s}{s_0}\right)^N \exp\left(-\frac{2\mu}{k_s} L\right) \simeq 1 \qquad \Longrightarrow \qquad L \simeq \frac{k_s}{2\mu} \ln \frac{s}{s_0}.$$

The contribution of the infinite tail of the series in (1.12) can be estimated using $f_{L+n} \sim f_L \exp(-2\mu\, n/k_s)$ and turns out to be subdominant:

$$\sum_{n=0}^{\infty} 2(L + n) \exp\left\{-\frac{2\mu}{k_s} n\right\} \simeq \frac{k_s}{\mu} L + \frac{k_s^2}{2\mu^2} \overset{s\to\infty}{\ll} L^2.$$

Thus,

$$\text{Im}\, A(s, t = 0) \propto L^2 \propto s \ln^2 \frac{s}{s_0}.$$

This is the Froissart theorem.

The magnitude of the partial wave as a function of ℓ is sketched here:

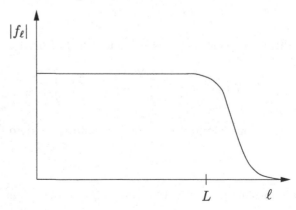

Since according to the optical theorem $\text{Im}\, A(s, t = 0) = s\sigma_{\text{tot}}(s)$, it follows from the Froissart theorem that the total cross section cannot grow with the centre of mass energy $\sqrt{s}$ faster than the squared logarithm of s, $\sigma_{\text{tot}}(s) \leq \sigma_0 \ln^2(s/s_0)$, and the interaction radius cannot grow faster than the logarithm of s.

An analogous consideration, together with the unitarity condition, leads to the similar inequality for the *real* part of the forward scattering amplitude, $|\text{Re}\, A(s, t = 0)| < \text{const} \cdot s \ln^2(s/s_0)$.

In order for the cross section not to decrease with increasing energy, the amplitude $A(s, t = 0)$ has to grow and, as a consequence, the number of partial waves contributing to the sum in (1.3a) has to be large. This allows us to replace the sum in (1.3a) by the integral over ℓ, using the well known approximate expression for the Legendre polynomials,

$$P_\ell(\cos\Theta) \simeq J_0\left[(2\ell + 1)\frac{\Theta}{2}\right], \qquad \ell \gg 1, \quad \theta \ll 1. \tag{1.13}$$

We obtain

$$A(s, t) \simeq \int f_\ell(s)\, J_0\left[(2\ell + 1)\frac{\Theta}{2}\right] (2\ell + 1)\mathrm{d}\ell.$$

It is convenient to replace ℓ by the impact parameter ρ, $\ell + 1/2 = k_s\rho$. Then, using $t \simeq -(k_s\Theta)^2$, we obtain

$$A(s, t) \simeq k_s^2 \int f(\rho, s)\, J_0\left(\rho\sqrt{-t}\right) 2\rho\, \mathrm{d}\rho. \tag{1.14}$$

If the values of ρ giving the dominant contribution to this integral do not depend on s (which is the case for the usual picture of diffractive scattering off a finite size object), then it is natural to expect that the amplitude takes the factorized form $A(s, t) \simeq a(s)F(t)$. If we additionally assume that the partial wave amplitudes $f(\rho, s)$ that are dominant in (1.11) approach constant values as $s \to \infty$, then $A(s, t) \sim sF(t)$ and the total cross section tends to a constant.

1.5 The Pomeranchuk theorem

In 1958 I.Ya. Pomeranchuk showed that if the total cross sections are constant at high energies, then the total cross sections of the scattering of a particle and its antiparticle off the same target should be asymptotically equal. The derivation of this result is based on the properties of the scattering amplitude in the s- and u-channels.

Let us identify the singularities of $A(s, t = 0)$ in the complex s plane. They are the right-hand cut $s \geq 4\mu^2$ and the left-hand cut $s \leq 0$. The latter cut corresponds to the right-hand cut $u \geq 4\mu^2$ in the complex u plane due to the relation $s + t + u = 4\mu^2$.

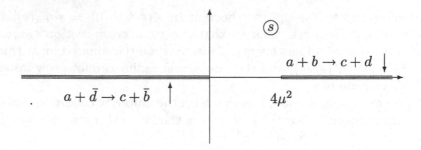

Fig. 1.3.   Amplitudes of two crossing reactions in the complex s plane

It is natural to assume that the amplitude of the reaction $a+b \to c+d$ is equal to the value $A(s,t)$ on the upper edge of the right cut in s, which corresponds to the usual definition of Feynman integrals in perturbation theory:

$$A(a + b \to c + d) \to \lim_{\varepsilon \to 0} A(s + i\varepsilon, t).$$

Similarly, the physical amplitude of the reaction $a + \bar{d} \to c + \bar{b}$ is given by the value of A on the upper edge of the right-hand cut in u, i.e.

$$A(a + \bar{d} \to c + \bar{b}) = \lim_{\varepsilon \to 0} A(u + i\varepsilon, t) = \lim_{\varepsilon \to 0} A(-(s - i\varepsilon) - t + 4\mu^2, t),$$

where the latter equality follows from the identity $s+t+u = 4\mu^2$ together with crossing symmetry. Thus the physical amplitude of the cross-channel reaction in the s plane is obtained by approaching the cut $s \leq 0$ *from below*, as shown in Fig. 1.3. Furthermore, since $A(s, t \leq 0)$ is real on the interval $0 < s < 4\mu^2$ which is free from singularities, the values of the amplitude on the two edges of the cut are complex conjugate. Therefore we may use the relation $A(s - i\varepsilon, t < 0) = A^*(s + i\varepsilon, t < 0)$ to finally arrive at

$$A_{a+\bar{d}\to c+\bar{b}}(s) \simeq [A_{a+b\to c+d}(-s)]^*, \qquad s \simeq -u. \tag{1.15}$$

Pomeranchuk proved the theorem under the assumption that the elastic scattering amplitude at large s has the form

$$A_{a+b\to a+b} = s\, F(t), \tag{1.16a}$$

so that the total cross section tends to a constant at $s \to \infty$. Using the relation (1.15) we then obtain

$$A_{a+\bar{b}\to a+\bar{b}} = -s\, F^*(t), \tag{1.16b}$$

yielding that the imaginary parts of the two amplitudes are equal whereas their real parts have opposite signs. (This implies that in such a model the part of the amplitude that is symmetric in s, u must be purely imaginary while the antisymmetric part must be real.) Since the total cross section is defined by the imaginary part of A, the Pomeranchuk theorem follows suit:

$$\sigma_{\text{tot}}(a + b) = \sigma_{\text{tot}}(a + \bar{b}).$$

If the total cross sections *increase* with energy, the asymptotic equality of σ_{ab} and $\sigma_{a\bar{b}}$ cannot, in general, be proved. The Pomeranchuk theorem, however, can be proved, assuming asymptotic factorization of the amplitude, $A(s,t) \simeq a(s)F(t)$, for a special class of the energy behaviour, namely, $a(s) = s(\ln s)^{\beta}$. To carry out the proof one must use the hypothesis that asymptotically the real part of the amplitude does not exceed its imaginary part

$$\lim_{s \to \infty} \frac{\text{Re}\, A(s,t)}{\text{Im}\, A(s,t)} < \text{const.} \tag{1.17}$$

It is supported by the observation that in general $\text{Re}\, f_{\ell}$ is a sign alternating function so that destructive interference in the series (1.3a) for $\text{Re}\, A(s,t)$ is possible. (Here it is important, once again, that at high energies the large values of ℓ are essential.)

We may illustrate the nature and significance of this hypothesis on a simple example. Consider an amplitude of the form

$$A(s,t) = s \ln \frac{-s}{s_0} \cdot c(t)$$

with $c(t)$ a real function. For $s > 0$ this amplitude is complex, and the cross section in the s-channel is constant, whereas at negative s (positive u) we have $\text{Im}\, A = 0$ and the u-channel cross section vanishes.

Did we manage to construct a counterexample to the Pomeranchuk theorem? Obviously not, since our model amplitude is not realistic. It gives rise to the elastic cross section exceeding the total cross section,

$$\sigma_{\text{el}} \sim \int \frac{dt}{s^2} \left| A(s,t) \right|^2 \propto \ln^2 s \gg \sigma_{\text{tot}} \sim \text{const},$$

which is a consequence of $\text{Re}\, A / \text{Im}\, A \sim \ln s \to \infty$, in contradiction with (1.17).

In this lecture we have demonstrated simple consequences of the analyticity and crossing symmetry of the scattering amplitude.

In the forthcoming lectures we will show how the t-channel unitarity can be used to study the asymptotics of the scattering amplitudes for $s \to \infty$. It is singularities of the amplitude in t (rather than those in u) that are located close to the physical region in the s-channel on the Mandelstam plane. This explains why the physics of the t-channel is important for large s.

2

Physics of the t-channel and complex angular momenta

In the previous lecture we have discussed analytic properties of the invariant amplitude $A(s,t,u)$ describing the scattering of neutral spinless particles (π^0 mesons).

As was already stressed, it is convenient to depict the kinematics of the reactions and the location of the singularities of $A(s,t)$ on the Mandelstam plane; see Fig. 2.1.

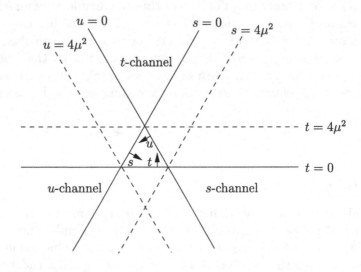

Fig. 2.1. Mandelstam plane

The same amplitude $A(s,t,u)$ describes the following three reactions

22

(see Fig. 1.1 for notation):

$$
\left.
\begin{aligned}
a + b &\to c + d && (s > 4\mu^2, t, u < 0), && s\text{-channel;} \\
a + \bar{c} &\to \bar{b} + d && (t > 4\mu^2, s, u < 0), && t\text{-channel;} \\
a + \bar{d} &\to \bar{b} + c && (u > 4\mu^2, t, s < 0), && u\text{-channel.}
\end{aligned}
\right\}
\qquad (2.1)
$$

As was shown in Lecture 1, the invariant amplitude has singularities at the thresholds of the corresponding reactions:

$$
s = 4\mu^2, 16\mu^2, \ldots; \quad t = 4\mu^2, 16\mu^2, \ldots; \quad u = 4\mu^2, 16\mu^2, \ldots .
$$

In the physical region of the s-channel the singularities of the amplitude $A(s,t)$ in s are related to the possibility of the transition of the initial state in the process of scattering into the intermediate states with two, four, etc. particles. Its imaginary part $A_1(s,t)$ is determined by the unitarity condition in the s channel, which is equivalent to the physical requirement that the sum of the probabilities of transitions into all possible states n should be equal to unity:

$$
\operatorname{Im}_s A \equiv A_1 = \frac{1}{2} \sum_n A_n A_n^*. \qquad (2.2a)
$$

The unitarity condition (2.2a) can be represented graphically as

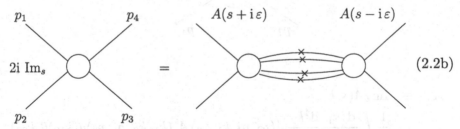

$$(2.2b)$$

To find the discontinuity (imaginary part) of the scattering amplitude corresponding to a given Feynman diagram, it suffices to cut the diagram in all possible ways into two connected parts, having incoming and outgoing particles on opposite sides of the cut. The cut lines must be identified as real particles in the intermediate state and we associate with each of them the factor $2\pi\theta(p_{i0})\delta(p_i^2 - m^2)$. These lines will be marked by crosses. Furthermore, one has to replace $i\varepsilon$ by $-i\varepsilon$ in all propagators of the block lying to the right of the cut, which corresponds to the conjugate amplitude $A^*(s + i\varepsilon) = A(s - i\varepsilon)$. Apart from these modifications, the Feynman rules remain unchanged.

2.1 Analytical continuation of the t-channel unitarity condition

Let us consider now the singularities with respect to momentum transfer. The first singularity is located at $t = 4\mu^2$. It is related to the possibility of exchanging two particles in the t-channel.

From the point view of the s-channel one might expect the existence of some relation expressing the physical requirement that the probability of the exchange of two particles in the cross-channel is restricted. To obtain such a relation it is necessary to continue the unitarity condition analytically from the physical region of the t-channel ($t > 4\mu^2$, $s < 0$) to the region of large positive s.

For $4\mu^2 < t < 16\mu^2$ it is sufficient to restrict ourselves to the two-particle unitarity condition:

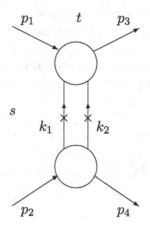

$$
\begin{aligned}
A_3 &\equiv \operatorname{Im}_t A(s,t) \\
&= \frac{1}{2} \int \frac{\mathrm{d}^4 k_1}{(2\pi)^3} \frac{\mathrm{d}^4 k_2}{(2\pi)^3} A(p_2, p_4, k_1, k_2) A^*(k_1, k_2, p_1, p_3) \theta(k_{10}) \theta(k_{20}) \\
&\quad \times \delta(k_1^2 - m^2) \delta(k_2^2 - m^2)(2\pi)^4 \delta(k_1 + k_2 - p_2 - p_4).
\end{aligned}
$$ (2.3)

The imaginary part of the amplitude $A_3(s,t)$ differs from zero only for $t > 4\mu^2$; therefore it is not analytic in t. Let us show, however, that for fixed $t > 4\mu^2$ it is an analytic function of s and thus can be analytically continued to the region of large positive s.

Since at large s the *relative distance* between $t > 4\mu^2$ and $t < 0$ is small, $4\mu^2/s \ll 1$, one may hope that the existence of a relation representing the analytical continuation of (2.3) ($t > 4\mu^2$) will impose a strong restriction on the amplitude $A(s,t)$ in the physical region of the s-channel ($t < 0$).

This important step – the continuation of the unitarity condition (2.3) to $s > 0$ – was done by Mandelstam. Let us briefly recall the main steps.

In the cms of the t-channel we have

$$A_3(z,t) = \frac{k_t}{\omega_t} \int \frac{\mathrm{d}\Omega}{64\pi^2}\, A(z_1,t)\, A^*(z_2,t)\,, \qquad (2.4)$$

with

$$k_t = \frac{\sqrt{t - 4\mu^2}}{2}\,, \qquad \omega_t = \frac{\sqrt{t}}{2}$$

the cms momentum and energy, respectively (cf. (1.4b)). In (2.4) we have traded the momentum transfer variable s for the cosine of the scattering angle, z:

$$s = -2k_t^2(1 - z)\,. \qquad (2.5a)$$

Analogously, $s_{1(2)}$ is the squared 4-momentum transfer between the intermediate and the initial (final) state; z_1, z_2 are the cosines of the corresponding scattering angles:

$$s_1 = -2k_t^2(1 - z_1), \qquad s_2 = -2k_t^2(1 - z_2). \qquad (2.5b)$$

It is convenient to replace integration over the azimuthal angle ϕ in $\mathrm{d}\Omega = \mathrm{d}z_1\mathrm{d}\phi$ by that over z_2 using the trigonometric relation

$$z_2 = zz_1 + \sqrt{(1 - z_1^2)(1 - z^2)}\cos\phi\,. \qquad (2.6)$$

Calculating the Jacobian of the transformation, $|\sin\phi|$, we obtain

$$\int \mathrm{d}\Omega \equiv \int \frac{2\,\mathrm{d}z_1\mathrm{d}z_2}{\sqrt{-K(z,z_1,z_2)}}\,, \qquad (2.7a)$$

where

$$K \equiv (z - z_1z_2)^2 - (1 - z_1^2)(1 - z_2^2)\,. \qquad (2.7b)$$

Finally,

$$A_3(z,t) = \frac{k_t}{32\pi^2\omega_t} \iint \frac{\mathrm{d}z_1\,\mathrm{d}z_2}{\sqrt{-K(z,z_1,z_2)}}\, A(z_1,t)\, A^*(z_2,t), \qquad (2.8)$$

with the integration domain determined by the condition $K(z,z_1,z_2) < 0$.

After the z-dependence has been explicitly extracted, it is clear that $A_3(z)$ is an analytic function of z that can be continued to $s > 0$ ($z > 1$, see (2.5a)). In the course of continuation we will have to deform integration contours in (2.8) appropriately, so as to avoid singularities. The singularity of the integral will appear at some $z = z_0(t) > 1$ when such contour deformation becomes no longer possible (singularities of the integrand 'pinch' the contour and immobilize it).

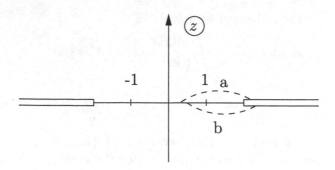

Fig. 2.2. Analytical continuation of $A_3(z, t)$ to $z > 1$

For $z > z_0(t)$ the amplitude is complex and double valued, i.e. two analytical continuations of (2.8), one along the path a and another along b, as shown in Fig. 2.2, will lead to two different expressions for $A_3(z)$.

One can calculate the discontinuity of $A_3(z)$ across the singularity at $z = z_0(t)$. This calculation yields

$$\rho(s, t) \equiv \mathrm{Im}\,_s A_3(z, t)$$

$$= \frac{k_t}{16\pi^2 \omega_t} \int \frac{\mathrm{d}z_1 \mathrm{d}z_2}{\sqrt{K(z, z_1, z_2)}} [A_1(z_1) A_1^*(z_2) + A_2(z_1) A_2^*(z_2)], \quad (2.9)$$

where the integration is performed over the region

$$z_1 > 1, \qquad z_2 > 1, \qquad z > z_1 z_2 + \sqrt{(z_1^2 - 1)(z_2^2 - 1)}. \qquad (2.10)$$

Here $A_1(z)$ is the imaginary part (absorptive part) of the amplitude in the s-channel:

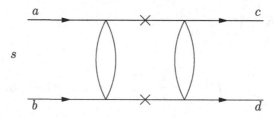

The region of negative z_1, z_2 is also taken into account in (2.9). Hence the contribution of the absorptive part of the amplitude in the u-channel:

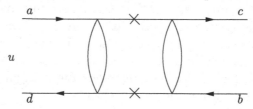

The real function $\rho(s,t)$ is 'the imaginary part of the imaginary part' – the double discontinuity of the amplitude $A(s,t)$ both in s and in t. In the Mandelstam plane, it differs from zero inside the region bounded by the Karplus curve $z = z_0(t)$. Inside this region $\rho(s,t)$ acquires contributions from each particle threshold, having a simple graphical meaning. For example, a four-particle s-channel state in A_1,

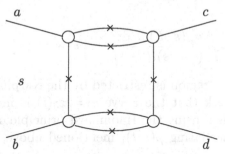

contributes to $\rho(s,t)$ inside the region determined by its own Karplus curve with the asymptotes* $t = (2\mu)^2$, $s = (4\mu)^2$:

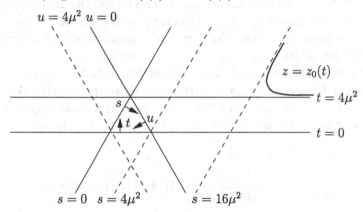

2.1.1 The Mandelstam representation

We have calculated $\rho(s,t)$ in two steps: we have found the imaginary part first with respect to t and then with respect to s. We could have done it differently: first find the imaginary part of the amplitude in the s-channel, $A_1(s,t)$, analytically continue it to the region $t > 4\mu^2$ and then calculate the imaginary part with respect to t.

In this way we would have got the same expression (2.9) for $\rho(s,t)$. This statement has been proved rigorously for Feynman graphs in a few

---

* This curve and also the complementary one with the asymptotes $t = (4\mu)^2$, $s = (2\mu)^2$ are the true first Karplus curves for $\pi\pi$ scattering, since the $\pi\pi \to \pi$ transition is forbidden by conservation of G-parity.

orders of perturbation theory. Although a general proof does not exist, one can give a simple argument for its validity.

Singularity described by the Karplus curve $z = z_0(t)$ is the singularity of the whole amplitude $A(s,t)$, and $\rho(s,t)$ is the discontinuity across this singularity. Its value cannot depend on the way in which it was calculated. For instance, suppose that the amplitude $A(s,t)$ admits the *Mandelstam representation*:

$$A(s,t) = \frac{1}{\pi^2} \int \frac{\rho_{st}(s',t')\mathrm{d}s'\mathrm{d}t'}{(s'-s)(t'-t)} + [s \to u] + [t \to u], \qquad (2.11)$$

where the integration region is restricted by the Karplus curve in the s'-t' plane. One can check that the curve $z = z_0(t)$ is indeed a singularity curve of $A(s,t)$. Then, using the Hadamard principle one can verify that the two ways of calculating $\rho(s,t)$, mentioned above, lead to the same result:

$$\rho_{st}(s,t) = \mathrm{Im}\,_s A_3(s,t) = \mathrm{Im}\,_t A_1(s,t).$$

Since 1958, when the representation (2.11) for the invariant amplitude $A(s,t)$ was suggested by Mandelstam, up to the present time no Feynman graph has been found for which it is not valid, provided all participating particles are stable (e.g., $m_a < m_b + m_c + m_d$).

Thus using the t-channel unitarity condition (2.8) we have obtained the expression for the imaginary part of the absorptive part $A_1(s,t)$ in the s-channel, analytically continued into the unphysical region $t > 4\mu^2$, in terms of the s- and u-channel absorptive parts A_1, A_2:

$$\rho(s,t) = \mathrm{Im}\,_t A_1(s,t)$$
$$= \frac{k_t}{16\pi^2\omega_t} \int \frac{\mathrm{d}z_1\mathrm{d}z_2}{\sqrt{K(z,z_1,z_2)}}[A_1(s_1,t)A_1^*(s_2,t) + A_2(s_1,t)A_2^*(s_2,t)].$$
$$(2.12)$$

2.1.2 Inconsistency of the 'black disk' model of diffraction

Using the relation (2.12) we can now show that the absorptive part of the s-channel amplitude A_1 at large s and fixed t *cannot* have the form

$$A_1(s,t) = s\,f(t), \qquad s \to \infty. \qquad (2.13)$$

Such a form arises within a simple model which corresponds to diffractive scattering off a black disk in quantum mechanics.

On the one hand, assuming (2.13) we can continue $A_1(s,t)$ into the unphysical region $t > 4\mu^2$ and determine $\rho(s,t)$ directly:

$$\rho(s,t) = s\,\mathrm{Im}\,f(t). \qquad (2.14)$$

On the other hand, we can calculate $\rho(s,t)$ by substituting the ansatz (2.13) into the r.h.s. of (2.12). To this end we first remark that $A_2(u,t)$ coincides (modulo sign) with $A_1(s,t)$ (see the proof of the Pomeranchuk theorem in Lecture 1). Therefore it suffices to consider only one term $A_1(s_1)A_1^*(s_2)$ in the integrand in (2.12).

For large $z \propto s$ the dominant contribution to the integral over z_1 and z_2 arises from $z_1 \sim z_2 \sim \sqrt{z} \gg 1$, since according to our assumption the integrand grows linearly with z_1 and z_2 $(A_1(z_i) \simeq z_i f(t))$. Therefore we may simplify the expression (2.7b) for $K(z_1, z_2, z)$:

$$K(z_1, z_2, z) \simeq z(z - 2z_1 z_2).$$

Substituting approximate asymptotic expressions into (2.12) and omitting irrelevant s-independent factors we obtain an estimate

$$\rho(s,t) \propto \int \frac{z_1 \mathrm{d}z_1\, z_2 \mathrm{d}z_2}{\sqrt{z(z - 2z_1 z_2)}}, \qquad z_1 > 1, \quad z_2 > 1, \quad z > 2z_1 z_2. \quad (2.15)$$

Introducing a convenient variable $x = 2z_1 z_2$ we arrive at

$$\rho(s,t) \propto \frac{1}{\sqrt{z}} \int_1^z \frac{\mathrm{d}z_1}{z_1} \int_{z_1/2}^z \frac{x\, \mathrm{d}x}{\sqrt{z-x}} \simeq z \ln z \int_0^1 \frac{y\, \mathrm{d}y}{\sqrt{1-y}}. \quad (2.16)$$

Comparing with (2.14) we observe that the r.h.s. of the relation (2.12) asymptotically exceeds its l.h.s. This disproves the black disk model (2.13): $A_1(s,t) \neq s f(t)$ for $s \to \infty$.

2.2 Complex angular momenta

We start now to consider the theory of complex angular momenta. This theory has a number of aspects. The first aspect can be already revealed within the framework of non-relativistic quantum mechanics. It is expressed by the fact that the bound states and resonances are grouped into families. In each family the binding energy is a continuous function of the orbital angular momentum ℓ.

Indeed, consider the Schrödinger equation with Hamiltonian

$$H = -\frac{1}{2m} \frac{1}{r^2} \frac{\mathrm{d}}{\mathrm{d}r} r^2 \frac{\mathrm{d}}{\mathrm{d}r} + U(r) + \frac{\ell(\ell+1)}{2mr^2}. \quad (2.17)$$

Then, if the potential $U(r)$ is attractive and has the form of a sufficiently deep and wide well, there will be in general several bound states with energies E_ℓ^n at every value of ℓ. With increasing ℓ the centrifugal barrier grows, so the energy of every level with definite n grows monotonically. (For example, for the Coulomb potential $E|_{n_r=0} = -e^2/2(\ell+1)^2$.)

At some sufficiently large ℓ one may encounter quasi-stationary states – resonances:

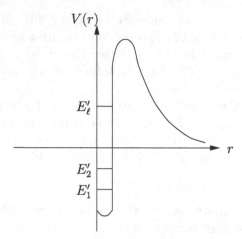

In the relativistic problem the potential depends on both the velocities and ℓ, and existence of a relationship between the states with different angular momenta seems unlikely. It is therefore even more surprising that elementary particles are grouped into families with the same internal quantum numbers (isospin, strangeness, spatial and charge parity) but different spins. In relativistic QFT the spins of particles play a rôle similar to that of the orbital angular momenta in the non-relativistic theory.

The second aspect is related to the asymptotics of the scattering amplitude $f(E, \cos \Theta)$ at large values of $\cos \Theta$. Within the framework of non-relativistic quantum mechanics this problem is only of academic interest.

In the relativistic theory, on the other hand, large $z_t = \cos \Theta_t$ at fixed E_t (in the t-channel) correspond to the transition to the physical region of another channel (s-channel), where $\cos \Theta_t$ is linearly related to the energy s (see (2.5a)).

Originally the idea that the asymptotics in the s-channel is related to the partial wave expansion in the t-channel was proposed by Mandelstam. Let us briefly recall his reasoning.

2.3 Partial wave expansion and Sommerfeld–Watson representation

Consider the partial wave expansion of the amplitude $A(s,t)$ in the physical region of the t-channel, i.e. for $t > 4\mu^2$, $s < 0$ (see Fig. 2.1),

$$A(s,t) = \sum_{n=0}^{\infty} (2n+1)\, f_n(t) P_n(z), \qquad z = 1 + \frac{2s}{t - 4\mu^2}. \qquad (2.18)$$

This series is convergent up to the first s plane singularity of $A(s,t)$, i.e. for $s < 4\mu^2$. It is natural to expect that the partial wave amplitudes $f_n(t)$

are large up to $n = kr_0$, where r_0 is the interaction radius (see Lecture 1).[†]

A question arises: what is the asymptotics of this series at large z? Although the series diverges at large z, one might expect that under an appropriate analytical continuation only partial wave amplitudes with $n \sim kr_0$ provide the dominant contributions to the series. Therefore we may truncate the series at $n \leq n_0 = kr_0$. Then at large s we would have $A(s,t) \sim z^{n_0}$. This implies that the asymptotics of $A(s,t)$ is related to the effective orbital angular momentum that is possible at given energy t.

To support this idea mathematically we use the method developed by Sommerfeld, Fock and Schwinger in solving the problem of the diffraction of radio waves around the Earth's surface and similar results obtained by Regge within the framework of quantum mechanics.

Our immediate goal is to perform the analytical continuation of (2.18) to large z.

Suppose that we can find an analytic function $f_\ell(t)$ that does not increase exponentially in any direction in the right half of the complex ℓ plane, and whose values coincide with the partial wave amplitudes at all integer ℓ:

$$f_\ell|_{\ell=n} = f_n. \tag{2.19}$$

Then the expansion (2.18) can be written in the form of a contour integral:

$$A = \frac{\mathrm{i}}{2} \int_L \frac{\mathrm{d}\ell}{\sin \pi \ell} f_\ell(t) P_\ell(-z)(2\ell + 1), \tag{2.20}$$

where the integration contour L is shown in Fig. 2.3. (In the derivation of (2.19) the property $P_n(-z) = (-1)^n P_n(z)$ has been used.)

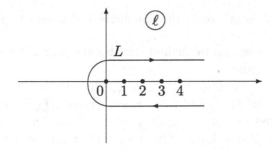

Fig. 2.3. Integration contour in the Sommerfeld–Watson representation

In (2.20), $P_\ell(z)$ is the Legendre function of the first kind, the solution of the differential Legendre equation that is regular at $z = 1$. Let us now

---

[†] We use here the letter n instead of ℓ to emphasize that the partial waves in this sum are defined only for integer values of the angular momentum.

deform the contour along the imaginary axis. For this to be possible the integrand in (2.20) should decrease at large $|\ell|$ in the right half-plane.

Recall the large-ℓ asymptotics of the Legendre function ($z > 0$):

$$P_\ell(z)|_{\ell \to \infty} \sim \exp(i\ell\Theta) + \exp(-i\ell\Theta), \qquad (2.21a)$$

$$P_\ell(-z)|_{\ell \to \infty} \sim \exp(i\ell(\pi - \Theta)) + \exp(-i\ell(\pi - \Theta)), \quad (2.21b)$$

where $z = \cos\Theta$. From (2.21) we conclude that the ratio $P_\ell(-z)/\sin\pi\ell$ falls exponentially with Im $\ell \to \infty$ in the physical region $0 < \Theta < \pi$. Therefore, the contour in the ℓ plane may be deformed, passing on the right all singularities of $f_\ell(t)$ that may appear at finite distances in the ℓ plane as shown in Fig. 2.4.

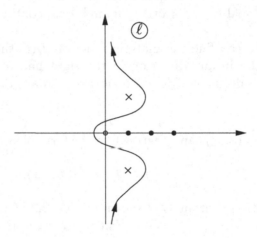

Fig. 2.4.   Contour deformation in the Sommerfeld–Watson integral

After the contour has been so deformed, the function (2.20) is defined in the entire complex z plane.

Indeed, consider first the region $z < -1$ in (2.20). Then, by virtue of (2.21a), the asymptotics of $P_\ell(z)$ will be of the form $\cosh(\ell\chi)$, where χ is a positive number related to z in a simple way. So, with ℓ moving along the imaginary axis, $P_\ell(-z)$ oscillates. This means that the integral (2.20) converges and defines an analytic function $A(t, z)$, free of singularities in the left half of the z plane.

If we consider the region of positive z, $z > 1$, then (2.21b) applies and as we move along the line Im $\ell \to +\infty$, $P_\ell(-z) \sim \exp(-i\ell\pi)\exp(\ell\chi)$. Taking into account the cancellation of the growing factor $\exp(-i\ell\pi)$ by $\sin\ell\pi$, the convergence of the contour integral (2.20) will depend on the behaviour of f_ℓ as Im $\ell \to \infty$. Thus, generally speaking, the amplitude $A(t, z)$ may have singularities on the right half of the z plane:

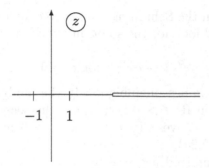

These singularities may occur only at real positive $z > 1$ since, as is easy to check, there are no singularities at complex values of z: taking $\mathrm{Im}\, z \neq 0$ immediately leads to a convergent integral.

Once the amplitude (2.20) is defined for all z, its asymptotics can be easily calculated using the property of the Legendre functions:

$$P_\ell(-z) \sim z^\ell \exp(-\mathrm{i}\,\ell\pi), \qquad z \to \infty. \tag{2.22}$$

From this equation one can see that the asymptotics of (2.20) is governed by the rightmost singularity of the partial wave amplitude in the ℓ plane. For instance if the rightmost singularity of $f_\ell(t)$ is a simple pole at $\ell = \alpha(t)$, then

$$A(s,t) \to \pi \frac{2\alpha + 1}{\sin \pi\alpha} P_\alpha(-z)\mathrm{Res}\, f_\alpha(t) + \int_{\mathrm{Re}\,\ell < \mathrm{Re}\,\alpha(t)} \mathrm{d}\ell \ldots , \tag{2.23}$$

where $\mathrm{Res}\, f_\alpha(t)$ is the residue of the partial wave amplitude at this pole. From (2.22) one can see that the 'background integral' (the second term on the r.h.s. of (2.23)) is asymptotically subdominant with respect to the first term.

Thus the ℓ plane singularities of $f_\ell(t)$ define the asymptotics of the amplitude $A(s,t)$, i.e. the asymptotics is defined by some ℓ_{eff}, in accordance with the intuitive picture of Mandelstam.

This is the second aspect of the theory of complex angular momenta.

2.4 Continuation of partial wave amplitudes to complex ℓ

Consider now the problem of analytical continuation of the partial wave amplitudes $f_n(t)$ into the complex ℓ plane.

2.4.1 Non-relativistic quantum mechanics

In non-relativistic quantum mechanics this problem is easily solved since for the analytical continuation of the radial wave function $\Psi_\ell(r)$ it is sufficient to consider in the Schrödinger equation the quantity $\ell(\ell+1)$ to be complex valued and look for the solution which at small r satisfies the condition

$$\Psi_\ell(r) \to c\,r^\ell, \quad \text{for } r \to 0. \tag{2.24}$$

Condition (2.24) determines unambiguously the analytic function at all complex ℓ in the region $\mathrm{Re}\,\ell > -1/2$, where the second solution to the Schrödinger equation behaves like $r^{-\ell-1}$ as $r \to 0$ and hence does not satisfy the condition (2.24).

The partial wave amplitudes

$$f_\ell(E) = \frac{1}{k} \sin \delta_\ell(E) \exp(\mathrm{i}\,\delta_\ell(E)) \tag{2.25}$$

can be found by studying the asymptotics of the radial wave function $\Psi_\ell(r)$ at large r:

$$\Psi_\ell(r) \sim \frac{1}{r} \sin\left(kr - \frac{\ell\pi}{2} + \delta_\ell(E) \right), \qquad r \to \infty, \tag{2.26}$$

where $\delta_\ell(E)$ is the scattering phase shift. The partial wave amplitudes $f_\ell(E)$ defined this way will decrease at large ℓ: from the general physical picture it is clear that the scattering at large impact parameters should be small, as long as we restrict ourselves to potentials with finite interaction radii.

Therefore, for the function f_ℓ analytically continued in this way, the representation (2.20), with the contour of integration deformed into the path going along the imaginary axis as shown in Fig. 2.4, will be valid.

2.4.2 Relativistic theory

Let us consider now the analytical continuation of the partial wave amplitudes in the relativistic theory.

It is worthwhile to mention that, generally speaking, there exists no method of constructing an analytic function f_ℓ that coincides with an arbitrary set of numbers f_n at integer values of $\ell = n$. In the mathematical literature this problem is solved using an infinite series which is not convergent however for all sets of f_n.

Let us consider the question of uniqueness of constructing an analytic function f_ℓ out of its values f_n at integer points. There exists a mathematical theorem (Carlson's theorem) which states that if one has found a function which is analytic at $\mathrm{Re}\,\ell > \ell_0$, and which grows in any direction in the right half-plane slower than $\exp(\pm\mathrm{i}\,\ell\pi)$, then this function is unique. Indeed if there were two such functions $f_{1,2}(\ell)$, then the function $\phi = (f_1(\ell) - f_2(\ell))/\sin\pi\ell$ would be analytic and exponentially falling along any beam in the right half-plane. It is next to obvious that such a function is identically equal to zero.

In the relativistic theory, a function $f_\ell(E)$ that satisfies all the above mentioned properties cannot exist. Indeed, otherwise $A(s,t)$ defined by (2.20) would have no singularities for $z < -1$, but this would contradict the fact that $A(s,t)$ does have a singularity at $u = -2p^2(1+z) > 4\mu^2$ (see the previous lecture).

2.5 Gribov–Froissart projection

We can show, however, that in the relativistic case there exist two functions f_ℓ^+ and f_ℓ^-, analytic in the ℓ plane, that are the analytical continuations of the partial wave amplitudes $f_n(t)$ with even and odd integer n, respectively.

Let us write down the formula which defines $f_n(t)$ in terms of $A(s,t)$:

$$f_n(t) = \frac{1}{2} \int_{-1}^{1} P_n(z) A(t,z) \, \mathrm{d}z. \qquad (2.27)$$

Consider also the Legendre function of the second kind for integer n:

$$Q_n(z) = \frac{1}{2} \int_{-1}^{1} \frac{P_n(z') \mathrm{d}z'}{z - z'}. \qquad (2.28)$$

At large z this function decreases,

$$Q_n(z) \sim \frac{c}{z^{n+1}}, \qquad |z| \to \infty, \qquad (2.29)$$

whereas for $-1 < z < 1$ it is complex valued and obeys the relation

$$Q_n(z + \mathrm{i}\,\varepsilon) - Q_n(z - \mathrm{i}\,\varepsilon) = -\mathrm{i}\,\pi P_n(z). \qquad (2.30)$$

Using this property we may rewrite (2.27) as

$$f_n(t) = \frac{1}{2\pi\mathrm{i}} \oint_{(a)} Q_n(z) \, A(t,z) \, \mathrm{d}z, \qquad (2.31)$$

where the integration contour (a), enclosing the interval $[-1,1]$ on the real axis in the z plane, is displayed in Fig. 2.5.

Away from the contour, the integrand has two branch cuts on the real axis which start at the points z_1 and $-z_2$ that correspond to the threshold s- and u-channel singularities of the amplitude $A(s,t)$; in our symmetric case, the thresholds are at $s = 4\mu^2$ and $u = 4\mu^2$, so that $z_1 = z_2 = 1 + 4\mu^2/2k_t^2 > 1$. The discontinuities across these cuts are, respectively, $2\mathrm{i}\,A_1(z,t)$ and $2\mathrm{i}\,A_2(z,t)$.

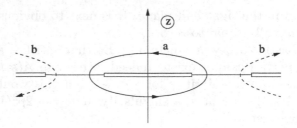

Fig. 2.5.   Original (a) and deformed contours (b) in the representation (2.31)

The integration contour (a) in (2.31) can be deformed for sufficiently large n by virtue of (2.29) and replaced by the contour (b) as shown in Fig. 2.5. Omitting the contribution of the circle at infinity we find

$$f_n = \frac{1}{\pi} \int_{z_1}^{\infty} Q_n(z)\, A_1(z,t)\, dz + \frac{1}{\pi} \int_{-z_2}^{-\infty} Q_n(z)\, A_2(z,t)\, dz. \qquad (2.32)$$

Using the relation

$$Q_n(-z) = (-1)^{n+1} Q_n(z), \qquad (2.33)$$

we may rewrite (2.32) in the following form ($z_u = -z$):

$$f_n = \frac{1}{\pi} \int_{z_1}^{\infty} Q_n(z)\, A_1(z,t)\, dz + \frac{(-1)^n}{\pi} \int_{z_2}^{\infty} Q_n(z_u)\, A_2(-z_u,t)\, dz_u. \qquad (2.34)$$

This representation is better suited for performing analytical continuation to complex ℓ than the expression (2.27) we started from.

The advantage of (2.34) becomes evident from the comparison of asymptotic behaviour of $P_n(z)$ (cf. (1.13)) and $Q_n(z)$ at large n:

$$P_n(z) \simeq J_0(n\Theta) \sim \cosh(n\chi); \qquad (2.35a)$$

$$Q_n(z) \sim \exp(-n\chi); \qquad (2.35b)$$

$$\cos\Theta = \cosh\chi = z \geq z_{1,2} > 1 \quad (\chi > 0).$$

From (2.35a) we see that if we had defined $f_n(z)$ by formula (2.27), partial waves would have grown exponentially with n, thus invalidating the Sommerfeld–Watson representation.

On the other hand, $Q_n(z)$ according to (2.35b) do not grow in the right half of the n plane. Therefore, had we omitted the factor $(-1)^n \sim \exp(\pi|\mathrm{Im}\, n|)$ in the second term of (2.34), the function $f_n(t)$ would have fallen off exponentially at positive n and would not have grown along the imaginary n axis.

To get rid of the oscillating factor $(-1)^n$, let us define two analytic functions f_ℓ^+ and f_ℓ^- which satisfy the condition

$$f_\ell^+ |_{\ell=n=2k} = f_n, \qquad f_\ell^- |_{\ell=n=2k+1} = f_n. \qquad (2.36)$$

For these functions we obtain from (2.34)

$$f_\ell^\pm = \frac{1}{\pi} \int_{z_1}^\infty Q_\ell(z) A_1(z,t)\, \mathrm{d}z \pm \frac{1}{\pi} \int_{z_2}^\infty Q_\ell(z_u) A_2(-z_u,t)\, \mathrm{d}z_u. \qquad (2.37)$$

We have shown that in relativistic theory no analytic function exists which extrapolates all partial wave amplitudes. At the same time, one can introduce two analytic functions $f_\ell^\pm(t)$, defined by (2.37), which are the continuations of the partial waves with even and odd n, respectively.

It is easy to show that the continuation (2.37) is unique. Indeed, if the function in the right half-plane grows slower than $\exp(\pm \mathrm{i}\,\ell\pi/2)$ and takes the given values f_n at even integer points then it is unique by virtue of the Carlson theorem. Note that our functions (2.37) do not grow at all for $\mathrm{Re}\,\ell > \ell_0$ (with ℓ_0 such that $|A_1(z,t)| < z^{\ell_0}$ for $z \to \infty$, so as to ensure convergence of (2.37)).

One can represent the scattering amplitude A given in (2.18) as a sum of odd and even parts in z: $\quad A = A^+ + A^-$,

$$A^+ = \sum_{n=2r} P_n(z)(2n+1)f_n^+, \quad A^- = \sum_{n=2r+1} P_n(z)(2n+1)f_n^-. \qquad (2.38)$$

For each of the amplitudes A^+ and A^- one may introduce analytic functions $f_\ell^\pm$ which are, generally speaking, different. In non-relativistic quantum mechanics a similar situation occurs in the presence of an exchange potential.

In conclusion we note that the transition from (2.31) to (2.37) is possible only for $n > n_0 \geq \ell_0$, when one can neglect the integral over the circle of infinite radius in the z plane. Therefore in the general case we must add to the contour integral a few of the first terms of the sum (2.38):[‡]

$$\left.\begin{aligned} A^+ &= \sum_{n=2r}^{n_0} P_n(z)(2n+1)f_n^+ + \frac{\mathrm{i}}{4} \int_{\ell_0-\mathrm{i}\infty}^{\ell_0+\mathrm{i}\infty} \frac{(2\ell+1)\,\mathrm{d}\ell}{\sin \pi\ell} f_\ell^+ \left[P_\ell(-z) + P_\ell(z)\right], \\ A^- &= \sum_{n=2r+1}^{n_0} P_n(z)(2n+1)f_n^- + \frac{\mathrm{i}}{4} \int_{\ell_0-\mathrm{i}\infty}^{\ell_0+\mathrm{i}\infty} \frac{(2\ell+1)\,\mathrm{d}\ell}{\sin \pi\ell} f_\ell^- \left[P_\ell(-z) - P_\ell(z)\right]. \end{aligned}\right\}$$
$$(2.39)$$

We can now find the expression for the absorptive part of the s-channel amplitude using

$$P_\ell(-(z+\mathrm{i}\varepsilon)) - P_\ell(z+\mathrm{i}\varepsilon) = \left[\mathrm{e}^{-\mathrm{i}\pi\ell} - \mathrm{e}^{\mathrm{i}\pi\ell}\right] P_\ell(z)$$
$$= -2\mathrm{i}\sin \pi\ell\, P_\ell(z). \qquad (2.40)$$

[‡] Recall that by virtue of the Froissart theorem $n_0 < 2$, so that each of the finite sums in (2.39) contains at most one term.

Applying (2.40) to (2.39) we derive

$$A_1^\pm(s,t) = \frac{1}{4\mathrm{i}} \int \mathrm{d}\ell\,(2\ell+1)\,f_\ell^\pm(t)\,P_\ell(z_s), \left.\begin{array}{c} \\ \\ \\ \\ \end{array}\right\}$$

$$z_s = 1 + \frac{2s}{t - 4\mu^2}. \qquad\qquad (2.41)$$

2.6 t-Channel partial waves and the black disk model

Let us show now that the t-channel unitarity condition (2.3) for the partial wave amplitudes with integer numbers n can be analytically continued onto the complex ℓ plane. To this end we rewrite (2.3) in the form

$$\frac{1}{2\mathrm{i}}\left[f_n(t+\mathrm{i}\varepsilon) - f_n(t-\mathrm{i}\varepsilon)\right] = \frac{k_t}{16\pi\omega_t} f_n(t+\mathrm{i}\varepsilon)f_n(t-\mathrm{i}\varepsilon). \qquad (2.42)$$

This equation is valid for even as well as for odd n. Substituting $n \to \ell$ and moving the r.h.s. of (2.42) to the l.h.s., we note that the function that we have thus formed in the l.h.s. of the equation is defined for arbitrary complex ℓ, equals zero at all even (odd) points and does not grow exponentially in the right half-plane. Therefore it is identically equal to zero:

$$\frac{1}{2\mathrm{i}}\left[f_\ell^\pm(t+\mathrm{i}\varepsilon) - f_\ell^\pm(t-\mathrm{i}\varepsilon)\right] = \frac{k_t}{16\pi\omega_t} f_\ell^\pm(t+\mathrm{i}\varepsilon)f_\ell^\pm(t-\mathrm{i}\varepsilon). \qquad (2.43)$$

It is clear that the unitarity condition (2.43) and the Sommerfeld-Watson representation (2.39) are together equivalent to the Mandelstam analytical continuation of the t-channel unitarity condition towards large z.

Let us demonstrate that (2.43) forbids the asymptotics of the absorptive part $sf(t)$ that is prescribed by the black disk model. Indeed, having assumed such an asymptotic behaviour of A_1 and A_2 we would get from (2.37)

$$f_\ell^+\big|_{\ell\to 1} \simeq \frac{r(t)}{\ell - 1}. \qquad (2.44)$$

The contradiction with unitarity is apparent, since the r.h.s. of (2.43) acquires a *double* pole, while its l.h.s. has only a simple pole at $\ell = 1$.

From this example it is clear that the t-channel unitarity condition forbids partial wave amplitudes to have any *fixed* (independent of t) singularities at real ℓ, such that $f_\ell = \infty$ at the singular point. 'Soft' fixed singularities (for example, $f_\ell \propto \sqrt{\ell - 1}$, or $f_\ell \propto \ln(\ell - 1)$) which correspond to σ_tot *falling* with energy are, generally speaking, possible.

We conclude that a seemingly natural picture of a hadron as an object with fixed, independent of the collision energy, interaction radius is inconsistent with unitarity in the cross-channel.

3

Singularities of partial waves and unitarity

In the previous lecture we have found a simple connection between ℓ plane singularities of $f_\ell^\pm(t)$ and the asymptotics of the scattering amplitude at fixed $t > 4\mu^2$ and $s \to \infty$, along the dashed line 1 in Fig. 3.1. It is

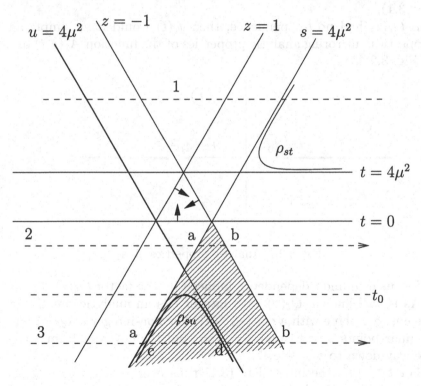

Fig. 3.1.   On continuation of the t-channel partial wave to $t < 0$

necessary to continue the formulæ (2.39) to the physical region of the *s*-channel, that is to $t < 0$.

3.1 Continuation of partial waves with complex ℓ to $t < 0$

Let us discuss analytic properties of the partial wave $f_\ell^\pm(t)$:

$$f_\ell^\pm(t) = \frac{1}{\pi} \int Q_\ell(z) A_1^\pm(z, t),$$

or

$$f_\ell^\pm(t) = \frac{1}{\pi} \int_{4\mu^2}^{\infty} Q_\ell \left(1 + \frac{2s}{t - 4\mu^2}\right) A_1^\pm(s, t) \, \frac{2 \, ds}{t - 4\mu^2}. \qquad (3.1)$$

As we have learned from the Mandelstam representation, $A_1(s, t)$ as a function of t has singularities both at $t > 4\mu^2$ where the *first spectral function* $\rho_{st} \neq 0$ and at negative t, due to the *third spectral function* ρ_{su}.

$A_1(s, t)$ is real, for arbitrary s, in the interval from $t = 4\mu^2$ down to $t = t_0$ (the line $t = t_0$ is tangential to the third Mandelstam curve, see Fig. 3.1).

If $Q_\ell(z)$ had no t-dependence, then $f_\ell(t)$ would have simple analytic properties, mirroring analytic properties of the function $A_1(s, t)$ as shown in Fig. 3.2.

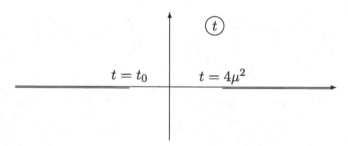

Fig. 3.2.   Analytic properties of $A_1(s, t)$

Let us examine t-dependence induced by the factor $Q_\ell(z)$.

As is well known, $Q_\ell(z)$ with integer ℓ is an analytic function of z in the entire z plane with a cut $[-1, +1]$. For non-integer ℓ, $Q_\ell(z)$ has three singular points $z = \pm 1$ and $z = -\infty$, and the cut goes from $z = +1$ all the way down to $z = -\infty$.

For $t > 4\mu^2$, the integral in (3.1) runs over $z > 1$ where $Q_\ell(z)$ does not have any singularities. In this region the imaginary part of f_ℓ is determined by that of A_1.

Now we take $t < 4\mu^2$. Here A_1 is real and we could expect partial waves to be real too. However, an additional complexity arises due to the fact that for $0 < t < 4\mu^2$ we have $z < -1$ and $Q_\ell(z)$ (with non-integer real ℓ) becomes complex valued. Let us show that this complexity of f_ℓ has a simple kinematical origin and can be eliminated.

3.1.1 Threshold singularity and partial waves ϕ_ℓ

Consider values of t close to the t-channel threshold. Here $|z| \to \infty$ and

$$Q_\ell(z) \propto \frac{1}{z^{\ell+1}} \simeq \left(\frac{t - 4\mu^2}{2s}\right)^{\ell+1}, \quad t - 4\mu^2 \to 0.$$

Hence

$$f_l^\pm(t) \propto (t - 4\mu^2)^\ell \int A_1^\mp(s,t) \frac{\mathrm{d}(2s)}{(2s)^{\ell+1}}, \tag{3.2}$$

which is nothing but the correct threshold behaviour of a partial wave with integer ℓ (recall that $(t - 4\mu^2)^\ell = (2k_t)^{2\ell}$).

Thus, the new complexity at $t - 4\mu^2 < 0$ is connected with the generalization of the standard threshold behaviour of partial wave amplitudes to non-integer ℓ.

Let us extract this singularity by introducing a new function

$$\phi_\ell(t) \equiv \frac{f_\ell(t)}{(t - 4\mu^2)^\ell} \tag{3.3}$$

and verifying that the redefined partial wave $\phi_\ell(t)$ has no singularity at $t = 4\mu^2$ and therefore is *real* in the interval $0 < t < 4\mu^2$.

To this end we invoke a representation of Q_ℓ in terms of the hypergeometric function,

$$Q_\ell(z) = \frac{\sqrt{\pi}\,\Gamma(\ell+1)}{\Gamma(\ell+3/2)(2z)^{\ell+1}} F\left(\frac{\ell}{2}+1, \frac{\ell+1}{2}, \ell + \frac{3}{2}, \frac{1}{z^2}\right), \tag{3.4}$$

to observe that for $z < -1$ there is no singularity but the cut due to the factor $1/z^{\ell+1}$.

Using (3.4) it is easy to check that

$$\begin{aligned}
\phi_\ell^\pm(t) &= \frac{1}{(t - 4\mu^2)^\ell}\frac{1}{\pi}\int Q_\ell\left(1 + \frac{2s}{t - 4\mu^2}\right) A_1^\pm(s,t)\frac{2\,\mathrm{d}s}{t - 4\mu^2} \\
&= \frac{2}{\pi(4\mu^2 - t)^{\ell+1}}\int_{4\mu^2}^\infty Q_\ell\left(\frac{2s}{4\mu^2 - t} - 1\right) A_1^\pm(s,t)\,\mathrm{d}s.
\end{aligned} \tag{3.5}$$

This expression remains real when t decreases from $4\mu^2$ down to the point where the argument of Q_ℓ in (3.5) becomes equal to $+1$:

$$\frac{2s}{4\mu^2 - t} - 1 = 1 \implies s = 4\mu^2 - t. \tag{3.6}$$

Since an integration over s is performed from $s_0 = 4\mu^2$, the greatest value of t for which the condition (3.6) is still satisfied is equal to zero.

Thus, the partial wave $\phi_\ell(t)$ defined in (3.3) is real in the interval $0 < t < 4\mu^2$.

3.1.2 $\phi_\ell(t)$ At $t < 0$ and its discontinuity

For $t < 0$ the function Q_ℓ in (3.5) is in general complex valued when s is varied within some finite interval (see the shaded area in Fig. 3.1). For example, along the line 2 in Fig. 3.1 it is the interval ab.

For $t < t_0$ the singularities of the function ϕ_ℓ are not only those related to the complexity of the Legendre function Q_ℓ, but also those coming from the discontinuity of $A_1(s, t)$ (where $\rho_{su} \neq 0$).

Finally, $\phi_\ell(t)$ has the singularities as shown below:

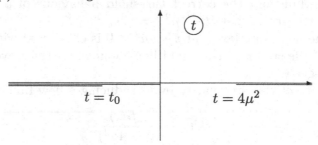

The discontinuity across the left-hand cut for $t > t_0$ can be easily found as it is related only with the complexity of $Q_\ell(z)$:

$$Q_\ell(z + i\varepsilon) - Q_\ell(z - i\varepsilon) = \frac{\pi}{i} P_\ell(z), \qquad -1 < z < 1.$$

For arbitrary $t < 0$ we have $z_t < 1$ satisfied automatically, while the condition $z_t > -1$ yields $s < s_b \equiv 4\mu^2 - t$. That is for $t_0 < t < 0$ we have

$$\operatorname{Im} \phi_\ell^\pm(t) = -\frac{1}{2} \int_{4\mu^2}^{s_b} P_\ell\left(\frac{2s}{4\mu^2 - t} - 1\right) A_1^\pm(s, t) \frac{2\,ds}{(4\mu^2 - t)^{\ell+1}}. \tag{3.7}$$

The integration is performed at $t =$ const along the interval ab on the line 2 inside the shaded area in Fig. 3.1.

For $t < t_0$, $\phi_\ell(t)$ develops an additional discontinuity which is related to the complexity of $A_1(s, t)$. Therefore for $t < t_0$ the expression (3.7) has a more complicated form:

$$\operatorname{Im} \phi_\ell^\pm(t) = -\frac{1}{2} \int_{4\mu^2}^{s_b} P_\ell\left(\frac{2s}{4\mu^2 - t} - 1\right) A_1^\pm(s, t + i\varepsilon) \frac{2\,ds}{(4\mu^2 - t)^{\ell+1}}$$
$$+ \int_{s_c}^{s_d} Q_\ell\left(\frac{2s}{4\mu^2 - t + i\varepsilon} - 1\right) \operatorname{Im}_t A_1^\pm(s, t) \frac{2\,ds}{(4\mu^2 - t)^{\ell+1}}. \tag{3.8}$$

The integration in the first term in (3.8) is performed along the interval ab of line 3 in Fig. 3.1, and that in the second term along the interval cd on the same line, inside the Karplus curve.

Note that the second integral vanishes for integer ℓ. Indeed, the integration region here is symmetric with respect to $z \to -z$ which is equivalent to interchanging $s \leftrightarrow u$. Let $\ell = 2n$. Remembering that $Q_{2n}(z)$ is an odd function of z and that the spectral function $\rho_{su}^{+}(s,t) \equiv \rho(s,t) + \rho(u,t)$ is an even function, we find that the second term is zero as the integral of an odd function over a symmetric interval. The same is valid for the function $\phi^{-}(t)$ at $\ell = 2n + 1$.

Thus we have shown that the discontinuity of the partial wave $\Delta\phi_{\ell}^{\pm}(t)$ across the left-hand cut in integer points of its proper *signature*,

$$\ell = 2n \quad \text{for} \quad \phi_{\ell}^{+}(t),$$
$$\ell = 2n + 1 \quad \text{for} \quad \phi_{\ell}^{-}(t),$$

is not related to $A_{1}^{\pm}(s,u)$ being complex.

Considering the scattering amplitude of non-identical particles the statement remains valid although in general the integration interval (s_c, s_d) in (3.8) loses its symmetry with respect to reflection $z \to -z$.

3.2 The unitarity condition for partial waves with complex ℓ

The discontinuity of $\phi_{\ell}^{\pm}(t)$ across the right-hand cut for t running from $4\mu^2$ to $16\mu^2$ can be determined by the two-particle unitarity condition, which was discussed in the second lecture:

$$\text{Im}\, f_n(t) = \frac{k_t}{16\pi\omega_t}\, f_n(t) f_n^{*}(t). \tag{3.9}$$

For $t > 16\mu^2$ the unitarity condition has a much more involved form.

In order to generalize (3.9) to complex ℓ let us write down this unitarity condition in a more convenient form:

$$\frac{1}{2i}[f_n(t + i\varepsilon) - f_n(t - i\varepsilon)] = \frac{k_t}{16\pi\omega_t}\, f_n(t + i\varepsilon) f_n(t - i\varepsilon).$$

Recall the integral representation

$$f_{\ell}^{\pm}(t \pm i\varepsilon) = \int_{4\mu^2}^{\infty} Q_{\ell}\left(1 + \frac{2s}{t - 4\mu^2}\right) A_{1}^{\pm}(s, t \pm i\varepsilon)\frac{2\,ds}{t - 4\mu^2}$$

that we invented to define partial waves with arbitrary non-integer $\ell > \ell_0$ (to ensure convergence of the integral over s). Note that Q_{ℓ} is real for $t > 4\mu^2$, so we don't need to assign a complex value to its argument t.

In terms of partial wave amplitudes ϕ_ℓ we have

$$\frac{1}{2i}[\phi_\ell(t+i\varepsilon) - \phi_\ell(t-i\varepsilon)] = C_\ell\,\phi_\ell(t+i\varepsilon)\phi_\ell(t-i\varepsilon), \qquad (3.10a)$$

where

$$C_\ell(t) = \frac{k_t}{16\pi\omega_t}(t-4\mu^2)^\ell. \qquad (3.10b)$$

This formula is more convenient since, as has been explained at the beginning of this lecture, for $0 < t < 4\mu^2$ the function $\phi_\ell(t)$ is real for any real ℓ, whereas $f_\ell(t)$ is a complex valued function for non-integer ℓ. As a result, the functions $\phi_\ell(t\pm i\varepsilon)$ for any ℓ are the values of the same analytic function on the upper (lower) edge of the right-hand cut at $4\mu^2 < t < 16\mu^2$.

The expression (3.10a) can be rewritten as

$$\frac{1}{2i}[\phi_\ell - \phi_{\ell^*}^*] = C_\ell(t)\,\phi_\ell\phi_{\ell^*}^*. \qquad (3.11)$$

The proof goes through a trivial calculation:

$$\phi_\ell^\pm(t-i\varepsilon) = \int Q_\ell\left(1+\frac{2s}{t-4\mu^2}\right)A^\pm(t-i\varepsilon,s)\frac{2\,ds}{(t-4\mu^2)^{\ell+1}}$$

$$= \int Q_{\ell^*}^*\left(1+\frac{2s}{t-4\mu^2}\right)A^{*\pm}(t+i\varepsilon,s)\frac{2\,ds}{(t-4\mu^2)^{\ell+1}} = \phi_{\ell^*}^*,$$

$$A^\pm(t-i\varepsilon) = (A^\pm(t+i\varepsilon))^*, \qquad 0 < t < 4\mu^2.$$

3.3 Singularities of the partial wave amplitude

Let us address the central question: what do we know about the singularities of the partial wave amplitudes? But first, a few words about the physical meaning of the formulæ obtained.

We have introduced the notion of partial wave amplitude $f_\ell(t)$. Singularities of $f_\ell(t)$ at $t > 4\mu^2$ (*right cut*) have a clear physical origin and are related to the masses of real particles in the intermediate states that can be created at given t.

What is, however, the meaning of the appearance of the *left cut* ($t < 0$) in the partial wave amplitude?

3.3.1 Left cut in non-relativistic theory

The left cut appears already in non-relativistic theory (though the discontinuity across it is given in this case only by the first term in (3.8)).

Where does the left cut come from in non-relativistic theory which, from the point of view of QFT, is equivalent to the ladder approximation to the scattering amplitude?

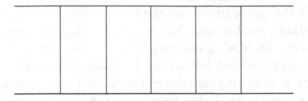

To answer this question, we have to consider singularities of the amplitudes with respect to momentum transfer. Let us show that it is these singularities, determined by the form of the potential, that build up the left cut.

Indeed, consider the Born amplitude for scattering with momentum transfer q, corresponding to the Yukawa potential $r^{-1}\exp(-\alpha r)$:

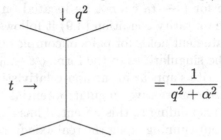

It is straightforward to calculate the partial wave amplitude corresponding to this graph. Constructing the cosine of the scattering angle, $|q| = k_t \sin\frac{\Theta}{2}$, $1 - z = 2q^2/(t - 4\mu^2)$, we obtain

$$f_\ell(t) = \frac{1}{2}\int_{-1}^{1}\frac{dz}{q^2+\alpha^2}P_\ell(z) = \frac{1}{t-4\mu^2}\int_{-1}^{1}\frac{P_\ell(z)dz}{1-z+2\alpha^2/(t-4\mu^2)}$$

$$= \frac{1}{t-4\mu^2}Q_\ell\left(1+\frac{2\alpha^2}{t-4\mu^2}\right). \qquad (3.12)$$

Let's take integer ℓ. Then the function $Q_\ell(z)$ has a cut only at $-1 < z < 1$, and so $f_\ell(t)$ develops a cut in t varying from $-\infty$ up to $t_0 = 4\mu^2 - \alpha^2$ (defined by the equation $1 + 2\alpha^2/(t_0 - 4\mu^2) = -1$).

In each successive order (n) in the scattering potential the amplitude will develop a new singularity $t_0^{(n)} < t_0$ on the negative t axis. Thus, the left cut plays in the theory a rôle akin to a potential. If the potential were defined, then one could calculate the discontinuity of $f_\ell(t)$ across the left cut, and using the unitarity condition find $f_\ell(t)$.

The difference between relativistic and non-relativistic theories is that while the form of the potential can be arbitrary in the latter, the restrictions arising from the unitarity conditions in the different channels have to be imposed in the former. There were numerous attempts to identify the discontinuity on the left cut with potential, i.e. to define the discontinuity rather than the potential. However, no interesting results have been obtained in this way, since the left cut discontinuity has some specific features that follow directly from expression (3.7).

What kind of singularities in ℓ can the partial wave amplitudes have? It is convenient to divide them into the two classes: fixed and moving singularities.

3.3.2 Fixed singularities

Location of 'fixed singularities' ($\ell = \ell_0$) does not depend on t. Actually, very little can be said about their properties. Fixed singularities such that $\phi_\ell \to \infty$ for $\ell \to \ell_0$ cannot be located on the real axis, $\mathrm{Im}\,\ell = 0$, since from the unitarity condition (3.9) it follows that $|\phi_\ell| \propto |f_\ell| < $ const. (The same statement holds for pairs of complex conjugate branch points.)

However, the singularities of the form $\sqrt{\ell - \ell_0}$ are not forbidden.*

Such singularities appear in the non-relativistic theory in the scattering problem with the attractive singular potential $-\gamma/r^2$. The partial wave amplitude corresponding to this potential has a fixed branch point associated with the cut running along the real axis from $\ell_1 = -1/2 + \sqrt{\gamma^2 + 1/4}$ down to $\ell_2 = -1/2 - \sqrt{\gamma^2 + 1/4}$.

From the point of view of high energy asymptotics such a weak singularity would lead to the cross section falling with increasing energy (even for $\ell_0 = 1$), since as we have discussed above, an asymptotically constant total cross section (which is translated according to the optical theorem into the asymptotics of the scattering amplitude at $t = 0$) corresponds to a stronger singularity, namely the simple pole $f_\ell(t) \propto 1/(\ell - 1)$.

3.3.3 Moving singularities

We will suppose that in relativistic theory there exist only *moving* singularities or, to be more precise, that only these are important for finding the asymptotics of the scattering amplitude. The argument in support of this hypothesis is twofold:

- in such a theory a beautiful self-consistent picture emerges, and

---

* Singularities whose *character* depends on t, for example $(\ell - \ell_0)^{\chi(t)}$ with $\chi(t \geq 4\mu^2) \geq 0$, are not forbidden by unitarity either.

• such singularities have to be present.

What can be said about them? It turns out, quite a lot.

Let us assume that the amplitude at large s is polynomially bounded uniformly in t, that is $A_1^{\pm}(s,t) < s^{\ell_0}$, $\ell_0 = \text{const}$ for any t.

Then for $\ell > \ell_0$ the integral in (3.1) is convergent and $f_\ell(t)$ has no singularities for any t.

Notice that up to this point we did not discuss the convergence properties of the integrals but rather only local properties of the Mandelstam representation (the location of the singularities).

Since a moving singularity in ℓ, $\ell = \ell(t)$ leads in turn to a singularity in $t = t(\ell)$, then for $\ell > \ell_0$, $f_\ell(t)$ has no moving singularities in t. The fact that $\phi_\ell(t)$ for $\ell > \ell_0$ has no singularities means that they are located on other sheets. When ℓ decreases, the singularities move from these sheets to the first sheet. So to learn what kind of singularities can appear on the physical sheet for $\ell < \ell_0$ it is necessary to know the content of other sheets.

First statement: a moving singularity cannot appear from beneath the left cut. To prove this one needs to show that for any ℓ there are no singularities on the unphysical sheets connected with the left cut:

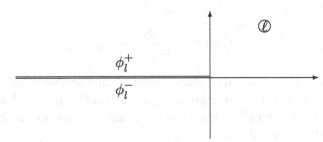

$$\phi_\ell^+(t) = \phi_\ell^-(t) + \Delta\phi_\ell(t), \qquad \phi_\ell^{\pm}(t) = \phi_\ell(t \pm \mathrm{i}\varepsilon).$$

Let us take here $t < 0$ and give it a negative imaginary part. Then $\phi_\ell^-(t)$ will stay on the physical sheet, while $\phi_\ell^+(t)$ will be moving to the unphysical sheet under the left cut, exploring its content. Since $\phi_\ell^-(t)$ is a regular function for complex t (physical sheet), we could encounter a moving singularity of $\phi_\ell(t)$ on the unphysical sheet ($\phi_\ell^+(t)$) only if $\Delta\phi_\ell(t)$ had such a singularity at some $t = t'(\ell)$.

But actually $\Delta\phi_\ell(t)$ has no moving singularity. This follows from the structure of the expression for the discontinuity:

$$\Delta\phi_\ell(t) = -\frac{1}{2}\int_{4\mu^2}^{s_b} P_\ell\left(\frac{2s}{4\mu^2 - t - \mathrm{i}\varepsilon} - 1\right) A_1^{\pm}(t + \mathrm{i}\varepsilon, s)\frac{2\,\mathrm{d}s}{(4\mu^2 - t)^{\ell+1}}$$

$$+ \int_{s_c}^{s_d} Q_\ell\left(\frac{2s}{4\mu^2 - t + \mathrm{i}\varepsilon} - 1\right) \rho_{su}(s,t)\frac{2\,\mathrm{d}s}{(4\mu^2 - t)^{\ell+1}}, \qquad (3.13)$$

The integration in (3.13) is performed over a finite region, so the integral converges and cannot generate singularities. As for an explicit ℓ-dependence, P_ℓ is an entire function of ℓ whereas the only singularities of Q_ℓ are poles at negative integer $\ell \leq -1$. Therefore all singularities of $\Delta\phi_\ell(t)$ in the ℓ plane are fixed poles. This completes the proof that moving singularities cannot emerge from under the left cut of the partial wave.

3.4 Moving poles and resonances

Hence all moving singularities of $f_\ell(t)$ come onto the physical sheet from the right cut which, as was argued above, is related to the physics of the t-channel.

Let us show now that for $16\mu^2 > t > 4\mu^2$ (where the elastic unitarity condition is applicable) only *poles* can appear.

Indeed, solving the equation

$$\phi_\ell^+ - \phi_\ell^- = 2\mathrm{i}\, C_\ell \phi_\ell^+ \phi_\ell^-, \qquad \phi_\ell^\pm = \phi_\ell(t \pm \mathrm{i}\,\varepsilon),$$

we obtain

$$\phi_\ell^+(t) = \frac{\phi_\ell^-}{1 - 2\mathrm{i}\, C_\ell \phi_\ell^-}. \tag{3.14}$$

Now we repeat the same procedure: move t in the lower half-plane, so that ϕ_ℓ^- (for $\ell > \ell_0$) stays regular (amplitude on the physical sheet) and $\phi_\ell^+(t)$ represents the amplitude on the first unphysical sheet related to the first, two-particle, threshold.

Hence the singularities of ϕ_ℓ on the first unphysical sheet appear when $\phi_\ell(t)$ on the physical sheet becomes equal to $1/(2\mathrm{i}\, C_\ell)$ at some point $t = t(\ell)$. These singularities are poles.

If such a pole on the unphysical sheet is close to the cut it can be identified as a resonance. Moreover, the trajectory of the pole $t = t(\ell)$ for all $\ell = 2n$ (if $\phi_\ell(t)$ was chosen with positive signature) describes a chain of resonances (provided they do not move too far away from the real axis).

Let us decrease ℓ below ℓ_0. The pole $t(\ell)$ may then move onto the physical sheet through the tip of the cut at $t = 4\mu^2$. If then ℓ should approach an integer ℓ (for instance $\ell = 2$), then in the real physical amplitude $f_\ell(t)$ the pole would appear, which corresponds to a pole in the scattering amplitude, i.e. to a bound state (with spin ℓ). If such bound states exist, then a single curve $t = t(\ell)$ can describe a whole set of physical particles – bound states and resonances.

To give an example, consider the partial wave of the electron scattering in the Coulomb field:[†]

$$f_\ell(t) = \frac{\Gamma(\ell + 1 + i/k)}{\Gamma(\ell + 1 - i/k)}. \qquad (3.15)$$

Its poles correspond to the poles of $\Gamma(\ell + 1 + i/k)$, which appear at integer negative values of its argument: $\ell + 1 + (i/k) = -n_r$, or $k = i/(\ell + n_r + 1)$. This equation gives the energy levels of the hydrogen atom. Thus, in complete agreement with the statements given above, the location of the pole of the partial wave amplitude defines the set of energy levels of the hydrogen atom with a given principal quantum number n_r.

If we tried to get onto unphysical sheets related to multi-particle thresholds using *inelastic* unitarity conditions with more than two particles in the intermediate state, we would encounter a lot of different things. Up to now this problem is not completely solved. Nevertheless the question what is the content of the other sheets can be answered for integer ℓ. It turns out that we encounter nothing unexpected there.

The cut from $4\mu^2$ to $16\mu^2$ is related to the Feynman graphs describing elastic scattering, with only two on-mass-shell particles in the intermediate state:

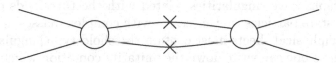

For $t > 16\mu^2$ apart from these diagrams there exist processes with four real particles in the intermediate state:

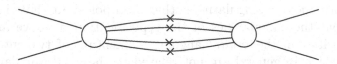

Among them there are Feynman graphs describing *rescattering* of two particles shown in Fig. 3.3.

[†] Note that the normalization of the non-relativistic amplitude f in (3.15) differs from the relativistic normalization that is being used elsewhere in the text.

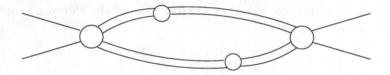

Fig. 3.3.   Two-particle rescattering in the four-particle intermediate state

Since the $2 \to 2$ block for $t_1 > 4\mu^2$ may contain a pole at some complex t (resonance), among the Feynman diagrams of Fig. 3.3 there will be the graphs of the type shown in Fig. 3.4.

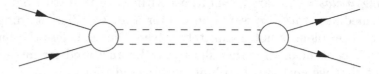

Fig. 3.4.   Two-resonance exchange generating complex threshold singularities on the unphysical sheet related with four-particle intermediate states $(t > 16\mu^2)$.

These graphs have singularities related with the thresholds of two intermediate state particles (in this case – with complex masses). Therefore the other unphysical sheets must contain threshold-type singularities.

For integer ℓ one can write down the unitarity condition which accounts for the four-particle intermediate state, and see that the new singularities are in fact *branch points*, generated by the poles. The word *generated* means that if poles were absent, then these branch points would disappear as well.

In contrast, for non-integer ℓ such a programme has not yet been accomplished. During 1961–1963 the hypothesis prevailed that $\phi_\ell(t)$ in general does not have moving singularities other than poles. At that time people believed that the singularities of the type mentioned above are due to the singularities in the *orbital* angular momentum of two intermediate particles, which in general are not related to the *total* angular momentum of the initial particles. When varying ℓ, the thresholds jump from one integer ℓ to another (together with masses of resonances) whereas at non-integer ℓ they turned out to be moving continuously.

The hypothesis is that for non-integer ℓ the structure of singularities is the same as for the integer case, namely moving poles and related branchings.

Up to 1969 no other singularities were found.

4

Properties of Regge poles

In the previous lecture we have shown that on the second sheet of the complex plane there could be only poles. One can show that on the third sheet there are only poles and branch points, corresponding to an exchange of two resonances.

4.1 Resonances

Let us discuss the properties of poles. Near the singularity a partial wave amplitude can be represented as a sum of a pole and a regular ('background') term:

$$f_\ell = \frac{r(t)}{\ell - \alpha(t)} + \tilde{f}_\ell.$$

The total amplitude reads

$$A^\pm(t, z) = \frac{\mathrm{i}}{4} \int \frac{\mathrm{d}\ell}{\sin \pi \ell} (2\ell + 1) f_\ell^\pm [P_\ell(-z) \pm P_\ell(z)]. \tag{4.1}$$

For $t < 4\mu^2$ the amplitude f_ℓ is real and therefore $r(t)$ and $\alpha(t)$ are also real.

Let us consider now $t > 4\mu^2$. Let us first show that there are no poles on the real axis. To do this we shall use the unitarity condition for the function

$$\phi_\ell(t) = \frac{\rho(t)}{\ell - \alpha(t)} + \tilde{\phi}_\ell, \qquad \rho(t) = \frac{r(t)}{(t - 4\mu^2)^{\alpha(t)}}. \tag{4.2}$$

This condition has the form

$$\frac{1}{2\mathrm{i}} \left[\phi_\ell^+(t) - \phi_\ell^-(t) \right] = C_\ell(t) \phi_\ell^+(t) \phi_\ell^-(t). \tag{4.3}$$

51

The superscripts $\pm$ in (4.3) denote that t has positive or negative imaginary parts, respectively; $\phi_\ell^-(t)$ is the complex conjugate of $\phi_\ell^+(t)$. Then we have

$$\phi_\ell^+(t) - \phi_\ell^-(t) \simeq \frac{\rho^+}{\ell - \alpha^+} - \frac{\rho^-}{\ell - \alpha^-} = \frac{\rho^+(\ell - \alpha^-) - \rho^-(\ell - \alpha^+)}{(\ell - \alpha^+)(\ell - \alpha^-)}$$

$$= 2\mathrm{i}\,C_\alpha \frac{\rho^+\rho^-}{(\ell - \alpha^+)(\ell - \alpha^-)}. \tag{4.4}$$

We have kept in $\phi_\ell^+(t)$ and $\phi_\ell^-(t)$ only the pole terms, and did not take into account finite corrections. This is legitimate only if ℓ is close to both α^+ and α^- simultaneously, i.e. if the imaginary part $\mathrm{Im}\,\alpha$ is small. The denominators on the l.h.s. and r.h.s. of the last equality in (4.4) are the same. The numerator on the r.h.s. does not depend on ℓ. Hence, the numerator on the l.h.s. should not depend on ℓ either. That means

$$\rho^+ = \rho^- = \rho,$$

and (4.4) takes the form

$$\rho(\alpha^+ - \alpha^-) \simeq 2\mathrm{i}\,\rho^2 C_\alpha, \qquad \mathrm{Im}\,\alpha = \rho C_\alpha. \tag{4.5}$$

We see that $\mathrm{Im}\,\alpha \neq 0$. So we have proved that there are no poles on the real axis in the ℓ plane for $t > 4\mu^2$.

Substituting ρ from (4.5) into the expression for the partial wave amplitude (4.2), we obtain

$$\phi_\ell = \frac{\mathrm{Im}\,\alpha}{\ell - \mathrm{Re}\,\alpha - \mathrm{i}\,\mathrm{Im}\,\alpha} \cdot \frac{1}{C_\ell}. \tag{4.6}$$

Let us assume that for some $t = t_1 \equiv M^2 > 4\mu^2$, $\mathrm{Re}\,\alpha(t_1) = n$; then, expanding $\mathrm{Re}\,\alpha$ in a series in $(t - t_1)$, we obtain

$$\phi_n(t) \simeq \frac{\mathrm{Im}\,\alpha}{n - \mathrm{Re}\,\alpha(t_1) - \mathrm{Re}\,\alpha'(t_1)(t - t_1) - \mathrm{i}\,\mathrm{Im}\,\alpha(t_1)} \cdot \frac{1}{C_n}$$

$$= \frac{\mathrm{Im}\,\alpha/\mathrm{Re}\,\alpha'(t_1)}{t_1 - t - \mathrm{i}\,\mathrm{Im}\,\alpha/\mathrm{Re}\,\alpha'(t_1)} \cdot \frac{1}{C_n}.$$

In spite of the fact that we are close to the pole, we do not get an infinity, because of the presence of $\mathrm{Im}\,\alpha$ in the denominator of $\phi_n(t)$.

In the vicinity of the pole we can approximate $t_1 - t \equiv M^2 - E^2 \simeq 2M(M - E)$ and write

$$\phi_n(E) \simeq \frac{\Gamma/2}{M - \mathrm{i}\Gamma/2 - E} \cdot \frac{1}{2MC_n}.$$

This is the Breit–Wigner formula.

So we see that a Regge pole $n = \alpha(t)$ at an integer angular momentum n corresponds to a resonance (unstable particle) with the mass $M = \sqrt{t_1}$, $\alpha(t_1) = n$, and the width

$$\Gamma = \frac{\operatorname{Im}\alpha(M^2)}{M\operatorname{Re}\alpha'(M^2)}.$$

4.2 Bound states

Partial wave $\phi_\ell(t)$ at $t < 4\mu^2$ is real. So if for some integer n it has a pole at $t = M^2 < 4\mu^2$ then ρ and α in the expression $\phi_\ell = \rho/(\ell - \alpha(t))$ are real. The partial wave amplitude then has the form

$$\phi_n = \frac{\rho}{\alpha'(M^2)\,(M^2 - t)}, \quad M^2 < 4\mu^2.$$

The partial wave amplitude ϕ_n having a pole at some energy below the threshold corresponds to a bound state (stable particle) with angular momentum n. That leads to the graph

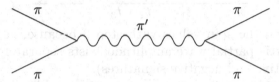

and on the Mandelstam plane there will be a pole at $t = t_2 = M_{\pi'}^2$ and thresholds in t corresponding to the possibility of creating two, three, etc. π' particles. If $M_{\pi'} > m_\pi$, then the new thresholds will be above the elastic one. (In reality, this bound state particle cannot be the π meson, because it has the quantum numbers of a system of two pions, which are different from the quantum numbers of a single pion.)

One more remark is needed: in the non-relativistic theory $\alpha'(t)$ should be positive, $\alpha'(t) > 0$, because there is the Schrödinger equation with a centrifugal barrier, and with increasing ℓ the barrier also increases and so does the mass of the bound state.

4.3 Elementary particle or bound state?

In the non-relativistic theory a pole in the scattering amplitude of scattering of *elementary particles* has a clear meaning of a bound state or a resonance. In the relativistic theory it is difficult to claim that some particle is a bound state of other elementary particles, because *any* real state contributes a pole to the scattering amplitude.

One may ask if it is possible to verify experimentally whether a given particle is an elementary one or a bound state.

4.3.1 Regge trajectories

Let us consider a set of particles (some of them may be unstable) with squared masses t_0, t_1, t_2, ... and spins n_0, $n_0 + 2$, $n_0 + 4$, ... ($n_0 = 0, 1$), respectively (we suppose that all other quantum numbers are the same). Then it is natural to assume that these particles are described by a single Regge pole trajectory $\ell = \alpha(t)$ that at $t = t_k$ passes (close to) the integer points $n_0 + 2k$ (the imaginary part of $\alpha(t)$ is supposed to be small).

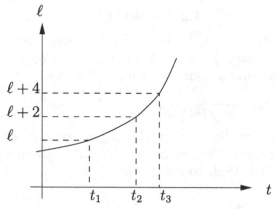

We have taken the spin values with step 2 because, as we have seen above, an analytic partial wave amplitude exists separately for even and odd spins (positive and negative signatures).

We saw that a Regge trajectory may contain stable particles – bound states – together with resonances. It is natural to assume conversely that any stable particle lying on a Regge trajectory is a bound state. In fact this is an operational definition of non-elementary (bound state) particles in relativistic theory.

4.3.2 Regge pole exchange and particle exchange (t > 0)

Where else does the Regge trajectory show itself? Let us consider a s-channel amplitude which is governed by some Regge pole. Substituting the pole term into expression (4.1) for the amplitude, we obtain

$$A^\pm = -\frac{\pi}{2} \frac{(2\alpha(t) + 1)r(t)}{\sin(\pi\alpha(t))} \left[P_{\alpha(t)}(-z) \pm P_{\alpha(t)}(z) \right].$$

What properties of the amplitude can we see in this formula?

We consider first the case of positive signature assuming that α passes through $2n$. At this point $P_\ell(-z)$ and $P_\ell(z)$ coincide, $P_{2n}(-z) = P_{2n}(z)$, and $\sin \pi\ell$ is zero. If we represent $\alpha(t)$ as $\alpha(t_1) + (t - t_1)\alpha'$, then

$$A^+ \simeq \frac{(2\alpha + 1)r(t)}{\alpha'(t_1 - t)} P_\alpha(z) \qquad (\alpha = 2n).$$

The amplitude A^+ has a pole at $t = t_1$. If we assume that $\alpha(t_1) = 0$, then $P_\alpha = 1$, and $A^+ = C/(t_1 - t)$. The Feynman graph corresponding to the exchange of a scalar particle in the t-channel (see Fig. 4.1) gives exactly

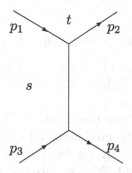

Fig. 4.1.   *t*-Channel particle exchange graph

the same expression. If, for example, $\alpha = 2$, then $A^+ = C_2 P_2(z)/(t_2 - t)$. It is easy to check that in the Feynman graph describing the exchange of a spin-2 particle, the tensor Green function of the particle, upon multiplication by the vertex functions, gives precisely $P_2(z)$.

If $\alpha = 2n+1$ and the signature is positive, then the sum $P_\ell(-z) + P_\ell(z)$ vanishes together with $\sin \pi \alpha$). In order to resolve this indeterminacy, we take into account that $P_\alpha(-(z + i\varepsilon)) = e^{-i\pi\alpha} P_\alpha(z)$. As a result we get

$$A^+ \sim -P_\alpha(z) \frac{e^{-i\pi\alpha} + 1}{\sin \pi \alpha} = P_\alpha(z)\left[i - \cot \frac{\pi\alpha}{2} \right].$$

We see that when α is an *odd* number, the positive signature amplitude $A^+(s, t)$ does not have any poles and is *purely imaginary*. Let us remark that for $\alpha = 1$ we would have $A^+ \propto i s$ at high energies.

We consider now a negative signature amplitude. Then it has the pole $A^- \sim P_\alpha(z)/(t_1 - t)$ near $\alpha(t_1) = 2n + 1$. For example, if $\alpha(t)$ passes through 1 at $t = t_1$, then the amplitude takes the form

$$A^- \propto \frac{P_1(z)}{t_1 - t} = \frac{z}{t_1 - t}, \qquad z \equiv z_t = 1 + \frac{2s}{t - 4\mu^2} = \frac{s - u}{t - 4\mu^2}, \qquad (4.7a)$$

so that for large energies $A^- \propto s/(t_1 - t)$.

Let us verify that this amplitude corresponds to vector particle exchange. A vector particle (with mass M) in the Feynman graph of Fig. 4.1 is described by the Green function proportional to $g_{\mu\nu}$, and there are conserved currents $(p_1 + p_2)_\mu$ and $(p_3 + p_4)_\nu$ at the vertices. In this case we obtain

$$A \propto \frac{(p_1 + p_2)_\mu (p_3 + p_4)_\mu}{M^2 - t} = \frac{s - u}{M^2 - t} \propto z \propto A^-[(4.7\text{a})]. \quad (4.7\text{b})$$

4.3.3 Regge exchange and elementary particles (t < 0)

Let us now pass to the region of negative t. If our pole $\alpha(t)$ is the rightmost singularity of $f_\ell(t)$ in the ℓ plane, then it will govern the asymptotics of the scattering amplitude.

As we have discussed above (see Lecture 2), analytic partial wave amplitudes $f_\ell^\pm(t)$ can be constructed using (the Gribov–Froissart projection) (2.37) only for $\operatorname{Re} \ell > \ell_0$, so that a few partial waves with $n \leq \ell_0$ are not included in the Sommerfeld–Watson integral and may not lie on Regge trajectories; see (2.39).

This is an important point, because it is related to the notion of elementarity of particles.

Let us assume that the amplitude of one of these *excluded* partial waves, f_0 for instance, has a pole $f_0 = C/(m^2 - t)$. This is a typical Born amplitude for particle exchange in the QFT framework. Therefore, in such a situation we would say that there is an elementary scalar particle in the spectrum of the theory.

The Born amplitude for the exchange of an elementary scalar particle reads

$$A = \frac{g^2}{m^2 - t}.$$

The corresponding partial wave amplitude is

$$f_n = \frac{1}{2} \int_{-1}^{1} A P_n(z) \, \mathrm{d}z = \frac{g^2}{m^2 - t} \delta_{n0}.$$

As we already know, the same amplitude can be obtained from the Regge pole, but only in the region close to $t = m^2$, whereas for other values of t the Regge amplitude would be different. Most importantly, its dependence on s will be different for different t while the elementary scalar gives the amplitude that is constant in s for all values of t.

The question arises: are there in nature truly elementary particles in the sense that in the partial wave amplitude they are represented by a term of the type $f_\ell \delta_{\ell n_0}$?

We are ready to show that there are no elementary particles with spin greater than 1. One can also almost always establish the fact that a given particle is *not* elementary. The converse statement is wrong.

4.3.4 There is no elementary particle with J > 1

Let us show that any partial wave amplitude with $n > 1$ lies on the Regge trajectory, that is all particles with spins $J > 1$ are necessarily non-elementary – bound states or resonances. To this end we shall use the fact that $A < Cs \ln^2 s$ for $t < 0$ by virtue of the Froissart theorem.

We have assumed that the amplitude is polynomially bounded, $A < s^N$, for $t > 4\mu^2$ and $s \to \infty$, and combined partial waves with sufficiently large angular momenta into an analytic function:

$$f_\ell^\pm(t) = \frac{1}{\pi} \int_{z_0}^{\infty} Q_\ell(z) A_1^\pm \, dz, \qquad \ell > N. \tag{4.8a}$$

For small angular momenta (non-reggeized partial waves) we have to use the standard definition

$$\bar{f}_n = \frac{1}{2} \int_{-1}^{1} A P_\ell(z) \, dz, \qquad \ell \leq N. \tag{4.8b}$$

We continue now $f_\ell(t)$ given in (4.8a) to the physical region of the s-channel, $t < 0$, where $A < Cs \ln^2 s$. Then for any $n > 1$ we may deform the contour and close it around the cut $[-1, 1]$ of Q_n:

$$f_\ell|_{\ell=n} = \frac{1}{\pi} \int_{z_0}^{\infty} Q_n(z) A_1(s,t) \, dz = \frac{1}{2} \int_{-1}^{1} P_\ell(z) A(s,t) \, dz \equiv \bar{f}_n.$$

Since $f_\ell|_{\ell=n}$ coincides with the actual partial wave amplitude $\bar{f}_n$, (4.8b), for $t < 0$ then, by virtue of analyticity in t, these functions are equal everywhere.

Non-elementarity of particles with spin $J > 1$ is in correspondence with the fact that one can construct a renormalizable QFT only containing particles with spins 0, 1/2 and 1; for fields with higher spins a renormalizable theory has not been constructed.

4.3.5 Asymptotics of s-channel amplitudes and reggeization

Let us take some $2 \to 2$ process and assume that experiment would give $A(s,t) \to 0$ with $s \to \infty$, at least for some (sufficiently) negative t. Then we would be able to perform the above calculation and identify f_ℓ with $\bar{f}_n$ for all values of ℓ and show that among the hadrons with corresponding quantum numbers that could have been exchanged in the t-channel there are no elementary particles at all. This is an experimental way to determine whether a particle is elementary or not.

For example, in the meson–proton scattering amplitude there is a graph corresponding to neutron exchange:

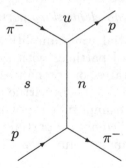

If the $\pi^- p$ scattering amplitude falls off with s for fixed u, then we can conclude that the neutron lies on a Regge trajectory.

Experiment shows indeed that there is no elementary particle among hadrons: spin-0 mesons (like π), spin-$\frac{1}{2}$ particles (nucleon) and vector mesons (like ρ) all lie on proper Regge trajectories.

4.4 Factorization

If there exist several states with the same quantum numbers, for instance $\pi\pi$ and $K\bar{K}$, then we have several reactions, corresponding to transitions between these states:

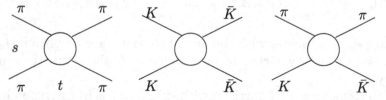

If there is a real particle with the same quantum numbers, then all three amplitudes will contain the pole corresponding to the exchange of this particle:

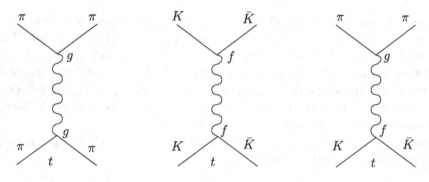

In the QFT framework, the following amplitudes correspond to these three graphs:

$$\frac{r_{11}}{t-t_1} = \frac{gg}{t-t_1}, \qquad \frac{r_{22}}{t-t_1} = \frac{ff}{t-t_1}, \qquad \frac{r_{12}}{t-t_1} = \frac{gf}{t-t_1}.$$

It is obvious from the above that $r_{11}r_{22} = r_{12}^2$. This property reflects the fact that the initial state knows nothing about the final one: the initial particles first transform into an intermediate state, which then gets converted into the final particles, with the amplitude independent of the properties of the initial state.

Let us show that Regge poles have the same property. If we write down the contribution of the Regge poles to our three amplitudes in the form

$$f_\ell^{11} = \frac{r_{11}(t)}{\ell - \alpha(t)}, \qquad f_\ell^{12} = \frac{r_{12}(t)}{\ell - \alpha(t)}, \qquad f_\ell^{22} = \frac{r_{22}(t)}{\ell - \alpha(t)},$$

then, by virtue of the unitarity condition for the residues of the poles, there exist the same relations, and it is natural to write

$$f_\ell^{11} = \frac{g_1^2(t)}{\ell - \alpha(t)}, \qquad f_\ell^{12} = \frac{g_1(t)g_2(t)}{\ell - \alpha(t)}, \qquad f_\ell^{22} = \frac{g_2^2(t)}{\ell - \alpha(t)}.$$

In order to show that, we will consider these three amplitudes in the region $4\mu^2 < t < 16\mu^2$. Then all amplitudes have a system of two pions as a real intermediate state. A real intermediate state gives rise to an imaginary part of the amplitudes, which can be derived from the unitarity condition. Note that the fact that the region of values of t is unphysical for the external state $K\bar{K}$ is not essential.

One can show, either by continuation from the region of higher t, which are physical for a $K\bar{K}$ system, or by straightforward evaluation of Feynman graphs, that the existence of imaginary parts in these reactions is related only to the properties of the intermediate states. Such a continuation of the unitarity condition yields the following expressions for the imaginary parts:

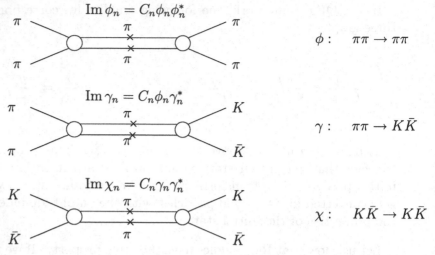

By continuation of the unitarity conditions from integer values n to complex ℓ we obtain

$$\frac{\phi_\ell^+ - \phi_\ell^-}{2\mathrm{i}} = C_\ell \phi_\ell^+ \phi_\ell^-, \qquad \frac{\gamma_\ell^+ - \gamma_\ell^-}{2\mathrm{i}} = C_\ell \phi_\ell^+ \gamma_\ell^-, \qquad \frac{\chi_\ell^+ - \chi_\ell^-}{2\mathrm{i}} = C_\ell \gamma_\ell^+ \gamma_\ell^-,$$

from where it follows that

$$\phi_\ell^+ = \frac{\phi_\ell^-}{1 - 2\mathrm{i}\,\phi_\ell^- C_\ell}, \qquad \gamma_\ell^+ = \frac{\gamma_\ell^-}{1 - 2\mathrm{i}\,\phi_\ell^- C_\ell}, \qquad \chi_\ell^+ = \chi_\ell^- + 2\mathrm{i}\,C_\ell \frac{(\gamma_\ell^-)^2}{1 - 2\mathrm{i}\,\phi_\ell^- C_\ell}.$$

Every one of the three amplitudes acquires a pole at the same point, where $1 - 2\mathrm{i}\,\phi_\ell^- C_\ell$ becomes zero. In the vicinity of this point we have

$$\phi_\ell^- \simeq \frac{1}{2\mathrm{i}\,C_\ell}, \qquad \frac{1}{1 - 2\mathrm{i}\,\phi_\ell^- C_\ell} \simeq \frac{1}{\beta(t - t_1)},$$

and

$$\phi_\ell^+ = \frac{1}{\beta(t - t_1)}\frac{1}{2\mathrm{i}\,C_\ell}, \qquad \gamma_\ell^+ = \frac{\gamma_\ell^-}{\beta(t - t_1)}, \qquad \chi_\ell^+ = 2\mathrm{i}\,C_\ell \frac{(\gamma_\ell^-)^2}{\beta(t - t_1)},$$

i.e. $r_{11}r_{22} = r_{12}^2$, which was required to be proved.

5

Regge poles in high energy scattering

The contribution of a Regge pole to the scattering amplitude at large s may be written in the following form:

$$A(s,t) = -r(t)\frac{(-s)^{\alpha(t)} \pm s^{\alpha(t)}}{\sin \pi\alpha(t)} = r(t)s^{\alpha(t)}\xi_\alpha, \qquad (5.1a)$$

where

$$\xi_\alpha = -(e^{-i\pi\alpha} \pm 1)/\sin \pi\alpha. \qquad (5.1b)$$

In the previous lecture we have proved that the residue $r(t)$ may be written as a product of two functions each of which is related only to one vertex:

$$r(t) = g_{ab}(t)\, g_{cd}(t).$$

This factorization property means that one may associate the amplitude to the diagram

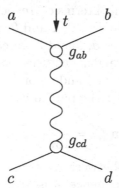

5.1 t-Channel dominance

Consider now the important property of Regge pole amplitudes, known as 't-channel dominance'.

61

Consider for instance the reaction involving π mesons,

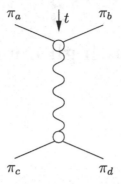

The π-meson exists in three charge states, so we could consider different amplitudes corresponding to various possible charge states of the initial and final pions. However, the charge independence of strong interactions means that some of the amplitudes may be expressed in terms of others.

Writing π in the form of a vector (π_1, π_2, π_3), we must demand invariance of the amplitude A under rotations in isospin space. From two pions one can construct three covariants: scalar, vector and second rank tensor. The amplitude A is the superposition of scalars built from these covariants:

$$A = (\boldsymbol{\pi}_c \boldsymbol{\pi}_d)(\boldsymbol{\pi}_a \boldsymbol{\pi}_b)\, a(s,t) + (\boldsymbol{\pi}_c \times \boldsymbol{\pi}_d)(\boldsymbol{\pi}_a \times \boldsymbol{\pi}_b)\, b(s,t)$$

$$+ \left[\pi_{a\alpha}\pi_{b\beta} + \pi_{a\beta}\pi_{b\alpha} - \frac{2}{3}\delta_{\alpha\beta}(\boldsymbol{\pi}_a \boldsymbol{\pi}_b) \right]$$

$$\times \left[\pi_{c\alpha}\pi_{d\beta} + \pi_{c\beta}\pi_{d\alpha} - \frac{2}{3}\delta_{\alpha\beta}(\boldsymbol{\pi}_c \boldsymbol{\pi}_d) \right] c(s,t). \qquad (5.2)$$

Here the first, second and third terms correspond to initial and final states in the t-channel with isospin $T = 0$, 1 and 2, respectively.

For each of the amplitudes a, b and c we can perform all the manipulations considered in the previous lectures: expansion into partial waves, continuation of f_ℓ^T to complex ℓ and so on.

The unitarity condition for the partial wave amplitudes takes the form

$$\text{Im}\, f_\ell^T = \frac{k}{16\pi\omega} f_\ell^T f_\ell^{*T}.$$

The absence of transitions between different values of isospin in the unitarity condition is the consequence of isospin conservation.

In terms of ϕ_ℓ^T one can obtain the familiar expression for ϕ_ℓ^{+T} on the unphysical sheet:

$$\phi_\ell^{+T} = \frac{\phi_\ell^{-T}}{1 - 2i\, c_\ell \phi_\ell^{-T}}.$$

As before, we see that ϕ_ℓ^{+T} has a pole at $\phi_\ell^{-T} = 1/(2\mathrm{i}\,C_\ell)$. The scattering amplitude, and thus the partial wave, depend on the value of isospin. So for different values of T the poles will be located at different points of the ℓ plane even at the same t. In other words the poles are described by different functions, the form of which depends on isospin.

In particular, as a result of Bose statistics of pions, the amplitude must be invariant under the replacement $\pi_a \leftrightarrow \pi_b$. Under such transposition z is changed to $-z$ ($s \leftrightarrow u$). Since the invariant coefficients of a, c (b) are even (odd) under this replacement, the partial wave expansion will involve only even values of ℓ for $T = 0, 2$ and only odd values of ℓ for $T = 1$.

The difference between partial wave amplitudes with $T = 0$ or $T = 2$ on one side, and $T = 1$ on the other side, is therefore related to the difference in their signatures.

States with $T = 0$ and $T = 2$ have the same signature; the partial wave amplitudes will be nevertheless different.

Thus if there is some conservation law, then the Regge pole must be characterized by the corresponding conserved quantum number. Note that signature is a new quantum number which arises from the concept of complex angular momenta.

We can always classify the amplitude by the quantum numbers in the s channel. At large s, provided the amplitude is determined by a single rightmost Regge pole, it can also be characterized by definite quantum numbers in the t-channel. This is the essence of 't-channel dominance'.

In our example the asymptotics of the scattering amplitude has the following form:

$$A = as^{\alpha_0(t)} + bs^{\alpha_1(t)} + cs^{\alpha_2(t)}.$$

If all the α_i, $i = 0, 1, 2$, are different, then at large s only one of the amplitudes with definite isospin in the t-channel survives.

Proceeding from empirical data on the existing particles, one can in some cases predict which of the amplitudes will survive asymptotically.

For example suppose that a vector (spin-1) resonance with isospin $T = 1$ and mass $t = m_\rho^2$ is discovered. Then one may expect that a reggeon exists with the trajectory $\alpha_T^\sigma = \alpha_1^-(t)$ (σ and T are its signature and isospin, respectively). So asymptotically the amplitude must contain a term of the form $b(t)s^{\alpha_1^-(t)}$.

If there is no resonance with $T = 2$, then presumably the corresponding reggeon $\ell_T^\sigma = \alpha_2^+(t)$ does not exist either.

Let us recapitulate our results:

1. If reggeons exist, then their trajectories are in general different for different quantum numbers of the t-channel (cross-channel);

2. The power of the asymptotic behaviour of the amplitude is connected with the quantum numbers of hadron resonances;

3. If the amplitude has no resonance in the t-channel ($t > 4\mu^2$), then it falls rapidly at large s in the physical region of the s channel.

5.2 Elastic scattering and the pomeron

Consider now elastic scattering of two arbitrary particles. If the collision were purely diffractive, then the amplitude would be of the form $A = \mathrm{i}\, s f(t)$, with $\sigma_{\text{tot}} = f(0)$. We saw that such an amplitude corresponds to a fixed pole at $\ell = 1$ whose existence contradicts the unitarity condition.

Nevertheless one can obtain a constant cross section. It suffices to postulate the existence of a single Regge pole $\alpha_P^+(t)$ satisfying the condition

$$\alpha_P^+(0) = 1. \tag{5.3}$$

Indeed, in this case the amplitude is purely imaginary at small t:

$$A = -r(t)s^{\alpha(t)} \frac{e^{-\mathrm{i}\pi\alpha} + 1}{\sin \pi\alpha} = \mathrm{i}\, r(t)s^{\alpha(t)},$$

$$\sigma_{\text{tot}} = \frac{\text{Im}\, A(0)}{s} = r(0) = \text{const}.$$

Such a pole has become known as a 'Pomeranchuk pole' (pomeron) because it automatically satisfies his theorem about the asymptotic equality of the total cross sections of particle and antiparticle scattering on the same target.

5.2.1 Quantum numbers of the pomeron

The Pomeranchuk pole is also sometimes called a *vacuum pole* since all its quantum numbers coincide with those of the vacuum. This in fact follows from the condition $\sigma_{\text{tot}} = (1/s)\text{Im}\, A(0) > 0$.

For instance, the isospin T of the Pomeranchuk pole cannot be equal to 2 since the imaginary part of an amplitude with $T = 2$ in the t-channel is in general a non-positive function (see (5.2)). By similar reasoning one can show that it is impossible to exchange a state with negative charge parity and so on. It is also obvious that the particle associated with the Pomeranchuk pole cannot carry electric charge.

Gell-Mann has noted that, if a vacuum pole exists, then it is natural to expect the existence of a particle with vacuum quantum numbers (charge, isospin, parity, etc.) and spin equal to 2. At present we know two such particles – tensor mesons. They have different masses. Analysing the energy dependence of the total cross sections one finds indeed two trajectories with vacuum quantum numbers: one of the trajectories has $\alpha_P(0) = 1$ and the other has $\alpha_{P'}(0) = 1/2$.

Furthermore, since $\alpha_P(0) = 1$ and $\alpha'_P(0) > 0$ (and the signature is positive), we could have a particle with the *imaginary mass* $t_1 = -m_p^2$ at $\alpha_P(t_1) = 0$. Gell-Mann has shown that a requirement of vanishing of the *residue* at this point is natural, which leads to the absence of a particle with $J = 0$ and imaginary mass (tachyon).

5.2.2 Slope of the pomeron trajectory

Let us prove now that the trajectory of the vacuum pole has a positive derivative, $\alpha'_P(t) > 0$, in the interval $0 < t < 4\mu^2$. In the non-relativistic theory the proof of the similar statement is based on the existence of a Hamiltonian.

In relativistic theory we will use analytic properties of the scattering amplitude in momentum transfer. We will assume that

$$A_1(s,t) = \text{Im}\, A(s,t) = r(t)s^{\alpha_P(t)}, \qquad A_1(s,0) = s\sigma_{\text{tot}}. \qquad (5.4)$$

The condition $r(0) > 0$ follows from $\sigma_{\text{tot}} > 0$.

Let us expand A into s-channel partial wave amplitudes:

$$A_1(s,t) = \sum_{n=0}^{\infty} \text{Im}\, f_n(s)(2n+1)P_n(z_s), \quad z_s = 1 + \frac{2t}{s - 4\mu^2}.$$

We know from the unitarity condition that $\text{Im}\, f_n(s) > 0$. Furthermore at $t > 0$ we have $P_n(z_s) > 1$ since the argument z_s exceeds 1. It can be shown that the derivatives of the Legendre polynomials are also positive:

$$\frac{\text{d}}{\text{d}t}P_n(z_s) \equiv P'_n(z_s) > 0,$$

from which it follows that $A'_1(s,t) > 0$. Calculating the derivative of $A_1(s,t)$ in the Regge pole form (5.4) we obtain

$$A'_1(t) = r(t) \cdot \alpha'_P(t)\ln s \cdot s^{\alpha_P(t)} + r'(t)\, s^{\alpha_P(t)} > 0.$$

The second term is asymptotically smaller than the first one, so taking into account the positiveness of $r(0)$ we obtain $\alpha'_P(t) > 0$.

For inelastic reactions, which can be dominated by non-vacuum poles, the partial wave amplitudes $\text{Im}\, f_n(s)$ have, in general, alternating signs. Therefore the condition $\alpha' > 0$ for non-vacuum poles cannot be proved.

5.3 Shrinkage of the diffractive cone

Let us elucidate now the properties of the scattering in the presence of a single vacuum pole. In the region of small t one can expand the trajectory to obtain

$$\frac{d\sigma_{el}}{dt} \propto r^2(t)\, e^{2\alpha' \xi t},$$

$$\alpha(t) \simeq 1 + \alpha' t, \qquad \xi \equiv \ln \frac{s}{\mu^2}. \tag{5.5}$$

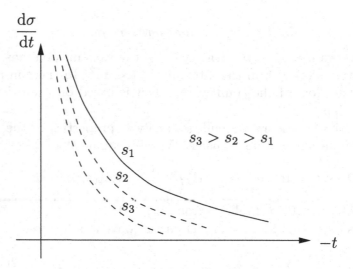

Fig. 5.1.   Shrinkage of the diffractive cone

From (5.5) and Fig. 5.1 we see that with increasing s the diffractive cone becomes *narrower* (in contrast with the ordinary diffraction picture) and the elastic cross section integrated over scattering angle, $-t \simeq k_s^2 \Theta^2$,

$$\sigma_{el} = \int \frac{d\sigma_{el}}{dt}\, dt \sim \frac{1}{\xi},$$

falls with energy as an inverse power of $\ln s$.

5.3.1 s-Channel partial waves in the impact parameter space

Let us estimate characteristic impact parameters that are essential for high energy scattering. Usually it is believed that scattering is due to exchange of virtual particles, and the essential impact parameters are limited by some finite value $1/\mu$ of the order of the Compton wavelength of the lightest hadron (π meson). For the vacuum pole we find, however, an essentially different result.

Indeed, the Pomeranchuk pole gives

$$A_1 = r(t)s^{\alpha(t)} \simeq r(0)\, s \exp(\alpha' \xi t). \qquad (5.6a)$$

On the other hand, from the s-channel partial wave expansion for the absorptive part of the amplitude we have (see Lecture 1):

$$A_1 = \sum_{\ell}(2\ell + 1)\mathrm{Im}\, f_{\ell}(s)P_{\ell}(\cos\theta)$$

$$\simeq 2k_s^2 \int \rho\, \mathrm{Im}\, f(\rho, s)\, J_0(\rho\sqrt{-t})\, \mathrm{d}\rho = \frac{s}{2}\int \mathrm{Im}\, f(\rho, s)\, e^{i\,\mathbf{q}\cdot\boldsymbol{\rho}}\, \frac{\mathrm{d}^2\rho}{2\pi}, \qquad (5.2)$$

where $\rho = \ell/k_s$, $t = -\mathbf{q}^2$. Performing the inverse Fourier transform and substituting (5.6a) for A_1 we obtain

$$\mathrm{Im}\, f(\rho, s) \simeq \frac{r(0)}{\pi}\int e^{-\alpha'\xi \mathbf{q}^2 - i\,\mathbf{q}\cdot\boldsymbol{\rho}}\, \mathrm{d}^2 q = \frac{r(0)}{\alpha'\xi}\exp\left(-\frac{\rho^2}{4\alpha'\xi}\right). \qquad (5.7)$$

One can easily show that under the approximations made, the region of validity of (5.7) is $\rho \lesssim \xi\alpha'(0)\mu$. The dependence of this function on ρ is shown in Fig. 5.2.

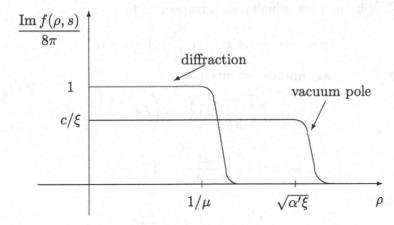

Fig. 5.2. Impact parameter dependence of $\mathrm{Im}\, f(\rho, s)$

Thus one can say that the particle swells and becomes grey. Let us discuss now how this may be reconciled with the old picture. The interaction potential may be obtained as the Fourier transform of the amplitude with a t-channel pole. So for the exchange of a spinless particle (pion) we have

$$\frac{1}{\mu^2 - t} \implies V(r) \sim \int \frac{\mathrm{d}^3 q}{\mu^2 + \mathbf{q}^2}\exp\left(i\,\mathbf{q}\cdot\mathbf{r}\right) \sim \frac{\exp\left(-\mu r\right)}{r}.$$

When the exchanged particle has spin 1, the singular amplitude has the form of $s/(\mu^2 - t)$, and the corresponding potential linearly increases with energy:

$$V(r) \sim s\frac{\exp(-\mu r)}{r} \propto E\frac{\exp(-\mu r)}{r},$$

with E the energy of one of the initial particles (projectile) in the rest frame of the other (target), $s = 2\mu E$. Similarly if the exchanged particle has spin ℓ, we obtain $V \sim s^\ell \exp(-\mu r)/r$. Given that the potential increases with energy, the range of distances in which the interaction remains strong, $rV(r) \sim 1$, grows with energy too: $r \lesssim r_0 \sim (\ell/\mu)\ln s = (\ell/\mu)\xi$.

Consider now the behaviour of the partial wave amplitude $f(\rho, s)$ in the region of very large impact parameters, $\rho \to \infty$, where the Gaussian expression (5.7) is no longer valid. Let us use the following expression for the s-channel partial wave amplitude:

$$f_\ell(s) = \frac{1}{\pi}\int_{4\mu^2}^\infty Q_\ell\left(1 + \frac{2t}{s - 4\mu^2}\right) A_3(s,t)\frac{2\,dt}{s - 4\mu^2}. \qquad (5.8)$$

This representation is analogous to (3.1) for the t-channel partial waves. A_3 in (5.8) is the discontinuity with respect to t:

$$A_3(s,t) \equiv \frac{1}{2i}\left[A(s,t + i\varepsilon) - A(s,t - i\varepsilon)\right].$$

At large ℓ, Q_ℓ is asymptotically given by

$$Q_\ell \simeq \frac{\sqrt{\pi}}{\sqrt{2\ell \sinh\theta}}\exp\left\{-\alpha\left(\ell + \frac{1}{2}\right)\right\},$$

where

$$\cosh\theta = 1 + \frac{2t}{s - 4\mu^2} \equiv 1 + \frac{t}{2k_s^2}.$$

If s is large then

$$\theta \simeq \frac{\sqrt{t}}{k_s}, \qquad e^{-\theta\ell} \simeq e^{-\rho\sqrt{t}}.$$

Inserting this asymptotic form of Q_ℓ into (5.8) we find that for $\rho \to \infty$ the main contribution to the integral comes from the region near the lower limit of integration. Therefore taking $A_3(s,t)$ outside the integral at $t = 4\mu^2$ we get

$$f(\rho, s) \simeq \frac{2\sqrt{\mu}}{\sqrt{\pi}}\frac{A_3(s, 4\mu^2)}{s}\frac{e^{-2\mu\rho}}{\rho^{3/2}}.$$

Evidently the imaginary part $A_3(s, 4\mu^2)$ is defined by the two-particle cut of the diagram

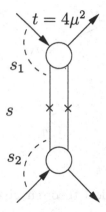

At large s the lower and upper blobs in this diagram must be taken in the asymptotic regime. Supposing that these blobs have Regge asymptotics of the form $s^{\alpha(t)}$ and taking into account that the energy invariants of the blobs, s_1 and s_2, obey the relation*

$$s_1 s_2 \simeq s \mu^2,$$

we finally obtain

$$f(\rho, s) \sim \frac{\exp(-2\mu\rho)}{\sqrt{\rho^3}} s^{\alpha(4\mu^2)-1} = \rho^{-\frac{3}{2}} \exp\left[-2\mu\left(\rho - \frac{\alpha(4\mu^2)-1}{2\mu}\xi\right)\right].$$

$$(5.9)$$

The region of validity of (5.9) is $\rho \gg \xi\alpha'(4\mu^2)\mu$. Thus, the fall-off of the partial wave with ρ changes from Gaussian, (5.7), to exponential, (5.9), at $\rho\mu \simeq \alpha'\mu^2 \cdot \ln s$.

5.4 Relation between total cross sections

By virtue of the factorization property, the vacuum pole leads to certain relations between the total cross sections. Consider for instance NN, πN and $\pi\pi$ scattering:

---

* This relation follows from the observation that in a renormalizable quantum field theory components of the exchanged particle momenta transversal to the direction of colliding particles are finite, $k_\perp = \mathcal{O}(\mu)$.

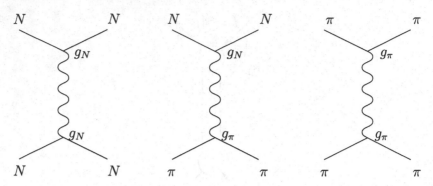

Factorization and the optical theorem give rise to the following relations:

$$\sigma^{\text{tot}}_{NN} = g^2_N, \qquad \sigma^{\text{tot}}_{\pi N} = g_N g_\pi, \qquad \sigma^{\text{tot}}_{\pi\pi} = g^2_\pi$$
$$\implies \quad \sigma^{\text{tot}}_{\pi\pi}\sigma^{\text{tot}}_{NN} = \left(\sigma^{\text{tot}}_{\pi N}\right)^2 .$$

The cross sections $\sigma_{\pi N}$ and σ_{NN} can be measured experimentally, whereas $\sigma_{\pi\pi}$ cannot be measured directly because there are no unstable targets. Nevertheless one can try to extract $\sigma_{\pi\pi}$ investigating the reaction $\pi + N \rightarrow 2\pi + N$ in the region of small angles, $|t| \lesssim \mu^2$, where the main contribution is expected to arise from the one-pion exchange diagram (Chew–Low process):

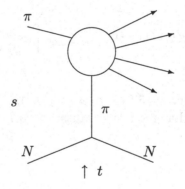

Then one can obtain $\sigma_{\pi\pi}$ from the factorization relation using the known values of $\sigma_{\pi N}$ and σ_{NN}.

6

Scattering of particles with spin

As was shown above, the definite quantum numbers of a reggeon in the t-channel lead to certain relations among amplitudes in the asymptotic region of the s-channel ($s \gg 4\mu^2$).

The most serious consequences arise from the assumption that there is a reggeon with vacuum quantum numbers whose trajectory $\alpha(t)$ passes through 1 at $t = 0$ ($\alpha(0) = 1$), so for this pole

$$\eta = P = +1, \qquad T = 0, \qquad C = +1$$

(η, P, T and C are the signature, parity, isospin and charge parity respectively).

The amplitude, corresponding to the exchange of a vacuum reggeon in the t-channel, has the form

$$A = r(t)\xi_\alpha^+ s^\alpha = -r(t)\frac{s^{\alpha(t)} + (-s)^{\alpha(t)}}{\sin \pi\alpha(t)}. \qquad (6.1)$$

Here $r(t)$ is the residue of the pole and ξ_α^+ is the signature factor:

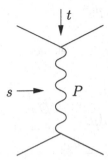

We saw that at small t the amplitude is purely imaginary and the angular distribution of the final particles becomes *narrower* with increasing energy: the diffractive cone shrinks; see Fig. 5.1.

71

Amplitude (6.1) can be written in terms of impact parameters in the following form:

$$A = \frac{s}{4\pi} \int f(\rho, s) \exp(\mathrm{i}\, \mathbf{q} \cdot \boldsymbol{\rho})\, \mathrm{d}^2\rho, \qquad (6.2)$$

$$f(\rho, s) \sim \frac{\mathrm{i}}{\alpha' \xi} \exp\left(-\frac{\rho^2}{4\alpha' \xi}\right), \qquad (6.3)$$

where $\xi = \ln(s/\mu^2)$, $\alpha(t) = 1 + \alpha' t$. This dependence of $f(\rho, s)$ on s leads both to the growth of the essential impact parameters with energy, $\rho \sim \sqrt{\alpha' \xi}$, and to an increase of the transparency as the absorption coefficient falls with increasing energy as $1/\alpha' \xi$.

Let us continue the discussion of consequences caused by the assumption of the dominance of a single Regge pole in the amplitude at $s \to \infty$.

First we show how one can write down the contribution of a Regge pole in the simplest way if the external particles have spin.

6.1 Vector particle exchange

Let us consider the amplitude, which corresponds to the exchange of a vector particle in the t-channel, see Fig. 6.1.

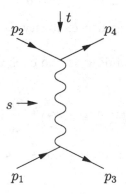

Fig. 6.1.   Exchange of an elementary particle with spin

$$A = \Gamma_\mu(p_2, q) \frac{D_{\mu\nu}}{m^2 - q^2} \Gamma_\nu(p_1, q), \qquad (6.4)$$

$$D_{\mu\nu} = g_{\mu\nu} - \frac{q_\mu q_\nu}{m^2}, \quad q = p_3 - p_1. \qquad (6.5)$$

We show now that at large energies only one contribution survives in (6.4), which corresponds to a definite spin state of the vector meson. To do this let us write down $g_{\mu\nu}$ in (6.5) in the form

$$g_{\mu\nu} = e^1_\mu e^1_\nu + e^2_\mu e^2_\nu + e^{\lambda\perp}_\mu e^{\lambda\perp}_\nu, \tag{6.6}$$

where e^1_μ and e^2_ν are unit vectors in the plane determined by the vectors p_1, p_2, and $e^{\lambda\perp}_{\mu,\nu}$ are the unit vectors lying in the transverse plane. Let us verify that at high energies one can neglect the contributions coming from $e^\perp_{\mu,\nu}$ and $q_\mu q_\nu / m^2$.

It is convenient to decompose the 4-vector q into two components: one in the plane p_1, p_2 and the other in the transverse subspace to the latter (Sudakov decomposition):

$$q = \alpha p'_2 + \beta p'_1 + q_\perp. \tag{6.7}$$

Following Sudakov, one chooses the vectors p'_1, p'_2 to be light-like, $p'^2_1 = p'^2_2 = 0$:

$$p_1 = p'_1 + \frac{m^2_1}{s}p'_2, \quad p_2 = p'_2 + \frac{m^2_2}{s}p'_1; \quad s = 2p'_1 p'_2 \simeq 2p_1 p_2. \tag{6.8}$$

According to (6.7) we have

$$\left.\begin{aligned}\alpha &= \frac{2p'_1 q}{s} \simeq \frac{m^2_3 - m^2_1 - q^2}{s}, \\[2mm] \beta &= \frac{2p'_2 q}{s} \simeq \frac{m^2_2 - m^2_4 + q^2}{s}, \\[2mm] q^2 &= \alpha\beta s + q^2_\perp \simeq q^2_\perp.\end{aligned}\right\} \tag{6.9}$$

One can see, inserting (6.5) and (6.6) into (6.4), that the quantities $\Gamma_\mu e^{\lambda\perp}_\mu$, $\Gamma_\nu e^{\lambda\perp}_\nu$ and $\Gamma_\mu q_\mu$ do not depend on energy. Therefore $e^{\lambda\perp}{}_{\mu,\nu}$ and $q_\mu q_\nu / m^2$ in (6.5) can be neglected as compared with e^1_μ and e^2_ν, whose contributions increase linearly with energy.

Indeed, let us introduce two normalized light-like vectors

$$n^1 = \frac{p'_1}{\sqrt{s/2}}, \quad n^2 = \frac{p'_2}{\sqrt{s/2}}, \tag{6.10}$$

which satisfy

$$(n^\mu_1)^2 = (n^\mu_2)^2 = 0, \quad n^\mu_1 n^\mu_2 = 1. \tag{6.11}$$

The first (in-the-scattering-plane) part of the unit tensor can be represented equivalently as

$$e^1_\mu e^1_\nu + e^2_\mu e^2_\nu = n_{1\mu} n_{2\nu} + n_{2\mu} n_{1\nu}. \tag{6.12}$$

Then the second term on the r.h.s. of (6.12) will give a small contribution to (6.4), and we obtain

$$A \simeq \Lambda_2 \frac{s}{m^2 - q^2} \Lambda_1, \qquad \Lambda_2 = \Gamma_\mu(p_2, q) \frac{\sqrt{2}p'_{1\mu}}{s}, \qquad \Lambda_1 = \Gamma_\mu(p_1, q) \frac{\sqrt{2}p'_{2\mu}}{s}.$$

So defined, the vertex factors $\Lambda_{1,2}$ asymptotically do not depend on s.

The vectors $n_{1,2}$ have a simple physical meaning in the t-channel: they are the vectors describing *circular polarizations* of the vector meson:

$$\mathbf{e}^\pm \equiv \frac{\mathbf{e}_x \pm i\,\mathbf{e}_y}{\sqrt{2}},$$

which obey the above relations (6.11). Solution of these equations (together with the condition $(eq) = 0$) leads to complex vectors for positive t, and to real vectors for negative t (s channel).

In the case when the spin of the exchanged particle equals 2,

$$A \propto \Gamma^{\mu\nu}(p_2, q) \times D^{\{\mu\nu; \rho\sigma\}}(q) \times \Gamma^{\rho\sigma}(p_1, q),$$

we can similarly derive that the dominant exchanged state is $n_{1\mu}n_{1\nu} \times n_{2\rho}n_{2\sigma}$, and hence the amplitude takes the form

$$A = \frac{s^2}{m^2 - t} \left(\Gamma^{\mu\nu} n_{1\mu} n_{1\nu} \right) \left(\Gamma^{\rho\sigma} n_{2\rho} n_{2\sigma} \right). \qquad (6.13)$$

This result can be easily generalized to the case of arbitrary spin of the exchanged particle in Fig. 6.1.

The values $\Lambda_{1,2}$ depend on $n_{1,2}$. This fact can be expressed differently, by stating that $\Lambda_{1,2}$ depend on the longitudinal,

$$q_\parallel^\mu = (qn_1)n_2^\mu + (qn_2)n_1^\mu,$$

and transverse, $q_\perp^\mu$, components of the momentum transfer, separately.

Then the general expression for the exchange of a particle of arbitrary spin n can be written at large $s \simeq 2p_1 p_2$ as follows:

$$A = \frac{s^n}{m^2 - t} \Lambda_2(p_2, q_\perp, q_\parallel) \Lambda_1(p_1, q_\perp, q_\parallel). \qquad (6.14)$$

Now we are in a position to write down the amplitude which corresponds to the t-channel exchange of a reggeon:

$$\frac{s^n}{t - m^2} \implies \xi_\alpha^\pm s^{\alpha(t)} = -\frac{s^{\alpha(t)} + (-s)^{\alpha(t)}}{\sin \pi\alpha(t)},$$

$$A \implies \xi_\alpha^\pm s^{\alpha(t)} \Gamma_1(p_1, q_\perp, q_\parallel) \Gamma_2(p_2, q_\perp, q_\parallel). \qquad (6.15)$$

If external particles are spinless, the vertices may only contain Lorentz products of momenta. Then, since $2p_1 q = q^2 - m_3^2 + m_1^2$, the vertex $\Gamma_i(p_i, q_\perp, q_\parallel)$ depends only on q^2 and the masses of the external particles in this vertex.

It is clear that the non-zero spins of external particles do not change the fact that the vertices are independent of the collision energy.

Let us discuss scattering of particles with non-zero spin.

6.2 Scattering of nucleons

Let particles 1 and 3 be nucleons. Then in the general case the vertex Γ_1 may be represented by the following sum of invariants:

$$\Gamma_1(p_1, q_\perp, q_\parallel) = \bar{u}(p_3)[g_1 + g_2\hat{q}_\parallel + g_3\gamma_5 + g_4\gamma_5\hat{q}_\parallel]u(p_1), \quad g_i = g_i(q^2). \quad (6.16)$$

Let us now take into account the quantum numbers of the t-channel states such as the parity and charge parity of the reggeon.

6.2.1 Reggeon quantum numbers and $N\bar{N} \to$ reggeon vertices

The parity of the t-channel state $1 + \bar{3}$ coincides with the parity of the reggeon. Replacing the $1 + \bar{3}$ system by its mirror state (i.e. the state obtained by spatial reflection of the initial state) affects both the vertex Γ_1 and the reggeon propagator. The latter acquires the multiplier $P_j = \pm 1$ equal to the *signature* of the reggeon, since under spatial reflection in the cms of the t-channel the vectors $\boldsymbol{p}_1$ and $-\boldsymbol{p}_3$ get permuted, $z_t \to -z_t$, which means the replacement $s \leftrightarrow u$ or, asymptotically, $s \to -s$.

Therefore the behaviour under spatial reflection of the *vertex* alone will be determined by the product of two factors, $P_r = PP_j$.

Now we turn to charge conjugation.

The charge parity operation on the system $1+\bar{3}$ leads (in addition to the matrix transformation of the spinor wave functions) to the permutation of p_1 with $-p_3$ in the initial amplitude, which, once again, implies the asymptotic replacement $s \to -s$. Repeating the arguments given above for the spatial reflection operation, we see that the behaviour of the vertex Γ_1 under charge conjugation is defined by the quantity $C_r = CP_j$ (with C the charge parity of the reggeon).

Each Regge pole can be characterized, apart from its signature P_j, by the quantum numbers P_r and C_r, instead of P, C. The former set of quantum numbers is more convenient, because the structure of the vertex coupled to a reggeon with given P_r, C_r does not depend on the signature of the reggeon.

1. The vacuum reggeon is classified by the following set of quantum numbers: $P_r = +1$, $C_r = +1$, $P_j = +1$. The corresponding vertex function Γ_1 has the form

$$\Gamma_1 = \bar{u}\left(f_1 + f_2 \hat{q}_\|\right) u, \qquad P_r = +1,\, C_r = +1,\, P_j = \pm 1. \quad (6.17a)$$

As we have already said, the vertex for the reggeons with $P_r = C_r = +1$ and the opposite signature, $P_j = -1$, has the same form.

2. For P-odd, C-even reggeons the vertex is

$$\Gamma_1 = f_3\left(\bar{u}\,\gamma_5\,u\right), \qquad P_r = -1,\, C_r = +1,\, P_j = \pm 1. \quad (6.17b)$$

3. For reggeons that are both spatial- and charge-parity-odd, one obtains

$$\Gamma_1 = f_3\left(\bar{u}\gamma_5\hat{q}_\| u\right), \qquad P_r = -1,\, C_r = -1,\, P_j = \pm 1. \quad (6.17c)$$

4. The state with quantum numbers $P_r = +1$, $C_r = -1$ cannot be realized in the $N\bar{N}$ system. It is interesting to note that experimentally meson resonances with such quantum numbers have not been observed.

6.2.2 Vacuum pole in πN and NN scattering

The contribution of the vacuum pole to the πN scattering amplitude has the form

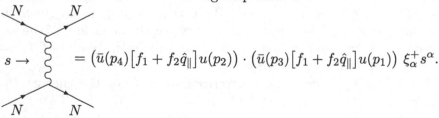

$$= g_\pi \cdot \left(\bar{u}(p_1)(f_1 + f_2\hat{q}_\|)u(\bar{p}_3)\right)\, \xi^+_{\alpha(t)} s^{\alpha(t)};$$

$$(6.18a)$$

its contribution to the NN scattering amplitude is

$$= \left(\bar{u}(p_4)\left[f_1 + f_2\hat{q}_\|\right]u(p_2)\right) \cdot \left(\bar{u}(p_3)\left[f_1 + f_2\hat{q}_\|\right]u(p_1)\right)\, \xi^+_\alpha s^\alpha.$$

$$(6.18b)$$

The form of the amplitudes (6.18) of πN and NN scattering completely determines the polarization properties of the scattered nucleons. As a consequence of factorization of residues in (6.18b), the polarization density matrix of the scattered nucleons 3 and 4 is also factorized. Reality of f_1 and f_2 at $q^2 < 0$ causes nucleons 3 and 4 to be unpolarized when a single Regge pole dominates the amplitude. In view of the fact that, in general, $f_1 \sim f_2$ there exists a large spin flip.

6.3 Conspiracy

We discuss now the situation which has recently been called *conspiracy of Regge poles*. We shall investigate the behaviour of the scattering amplitude at $q^2 \to 0$. In two-component form, the contribution of the vacuum pole to the NN scattering amplitude is

$$A_V \sim \left(f_1' + f_2'[\sigma^{(1)} \times \mathbf{q}_\perp]_z \right) \left(f_1' + f_2'[\sigma^{(2)} \times \mathbf{q}_\perp]_z \right) \xi_\alpha s^{\alpha(t)}, \qquad (6.19)$$

where z marks the collision axis in the direction of $\boldsymbol{p}_1$. Correspondingly the contributions of the other poles become

$$A_{P_r=-1,C_r=+1} \sim f_3^2 \left(\sigma^{(1)} \cdot \mathbf{q}_\perp \right) \left(\sigma^{(2)} \cdot \mathbf{q}_\perp \right) \xi_\alpha s^{\alpha(t)}, \qquad (6.20)$$

$$A_{P_r=-1,C_r=-1} \sim f_4^2 \sigma_z^{(1)} \sigma_z^{(2)}. \qquad (6.21)$$

In the general case due to angular momentum conservation the amplitude can have the following form at $q^2 = 0$:

$$C_1 + C_2 \, \sigma_\perp^{(1)} \cdot \sigma_\perp^{(2)} + C_3 \, \sigma_3^{(1)} \sigma_3^{(2)}. \qquad (6.22)$$

However, the second term does not correspond to the contribution of a Regge pole with definite parity, provided the form factors f_2', f_3 are regular at $q^2 = 0$. The structure $(\sigma_\perp^{(1)} \cdot \sigma_\perp^{(2)})$ does not appear from the contribution of Regge poles (the so-called *evasive solution*).

Nevertheless one can try to keep the contribution of such a form to the scattering amplitude. If we were to assume that $f_2' \sim \mathbf{q}_\perp^{-2}$, then we would get an expression of the following form from the contribution of vacuum-type Regge poles in (6.19):

$$\frac{[\sigma^{(1)} \times \mathbf{q}_\perp]_z [\sigma^{(2)} \times \mathbf{q}_\perp]_z}{\mathbf{q}_\perp^2}. \qquad (6.23a)$$

This expression has no meaning as $\mathbf{q}_\perp \to 0$ since the result depends on the direction along which $\boldsymbol{q}_\perp$ tends to zero. For a short range potential such a behaviour is impossible because the amplitude has no singularity at $q^2 = 0$.

The same could be said about the expression

$$\frac{(\boldsymbol{\sigma}^{(1)} \cdot \boldsymbol{q}_\perp)(\boldsymbol{\sigma}^{(2)} \cdot \boldsymbol{q}_\perp)}{\boldsymbol{q}_\perp^2} \tag{6.23b}$$

which arises from the contribution (6.20) of the pole with $P_r = -1, C_r = +1$ if the form factor f_3 is also singular: $f_3^2 \sim \boldsymbol{q}_\perp^{-2}$.

However, if the positions of the vacuum poles $P_r = C_r = +1$ and the pole with $P_r = -1, C_r = 1$ coincide at $q^2 = 0$,

$$\alpha_{P_r=1,C_r=1}(0) = \alpha_{P_r=-1,C_r=1}(0), \tag{6.24}$$

the so-called *conspiracy condition*, then the sum of contributions from (6.19) and (6.20) can be equal to $\boldsymbol{\sigma}_\perp^{(1)} \cdot \boldsymbol{\sigma}_\perp^{(2)}$, which remains meaningful in the $\boldsymbol{q}_\perp \to 0$ limit.

From the point of view of the t-channel, which plays the rôle of a *potential* for the s-channel, the appearance of additional relations in the amplitude at $t = 0$ means that the potential has an additional symmetry at this point.

7
Fermion Regge poles

The purpose of this lecture is to discuss the properties of fermion Regge poles. Let us consider the two graphs shown in Fig. 7.1.

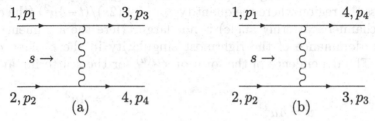

Fig. 7.1. s- And u-channel reggeon exchange graphs

Recall that the corresponding Mandelstam variables are defined as

$$s = (p_1 + p_2)^2, \ t = (p_1 - p_3)^2, \ u = (p_1 - p_4)^2, \qquad (7.1)$$

and we suppose, as always, that s is much larger than the squared masses of the external particles. The cosine of the angle between the directions p_1 and p_3 in the cms of the s-channel is

$$\left. \begin{aligned} z_{1,3} &= \frac{t-u}{s-4\mu^2} = 1 + \frac{2t}{s-4\mu^2}, \\ s+t+u &= 4\mu^2 \end{aligned} \right\} \qquad (7.2)$$

(for simplicity, we assume that all masses are equal).

The wavy lines in Fig. 7.1 represent reggeons. One can expect that at $s \to \infty$ the first graph dominates for small t, whereas the second one dominates at small u.

From (7.2) it is seen that small angles between p_1 and p_3 correspond to small $-t$, $-t \ll s$, and hence the first graph describes *forward* scattering in the cms of the s channel (for identical particles 1 and 3). The case of small u, where the second graph dominates, corresponds to *backward* scattering.

The order of magnitude of the contribution of the first graph is $s^{\alpha_1(t)}$ and that of the second one is $s^{\alpha_2(u)}$. The difference in quantum numbers of the systems 1–3 and 1–4, in general, leads to essentially different trajectories α_1 and α_2 of the t-channel and of the u-channel reggeons, respectively.

The fact that the asymptotic expression for the scattering amplitude contains two terms is important. Note that even a detailed knowledge of the t-channel trajectory $\alpha_1(t)$ at *all* t does not make it possible to calculate the asymptotics of the scattering amplitude at all t, and at small u in particular. This is so because the theory of complex angular momenta is valid when $|t|/s \ll 1$, or $|u|/s \ll 1$. To go to the region of small $-u$ from the region of small $-t$ through the physical region of the s-channel, it is necessary to move along the path shown in Fig. 7.2 by a dashed line, i.e. to cross the region where the quantity $z_t = 1+(2s)/(t-4\mu^2)$ (the cosine of the t-channel scattering angle) is not large. Here the arguments relying on the dominance of the rightmost singularity in the ℓ plane cease to work. The dependence of the form of $s^{\alpha_2(u)}$ for the scattering amplitude

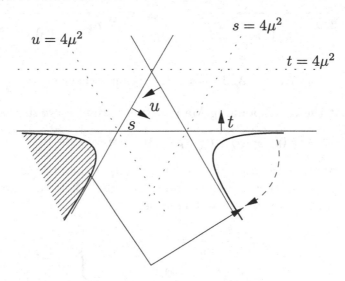

Fig. 7.2.   Continuation path from u-channel to high energy large angle s-channel scattering – solid arrow – as opposed to going from small to large scattering angles – dashed

at small $-u$ (large angles in the s-channel) arises when one makes the partial wave expansion in the u-channel (the shaded area in Fig. 7.2) and subsequently continues the series in the form of an integral over ℓ (see previous lectures) along the path shown by the solid arrow, along which $z_u \sim s/u$ stays large.

7.1 Backward scattering as a relativistic effect

From the assumption of dominance of the amplitude of Fig. 7.1(a) in the region of small angles, $\Theta \ll 1$, and of Fig. 7.1(b) for large angles, $\pi - \Theta \ll 1$, we obtain the angular dependence of the differential cross section, sketched in Fig. 7.3.

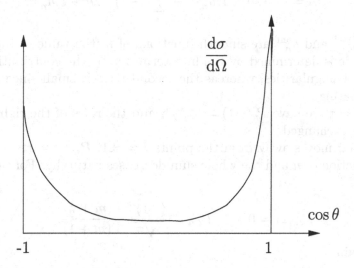

Fig. 7.3. Forward and backward peaks in the differential scattering cross section

This prediction does not contradict experimental data. The following simple but rather general arguments can be given in favour of the naturalness of this peculiar angular dependence.

We present the amplitude in the form of a sum over partial wave amplitudes:

$$A(s, z) = \sum_n (2n + 1) a_n(s) P_n(z), \quad z = \cos \Theta_{\widehat{13}}. \tag{7.3}$$

From the single dispersion relation for the amplitude at fixed large s we have

$$a_n(s) = \frac{1}{\pi} \int_{\text{right cut}} \frac{2dt}{s - 4\mu^2} A_3(s, t) \, Q_n \left(1 + \frac{2t}{s - 4\mu^2} \right)$$
$$+ \frac{(-1)^n}{\pi} \int_{\text{left cut}} \frac{2dt}{s - 4\mu^2} A_2(u, t) \, Q_n \left(1 + \frac{2t}{s - 4\mu^2} \right), \tag{7.4}$$

where A_3 is the discontinuity of the amplitude across the right cut in the t plane and its value is determined by the t-channel singularities; A_2 is the discontinuity across the left cut, whose value is restricted by the u-channel unitarity condition and is determined by the u-channel singularities.

Thus $a_n(s)$ can be represented as the sum of two contributions arising from the right and left cuts, respectively:

$$a_n(s) = a_n^{\text{right}}(s) + (-1)^n a_n^{\text{left}}(s). \tag{7.5}$$

Consider the case $\theta = 0$. Then $P_n(1) = 1$ and we have

$$A = \sum_n (2n+1) a_n^{\text{right}} + \sum_n (-1)^n (2n+1) a_n^{\text{left}}. \tag{7.6}$$

Since a_n^{right} and a_n^{left} are smooth functions of n, the value of the forward amplitude is determined by the first term, i.e. by the contribution of the t-channel singularities, whereas the second term is small since the series is alternating.

At $\Theta = \pi$ we have $P_n(-1) = (-1)^n$, and the rôles of the right and left cuts are exchanged.

When z moves away from the points $z = \pm 1$, $P_n(z)$ starts to oscillate as a function of n and the whole sum decreases naturally. For instance at $\Theta = \pi/2$,

$$P_{2m+1} = 0, \qquad P_{2m} = \frac{(-1)^m}{\sqrt{\pi}} \frac{\Gamma(m + \frac{1}{2})}{\Gamma(m+1)}.$$

We obtain

$$A = \sum_m (4m+1)(a_{2m}^{\text{right}} + a_{2m}^{\text{left}}) \frac{\Gamma(m + \frac{1}{2})}{\sqrt{\pi}\Gamma(m+1)} \cdot (-1)^m. \tag{7.7}$$

Since the series is alternating, and a_n is smooth in n, the amplitude turns out to be very small.

It should be stressed that the enhancement of the amplitude near $z = -1$ is a purely relativistic effect since in this phenomenon the existence of the left cut, generated by the u-channel singularities, is crucial. Physically, large backward scattering is due to the fact that in a field theory a relativistic particle tries to keep the direction of its motion, so it is preferable to change the particle's identity after collision rather than to have a large momentum transfer $\mathcal{O}(s)$.

7.2 Pion–nucleon scattering

Consider now the $\pi N \to \pi N$ scattering process. The notions of *forward* and *backward* scattering are clearly defined here, and these two kinematical regions have their special features. Already in perturbation theory there is a Feynman graph, shown in Fig. 7.4, whose magnitude is given by the expression

$$\bar{u}(p_4)\,\frac{\hat{q}+m}{m^2-q^2}\,u(p_1)\,,\qquad q = p_1 - p_3 = p_4 - p_2\,,\qquad (7.8)$$

with m the nucleon mass.

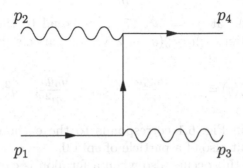

Fig. 7.4. πN scattering. Nucleon shown with the solid line, pion wavy

This diagram has the physical meaning of nucleon exchange and gives a big contribution at low $q^2 = u$ (backward scattering). Within the framework of quantum mechanics it is unusual for colliding particles to exchange a fermion.

Let us see how to generalize this diagram to the exchange of a reggeon.

What is the order of magnitude of the graph in Fig. 7.4 at large s? Choosing the normalization which will be convenient in the following,

$$\bar{u}u = 2m,\qquad (7.9)$$

we have

$$u(p) = \left(\begin{array}{c} \sqrt{p_0+m} \\ \sigma_p\sqrt{p_0-m} \end{array} \right) \phi = \frac{1}{\sqrt{p_0+m}} \left(\begin{array}{c} p_0+m \\ \boldsymbol{\sigma}\cdot\boldsymbol{p} \end{array} \right) \phi. \qquad (7.10)$$

Here ϕ is the usual two-component spinor, normalized to unity, $\phi^\dagger\phi = 1$, $\sigma_p = (\boldsymbol{\sigma}\cdot\boldsymbol{p})/|\boldsymbol{p}|$ is the helicity matrix and p_0 is the energy of the particle. In the laboratory frame we have $p_1 = (m,0,0,0)$, $p_4 \simeq (s/2m, s/2m, 0, 0)$ and hence the magnitude of the amplitude corresponding to this graph, $A \sim \mathcal{O}(u(p_4))$, is of the order of $s^{1/2}$, for the chosen normalization of the spinors. This should have been expected for the exchanged particle with spin $J = 1/2$.

However, the expression (7.8) cannot correspond to the exchange of a reggeon since its spin structure $\hat{q}+m$ does not have definite spatial parity.

7.2.1 Parity in the u-channel

To extract from (7.8) the reggeon contribution we shall advance the following argument. We recall that for the exchange of a *vector* particle (see Fig. 6.1) the following expression has been attributed to the internal line:

$$g_{\mu\nu} - \frac{q_\mu q_\nu}{m^2}. \tag{7.11}$$

On the mass shell, at $q^2 = m^2$, (7.11) is equal to the sum over three physical polarizations $e_\mu^\lambda(q)e_\nu^\lambda(q)$ of a vector particle, but off mass shell the equality is violated:

$$\sum_\lambda e_\mu^\lambda(q)e_\nu^\lambda(q) = g_{\mu\nu} - \frac{q_\mu q_\nu}{q^2} \neq g_{\mu\nu} - \frac{q_\mu q_\nu}{m^2}; \quad q^2 \neq m^2. \tag{7.12}$$

Thus the graph in Fig. 6.1 corresponds to the exchange not only of a vector particle, but also of a particle of spin 0.

Something like this occurs also when a fermion is exchanged. Turning now to half-integer spin, at $q^2 = m^2$ we can represent the fermion wave function (7.10) as

$$\left.\begin{aligned} u_\alpha^\lambda &= \frac{\hat{q} + \sqrt{q^2}}{\sqrt{q_0 + \sqrt{q^2}}}\phi^\lambda, \\ u_\alpha^\lambda(q)\bar{u}_\beta^\lambda(q) &= (\hat{q} + \sqrt{q^2})_{\alpha\beta}. \end{aligned}\right\} \tag{7.13}$$

A similar construction can be used to describe the exchange of a particle with an arbitrary half-integer spin. For example, for spin 3/2 one may write the propagator as follows:

$$e_\mu^\nu u_\alpha^\sigma \otimes e_\nu^\lambda \bar{u}_\beta^\sigma \sim u_\alpha^\sigma \bar{u}_\beta^\sigma \frac{q_\mu^\| q_\nu^\|}{q^{\|2}} \sim s \cdot (q_\mu^\| u_\alpha^\sigma) \otimes (q_\nu^\| \bar{u}_\beta^\sigma). \tag{7.14}$$

Thus we see that an additional power of s appears due to the presence of the vector representation in the wave function of a particle with spin 3/2 (cf. (6.7)–(6.14)).

What is the difference between $\hat{q} + \sqrt{q^2}$ in (??) and $(\hat{q} + m)$ of (7.8)? Recall that the Dirac equation

$$(\hat{q} - m)\psi = 0 \tag{7.15}$$

can be recast into another form, namely

$$(\hat{q} + m)\psi' = 0, \tag{7.16}$$

if we replace the Dirac spinor ψ by

$$\psi \to \psi' = \gamma_5 \psi.$$

The matrix γ_5 interchanges the upper and lower components of a spinor which differ from each other by a factor of σ_p (helicity) that changes sign under parity transformation. It makes it clear that when $q^2 \neq m^2$, the operator $\hat{q} + m$ corresponds to the exchange of a state of indefinite parity.

At the same time, the operator $\hat{q} + \sqrt{q^2}$ in the cms of the u-channel, where $q = (q_0, 0, 0, 0)$, is equal to

$$\sqrt{q^2}\,(\gamma_0 + 1),$$

i.e. is proportional to the projection operator that extracts the upper spinor components, which do not change sign under the parity transformation. It is this state that couples to an *even* fermionic reggeon.

7.2.2 Fermion poles with definite parity and singularity at $u = 0$

The contribution to the amplitude of the P-even reggeon exchange can be written therefore in the following form:

$$A_+ = \xi_{\alpha_+ - \frac{1}{2}}\, s^{\alpha_+ - \frac{1}{2}}\, \bar{u}(p_4)\hat{\Gamma}_1^+ \left(\hat{q} + \sqrt{q^2}\right)\hat{\Gamma}_2^+ u(p_1). \qquad (7.17\text{a})$$

The fact that $\alpha_+ - \frac{1}{2}$ appears in the exponent of s and in the signature is related to the requirement of the existence of a pole of positive signature and angular momentum $\frac{1}{2}$ (nucleon). At $\alpha_+(q) \to \frac{1}{2}$ the signature factor in (7.17a) develops a pole; taking account of the factor $s^{\frac{1}{2}}$ originating from the spinors, we get $A_+ \propto s^{\frac{1}{2}}$, as one should expect in the case of the exchange of a spin-$\frac{1}{2}$ particle.

Analogously, the contribution of a P-*odd* reggeon to the scattering amplitude has the form

$$A_- = \xi_{\alpha_- - \frac{1}{2}}\, s^{\alpha_- - \frac{1}{2}}\, \bar{u}(p_4)\hat{\Gamma}_1^- \left(\hat{q} - \sqrt{q^2}\right)\hat{\Gamma}_2^- u(p_1). \qquad (7.17\text{b})$$

The vertices $\hat{\Gamma}_i^\pm$ in (7.17) describe the couplings of the reggeons to the external particles and are built from Dirac γ-matrices, $q_\perp$ and $q_\parallel$. Each of the expressions (7.17) separately has a branch point at $u = q^2 = 0$, which contradicts the requirement that the amplitude be analytic at $u = 0$. To remove this fake singularity, the following relation between the two trajectories and corresponding vertices is needed:

$$\left.\begin{aligned}
\alpha_+(\sqrt{q^2}) &= \alpha_-(\sqrt{q^2}), \\
\Gamma_+(\sqrt{q^2}) &= \Gamma_-(\sqrt{q^2}).
\end{aligned}\right\} \qquad (7.18)$$

It is easy to see that (7.18) ensures regularity of the full amplitude at $u = q^2 = 0$.

Relation (7.18) follows from the fact that any Feynman graph becomes a function of $\sqrt{q^2}$ when one extracts an amplitude with definite parity. If we were to introduce corresponding partial wave amplitudes of definite parity, $f_{\lambda\lambda'}^{j\pm}(\sqrt{q^2})$, then the relation between them similar to (7.18), i.e.

$$f_{\lambda\lambda'}^{j+}(\sqrt{q^2}) = f_{\lambda\lambda'}^{j-}(-\sqrt{q^2}), \qquad (7.19)$$

would become vague. This happens because of an additional left cut of partial wave amplitudes, related to $Q_\ell(z)$. In the case of equal masses this cut started at $u = 0$ (see previous lectures). In the case of unequal masses there is an additional cut in the form of a circle:

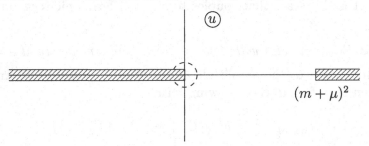

Given this structure of the u plane, the point $u = 0$ cannot be reached and this makes the assertion (7.18) indefinite.

For the residues and trajectories, however, the relations (7.17), (7.18) are entirely reasonable. This is due to the fact that, as we know from Lecture 3 (see (3.13)), the discontinuity across the left cut of the partial wave amplitude has no moving singularities. Therefore, conversely, the contribution of the Regge pole to this amplitude cannot have a left cut.

In the spinless case, from this consideration it immediately follows that the residue and trajectory have no left cuts. Indeed, if Δr and $\Delta\alpha$ were not equal to zero on the left cut,

$$f_\ell = \frac{r(t)}{\ell - \alpha(t)} \implies \Delta f_\ell^{\text{left}} = \frac{\Delta r}{\ell - \alpha(+)} + \frac{r(-)\,\Delta\alpha}{[\ell - \alpha(+)][\ell - \alpha(-)]}, \qquad (7.20)$$

then Δf_ℓ would tend to infinity at some ℓ, which is impossible (see previous lectures).

In the $J = \frac{1}{2}$ case we see that the residues and trajectories of each of the two Regge poles with opposite parity may have root singularities at $u = 0$, though the full amplitude remains regular at $u \leq 0$.

The structure of singularities of the amplitude $f_\ell(u)$ in the ℓ plane is shown in Fig. 7.5. For $0 < u < (m + \mu)^2$ (unphysical region) there are two poles on the real axis, situated, generally speaking, arbitrarily with respect to each other. With u decreasing they come closer and they collide at $u = 0$. When we pass to the region $u < 0$ they become mutually

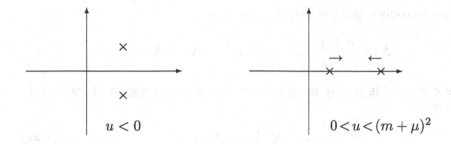

$u < 0$

$0 < u < (m + \mu)^2$

Fig. 7.5. Fermion poles in the ℓ plane

complex conjugate: $\alpha_+ = \alpha_-^*$. Thus in the physical region of the s-channel both poles are equally important, $\operatorname{Re}\alpha_+ = \operatorname{Re}\alpha_-$.

At sufficiently small u the trajectory of the pole can be approximately written as

$$\alpha^{\pm}(u) \simeq \alpha_0 \pm \gamma\sqrt{u} + \alpha' u. \qquad (7.21)$$

The coincidence of α^+ and α^- ($\gamma \equiv 0$) would correspond to parity degeneracy and demand the existence of fermion resonances of equal masses and spins but opposite parities (if at the physical points, where $\alpha(u)$ is half-integer, the residues do not vanish). The value $\gamma \lesssim 0.2$ does not contradict experiment.

7.2.3 Oscillations in the fermion pole amplitude

The presence of $\gamma \neq 0$ in the trajectory leads to the amplitude oscillating with energy, which is characteristic for fermion poles. In the cross section, however, these oscillations cancel out. Let us see how it happens.

First we note that in the region of large s, the operator $\hat{q}$, which is present in both the reggeon propagator and the vertices $\hat{\Gamma}$, can be replaced by $\hat{q}_{\perp}$. Indeed, one can decompose q in terms of the Sudakov parameters into a part that lies in the scattering plane of the u-channel cms ($(\boldsymbol{p}_1, \boldsymbol{p}_4)$ plane), and a part perpendicular to it:

$$q = \alpha_q p_1 + \beta_q p_4 + q^{\perp}, \quad |\alpha_q| \sim |\beta_q| \sim \left|\frac{q^2 - m^2 + \mu^2}{s}\right| \ll 1.$$

Recalling that $\hat{p}u = mu$ by virtue of the Dirac equation, we see that

$$\bar{u}(p_4)\,\hat{q}\,u(p_1) = \bar{u}(p_4)\hat{q}^{\perp}u(p_1)\left[1 + \mathcal{O}\!\left(\frac{m^2}{s}\right)\right] \simeq \bar{u}(p_4)\,\hat{q}^{\perp}\,u(p_1).$$

Now we introduce two projection operators

$$\hat{\Lambda}_\eta = \frac{q + \eta \hat{q}^\perp}{2q}, \quad \eta = \pm 1, \quad \Lambda_+ + \Lambda_- = 1,$$

where $q \equiv \sqrt{q^2}$ is purely imaginary in the physical region. It is easy to see that

$$\Lambda_\eta^+ = \Lambda_\eta, \quad \Lambda_{\eta_1}\Lambda_{\eta_2} = \delta_{\eta_1 \eta_2}\Lambda_\eta. \tag{7.22}$$

Λ_η with $\eta = \pm 1$ corresponds to the propagator of a reggeon of positive (negative) parity. Now the total amplitude can be expressed in the form

$$A = A_+ \bar{u}(p_4)\Lambda_+ u(p_1) + A_- \bar{u}(p_4)\Lambda_- u(p_1).$$

Since $A_\pm \propto s^{\alpha_\pm - \frac{1}{2}}$, oscillations in the cross section might occur in terms of the type Re $A_+^* A_-$ or Im $A_+^* A_-$ (provided Im $\alpha_+ = -$Im $\alpha_- \neq 0$).

The differential cross section $d\sigma/dt$ is proportional to

$$\begin{aligned}
|M^2| &= \sum_{\eta_1 \eta_2} \sum_{\lambda \sigma} [\bar{u}^\lambda(p_4) A_{\eta_1} \Lambda_{\eta_1} u^\sigma(p_1)]^+ [\bar{u}^\lambda(p_4) A_{\eta_2} \Lambda_{\eta_2} u^\sigma(p_1)] \\
&= \sum_{\eta_1 \eta_2} \bar{u}(p_1)\Lambda_{\eta_1} u(p_4)\bar{u}(p_4)\Lambda_{\eta_2} u(p_1) A_{\eta_1}^* A_{\eta_2} \\
&\simeq \sum_{\eta_1 \eta_2} \text{Tr} \left((\hat{p}_1 + m)\Lambda_{-\eta_1} \hat{p}_4 \Lambda_{\eta_2} \right) A_{\eta_1}^* A_{\eta_2}. \tag{7.23}
\end{aligned}$$

The sum over λ, σ means summing and averaging correspondingly over the final and initial polarizations of the nucleon. The operator $\Lambda_{-\eta_1}$ appeared as a result of the anticommutation between γ_0 (recall, $\bar{u} = u^\dagger \gamma_0$) and $\hat{q}^\perp$ in Λ_{η_1}. Here it was also used that

$$\sum_\lambda u^\lambda(p_4)\bar{u}^\lambda(p_4) = \hat{p}_4 + m \simeq \hat{p}_4 \quad \text{at } s \gg m^2.$$

$\hat{p}_4$ Also anticommutes with $\hat{q}^\perp$, hence after shifting $\hat{p}_4$ to the left, one translates $\Lambda_{-\eta}$ back into Λ_η. Thus in the middle there appears the product in (7.22),

$$\Lambda_{\eta_1}\Lambda_{\eta_2} = \delta_{\eta_1 \eta_2}\Lambda_{\eta_1},$$

and this leads the differential cross section to be proportional to

$$\frac{d\sigma}{dt} \sim a|A_+|^2 + b|A_-|^2. \tag{7.24}$$

Neither term in (7.24) does oscillate. Analogous cancellations of the oscillating terms happen in the polarization vector. These oscillations can manifest themselves only in more intricate correlation effects.

7.3 Reggeization of a neutron

The second question I would like to discuss in this lecture is whether a neutron is a reggeon? How can this be established experimentally? The easiest and most exact way is to study the differential cross section of backward scattering, say for reactions of the type $\pi^+ + p \to p + \pi^+$, in which the neutronic reggeon gives a contribution corresponding to Fig. 7.6.

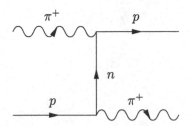

Fig. 7.6. *t*-Channel neutron exchange in $\pi^+ + p \to p + \pi^+$ scattering

The normal energy dependence of the cross section, arising as a result of the Regge pole exchange, is

$$\frac{d\sigma}{du} \propto s^{2[j(u)-1]}.$$

We have $j(u = m_n^2) = \frac{1}{2}$, since the neutron belongs to this Regge trajectory. By analogy with quantum mechanics, it is natural to expect that $j(u)$ decreases with decreasing u, i.e. $j(u = 0) < \frac{1}{2}$ in the physical region of the *s*-channel. Thus the differential cross section, caused by neutronic reggeon exchange, must fall faster than $1/s$. That is indeed observed experimentally for the reaction $\pi^+ p \to p\pi^+$.

The inverse statement is also true, as we have proved in Lecture 4: if a backward peak falls faster than $1/s$, then a neutron must be a reggeon.

Moreover, since the neutron is reggeized, a very fine prediction arises: the neutron is expected to have partners, belonging to the same trajectory $j(u)$ and having identical quantum numbers, except for their spins which differ by 2. At present two particles are known which lie on the neutron trajectory, whereas on the Δ trajectory there are three particles. If we assume that the trajectories are approximately linear in u (for which there are experimental indications), then one can predict the positions of other partners and, using the same parameters, the energy dependence of the backward cross sections for a number of reactions, in reasonable agreement with experiment.

8

Regge poles in perturbation theory

Let us now consider how the Regge poles arise in perturbation theory. There are many field theories within the framework of which one can trace the origin and behaviour of moving singularities in the ℓ plane. We shall investigate some of them.

8.1 Scattering of a particle in an external field

The process of non-relativistic particle scattering in an external field is described by a set of graphs

where $\times$ denotes the vertex of interaction with an external field. We will not consider this theory, because it is completely equivalent to the quantum-mechanical problem of potential scattering and, therefore, leads to the appearance of Regge poles.

8.2 Scalar field theory $g\phi^3$

Relativistic theories seem to be more interesting. Let us investigate the easiest of them – the $g\phi^3$ model with ϕ the field of scalar neutral particles. Its Lagrangian has the form

$$\mathcal{L} = \frac{1}{2}\left(\frac{\partial \phi}{\partial x_\mu}\right)^2 - \frac{1}{2}m^2\phi^2 - g\phi^3, \tag{8.1}$$

where the term $g\phi^3$ describes an interaction of these particles. In the Feynman graph language it leads to a vertex which looks as follows:

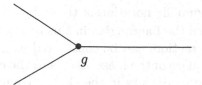

This theory is obviously unsatisfactory from a physical point of view, because the energy density derived from it is not a positive definite quantity. Indeed, the energy–momentum tensor of the field ϕ is

$$T_{\mu\nu} = \frac{\partial\phi}{\partial x_\mu}\frac{\partial\mathcal{L}}{\partial\left(\frac{\partial\phi}{\partial x_\nu}\right)} - \mathcal{L}g_{\mu\nu}. \tag{8.2}$$

From this formula we get the energy density

$$T_{00} = \left(\frac{\partial\phi}{\partial x_0}\right)^2 - \mathcal{L} = \frac{1}{2}\sum_{i=0}^{3}\left(\frac{\partial\phi}{\partial x_i}\right)^2 + \frac{1}{2}m^2\phi^2 + g\phi^3. \tag{8.3}$$

Due to the term $g\phi^3$ the energy density is not positive definite, independently of the sign of the coupling constant g.

8.2.1 $g\phi^3$ Theory in the Duffin–Kemmer formalism

This trouble can be avoided using the Duffin–Kemmer formalism, within which a wave function of a scalar particle satisfies the first order differential equation $\hat{p}u = mu$, where u is a column of five rows:

$$u = \begin{pmatrix} m\phi \\ p_\mu\phi \end{pmatrix}. \tag{8.4}$$

The 5×5 matrix $\hat{p}$ has the form

$$\hat{p} = \begin{pmatrix} 0 & \tilde{p}_\mu \\ p_\mu & 0 \end{pmatrix} = p_\mu\beta_\mu, \tag{8.5}$$

where

$$p_\mu = \begin{pmatrix} p_0 \\ p_1 \\ p_2 \\ p_3 \end{pmatrix}, \quad \tilde{p}_\mu = (p_0, -p_1, -p_2, -p_3).$$

In the Duffin–Kemmer formalism one can introduce an interaction between particles of type $g\phi^3$, if one writes down the full Lagrangian in the following form:

$$\mathcal{L} = \frac{1}{2i}\left\{\bar{u}\beta_\mu\partial_\mu u - (\partial_\mu\bar{u})\beta_\mu u\right\} - m\bar{u}u + \lambda u^3, \tag{8.6}$$

where λu^3 is a symbolic notation of the cubic combination of field u.

If we write down the Lagrangian in this form, then the term λu^3, which describes the interaction, can be considered as a supplement to the mass (and not to the square of the mass, as it is in the conventional $g\phi^3$ theory). Therefore the energy density in the Duffin–Kemmer formalism turns out to be positive definite.

Let us explain this by the example of a particle in a constant external field ψ. The wave function of the particle satisfies the equation

$$(\hat{p} - m - \lambda\psi)u = 0. \tag{8.7}$$

It follows from this equation that the energy of the particle,

$$p_0 = \sqrt{\boldsymbol{p}^2 + (m + \lambda\psi)^2}, \tag{8.8}$$

has a quite reasonable form, independent of the sign of $\lambda\psi$.

Note that the linear equation (8.7) leads to a second order equation for the function ϕ:

$$p^2\phi = (m + \lambda\psi)^2\phi = (m^2 + 2m\lambda\psi + \lambda^2\psi^2)\phi. \tag{8.9}$$

In the conventional formalism this means that apart from the vertex in Fig. 8.1a there is the four-particle vertex shown in Fig. 8.1b. Thus a simple

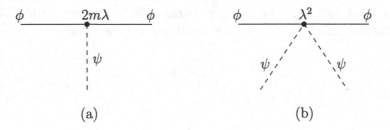

(a)   (b)

Fig. 8.1.   Interaction of a scalar particle with an external field

interaction in the Duffin–Kemmer formalism is reduced to a complicated one in the conventional formalism, and this provides a positive definite energy.

This peculiarity of the Duffin–Kemmer theory happens to have its consequence in the fact that the propagator of the u field does not decrease at large momenta $\hat{p}$ but behaves like a constant. Because of this, the properties of the usual pole graph in Fig. 8.2a of the Duffin–Kemmer theory are reminiscent of those of the point-like four-particle interaction of Fig. 8.2b.

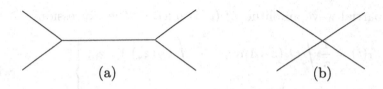

Fig. 8.2. Particle exchange and point interaction graphs

8.2.2 Analytic properties of the amplitudes

The drawback of the $g\phi^3$ theory remarked above does not prevent us, however, from studying the analytic properties of the amplitude in the ℓ plane. Therefore, we consider the usual formalism of the $g\phi^3$ theory, where the constant g is assumed to be small, and make use of perturbation theory. In the lowest order the scattering amplitude is described by three pole graphs:

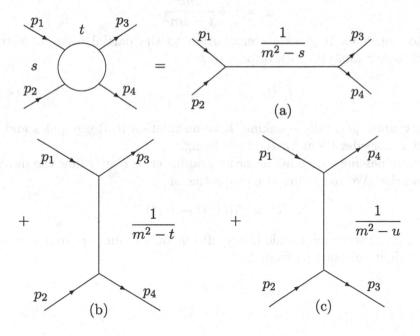

Fig. 8.3. Born diagrams for $2 \to 2$ scattering in $g\phi^3$ theory

As the particles are identical, there is only a positive signature. The

t-channel partial wave amplitude $f_\ell(t)$ is fixed by the expression

$$\left.\begin{aligned} f_\ell(t) &= \frac{1}{\pi} \int Q_\ell(z_s) A_1 \mathrm{d}z_s + \frac{1}{\pi} \int Q_\ell(z_u) A_2 \mathrm{d}z_u, \\ z_s &= -z_u = 1 + \frac{2s}{t - 4m^2} = \frac{s - u}{t - 4m^2}. \end{aligned}\right\} \tag{8.10}$$

In the lowest order in g we have

$$A_1 = \pi g^2 \delta(s - m^2) \qquad \text{from Fig. 8.3a,} \tag{8.11a}$$
$$A_2 = \pi g^2 \delta(u - m^2) \qquad \text{from Fig. 8.3c.} \tag{8.11b}$$

Therefore the total contribution of Figs. 8.3a and 8.3c to the amplitude $f_\ell(t)$ has the form

$$f_\ell^{(1)}(t) = Q_\ell(z_0) \frac{4g^2}{t - 4m^2}, \tag{8.12}$$

where

$$z_0 = 1 + \frac{2m^2}{t - 4m^2}.$$

As to Fig. 8.3b, it gives a contribution to the partial wave amplitude which is non-analytic with respect to ℓ:

$$f_\ell^{(1)}(t) \propto \delta_{l0} \frac{1}{m^2 - t}. \tag{8.13}$$

Such graphs, generally speaking, have no relation to Regge poles and we will not consider them for the time being.

Before passing to the higher order graphs let us clarify how the moving poles arise. We recall analytic properties of

$$\phi_\ell(t) = f_\ell(t) \cdot (t - 4m^2)^{-\ell}.$$

This partial wave amplitude has a left cut over t running from $-\infty$ to 0 and a right cut starting from $4m^2$:

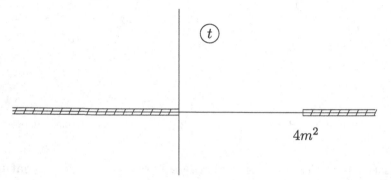

$$\phi_\ell(t) = \frac{1}{\pi} \int\limits_{-\infty}^{0} \frac{\mathrm{d}t'}{t'-t} \Delta\phi_\ell(t') + \frac{1}{\pi} \int\limits_{4m^2}^{\infty} \frac{\mathrm{d}t'}{t'-t} \delta\phi_\ell(t'). \tag{8.14}$$

The discontinuity across the left cut is given by the expression (3.13):

$$\Delta\phi_\ell(t) = -\frac{1}{2} \int\limits_{z<1} \frac{A_1(z,t)}{(4m^2-t)^\ell} P_\ell(-z)\mathrm{d}z + \int \frac{\rho_{su}(z,t)}{(4m^2-t)^\ell} Q_\ell(-z)\mathrm{d}z, \tag{8.15}$$

where the integration intervals in both terms are finite (see Lecture 3).

On the right cut, from $t = 4m^2$ to $t = 16m^2$, the discontinuity is derived from the elastic unitarity condition, see (3.10):

$$\delta\phi_\ell = \frac{2^{2\ell-4}}{\pi} \frac{k^{2\ell+1}}{\omega} \phi_\ell^+ \phi_\ell^-; \quad k = \frac{\sqrt{t-4m^2}}{2}, \quad \omega = \frac{\sqrt{t}}{2}. \tag{8.16}$$

From the unitarity condition it follows that the discontinuity on the right cut is $\delta\phi_\ell = \mathcal{O}(g^4)$, therefore in the lowest order in g^2 the amplitude ϕ_ℓ cannot develop a singularity in ℓ due to the right cut.

For a singularity to appear it is necessary to have a divergence in the integral of the left cut discontinuity $\Delta\phi_\ell$ in (8.14), starting from some ℓ.

To see that this indeed happens, we substitute the lowest order A_1 from (8.11a) into (8.15) to obtain

$$\Delta\phi_\ell^{(1)} = -\pi P_\ell(-\tilde{z}_0) \frac{g^2}{(4m^2-t)^{\ell+1}}, \quad \tilde{z}_0 = 1 - \frac{2m^2}{4m^2-t}. \tag{8.17}$$

Near $\ell = -1$ this expression tends to a constant at $t \to \infty$, therefore the first integral in (8.14) diverges at $\ell = -1$. Substituting (8.17) into (8.14) we find that near $\ell = -1$

$$\phi_\ell^{(1)} \propto \frac{g^2}{\ell+1}. \tag{8.18}$$

This result can be obtained directly from (8.12), if one takes into account that the function $Q_\ell(z_0)$ has poles at negative integer ℓ.

Thus the leading singularity is situated at $\ell = -1$. This is related to the fact that the s-channel pole term $1/(m^2 - s)$ decreases as $1/s$ at large s.

Let us show that this pole *reggeizes* in higher orders of perturbation theory. To this end we consider the unitarity condition (8.16), rewriting it as follows:

$$\delta\left(-\frac{1}{\phi_\ell}\right) = \frac{1}{2i}\left(-\frac{1}{\phi_\ell^+} + \frac{1}{\phi_\ell^-}\right) = \frac{\delta\phi_\ell}{\phi_\ell^+ \phi_\ell^-} = \frac{2^{2\ell-4}}{\pi} \frac{k^{2\ell+1}}{\omega}. \tag{8.19}$$

It follows from this relation that

$$-\frac{1}{\phi_\ell(t)} = c(\ell) + \frac{2^{2\ell-4}}{\pi^2} \int\limits_{4m^2}^{\infty} \frac{dt'}{t'-t} \frac{k^{2\ell+1}(t')}{\omega(t')}. \tag{8.20}$$

At small g^2 and ℓ near -1 we had for ϕ_ℓ the expression (8.18), which permits us to find the constant $c(\ell) = -(\ell+1)/4g^2$. Therefore, the amplitude $\phi_\ell(t)$, satisfying two-particle unitarity, should have at ℓ near -1 the form

$$\phi_\ell(t) = g^2 \left[\ell + 1 - \frac{g^2}{16\pi^2} \int_{4m^2}^{\infty} \frac{dt'}{t'-t} \frac{1}{\omega(t')k(t')} \right]^{-1}, \tag{8.21}$$

where we have put $\ell = -1$ in the integral term ($k^{2\ell+1} = k^{-1}$).

From (8.21) we see that the amplitude $\phi_\ell(t)$ has a moving pole $\ell = \alpha(t)$ with the trajectory

$$\alpha(t) = -1 + \frac{g^2}{16\pi^2} \int_{4m^2}^{\infty} \frac{dt'}{t'-t} \frac{1}{\omega(t')k(t')}$$

$$= -1 + \frac{g^2}{4\pi^2} \int_{4m^2}^{\infty} \frac{dt'}{(t'-t)\sqrt{t'(t'-4m^2)}}. \tag{8.22}$$

Thus, if in the lowest order in g^2 there was a fixed pole at $\ell = -1$, then in higher orders of perturbation theory it starts to move.

Our derivation of (8.22) was not rigorous for a number of reasons:

1. Only two-particle unitarity was considered;

2. The discontinuity on the left cut was calculated in the lowest order in g^2 (this is not too bad an approximation, since accounting for higher order contributions to $\Delta\phi_\ell$ leads to relatively small corrections $\mathcal{O}(g^4/(\ell+1))$, and does not contain contributions of the type of $g^4/(\ell+1)^2$);

3. We have chosen the simplest expression for $c(\ell)$ independent of t, while the condition (8.19) does not forbid us to add an arbitrary polynomial in t to $c(\ell)$. With the chosen $c(\ell)$ the expression (8.21) for $\phi_\ell(t)$ coincides with the one that emerges from the iteration of the two-particle unitarity condition.

The positive lesson obtained from the derivation of (8.21) carried out is a clear demonstration that the movement of a pole arises due to the two-particle unitarity condition, that is due to the graphs having two-particle intermediate states.

8.2.3 Order $g^4 \ln s$

Let us demonstrate that the direct consideration of Feynman graphs in perturbation theory leads to the same result. The graphs of order g^4 are shown in Fig. 8.4.

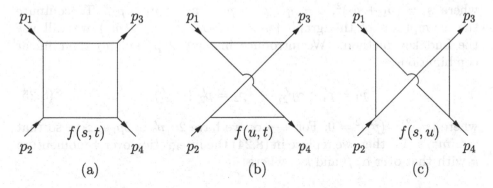

Fig. 8.4. Feynman graphs of order g^4

The asymptotics of the amplitude at large s and small t related to the Regge pole has the form

$$s^{\alpha(t)} = \frac{1}{s} s^{g^2 f(t)} = \frac{1}{s} e^{g^2 f(t) \ln s} = \frac{1}{s} \sum \frac{1}{n!} [g^2 f(t) \ln s]^n. \qquad (8.23)$$

Therefore to study the Regge poles in perturbation theory it suffices to consider only graphs behaving at large s as g^2/s, $g^4 \ln s/s$ and so forth, i.e. to investigate the asymptotics of graphs under conditions $g^2 \ll 1$, $g^2 \ln s \sim 1$. In this approximation – the leading logarithmic approximation – the problem can be solved exactly.

First let us calculate the asymptotics of the box diagram of Fig. 8.5.

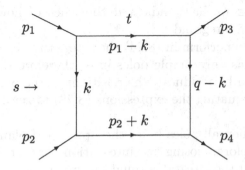

Fig. 8.5. The box diagram in $g\phi^3$ theory

The amplitude corresponding to this graph is given by

$$f(s,t) = g^4 \int \frac{\mathrm{d}^4 k}{(2\pi)^4 \mathrm{i}} \frac{1}{(m^2 - k^2)[m^2 - (q-k)^2][m^2 - (p_1-k)^2][m^2 - (p_2+k)^2]},$$
(8.24)

where $s = (p_1 + p_2)^2$, $t = q^2$, $q = p_1 - p_3 = p_4 - p_2$. To compute the asymptotes of this graph for $s \to \infty$, $|q^2| = \mathcal{O}(m^2)$ we will use the Sudakov method. We introduce instead of p_1 and p_2 their linear combinations,

$$p_1 = p'_1 + \gamma p'_2, \qquad p_2 = p'_2 + \gamma p'_1,$$
(8.25)

where $(p'_1)^2 = (p'_2)^2 = 0$. For $s \to \infty$ we have $2p'_1 p'_2 \simeq 2p_1 p_2 \simeq s$, so that $\gamma \simeq m^2/s$. Further, we replace in (8.24) the integration over 4-momentum k with that over α, β and $k_\perp$, where

$$k = \alpha p'_2 + \beta p'_1 + k_\perp, \quad k_\perp^2 < 0, \quad \mathrm{d}^4 k = \frac{s}{2} \mathrm{d}\alpha \, \mathrm{d}\beta \, \mathrm{d}^2 k_\perp.$$
(8.26)

Similarly the vector q is represented as

$$q = \alpha_q p'_2 + \beta_q p'_1 + q_\perp, \quad \alpha_q \sim -\beta_q \sim \frac{q^2}{s}, \quad q^2 \simeq q_\perp^2 < 0.$$
(8.27)

The denominator in (8.24) is expressed through the new variables $\alpha, \beta, k_\perp$ according to the formulæ

$$\left.\begin{aligned}
m^2 - k^2 &\simeq m^2 - s\alpha\beta - k_\perp^2 - \mathrm{i}\varepsilon, \\
m^2 - (q-k)^2 &\simeq m^2 - (\alpha - \alpha_q)(\beta - \beta_q)s - (q_\perp - k_\perp)^2 - \mathrm{i}\varepsilon, \\
m^2 - (p_1-k)^2 &\simeq m^2 - (\gamma - \alpha)(1 - \beta)s - k_\perp^2 - \mathrm{i}\varepsilon, \\
m^2 - (p_2+k)^2 &\simeq m^2 - (1 + \alpha)(\gamma + \beta)s - k_\perp^2 - \mathrm{i}\varepsilon.
\end{aligned}\right\}$$
(8.28)

By introducing $\mathrm{i}\varepsilon$ we have indicated the rules of how to pass by the singularities of the integrand.

First of all let us perform in (8.24) the integration over α. The denominator in (8.24) has only simple poles in α, therefore the integral over α is easily calculated by residues. The positions of the poles in the α plane can be found by equating the expressions (8.28) to zero. They depend on the value of β.

If $\gamma + \beta < 0$, then all poles lie in the upper half-plane of the complex variable α. Therefore, closing the integration contour in the lower half-plane, we find that the integral is equal to zero. In the same way, at $\beta > 1$ all poles in α lie in the lower half-plane of the variable s, therefore the integral is equal to zero as well.

Thus, it suffices to integrate over β within the limits $-\gamma < \beta < 1$. We divide this region of integration into two parts:

$$\beta_q < \quad \beta \quad < 1, \tag{8.29a}$$
$$-\gamma < \quad \beta \quad < \beta_q \qquad \left(\gamma \sim \beta_q = \mathcal{O}\left(s^{-1}\right) \ll 1\right). \tag{8.29b}$$

(Note that $\beta_q > 0$ when $q^2 < 0$.) These regions give essentially different contributions to the asymptotics of the amplitude. It is easy to show that large contributions proportional to $\ln s/s$ arise only when we integrate over the broad region (8.29a) whereas the integral over the small region (8.29b) gives a subleading correction $\mathcal{O}(1/s)$.

Because of that, let us consider the region $\beta_q < \beta < 1$. At these values of β the pole in α, corresponding to $m^2 - (p_1 - k)^2 = 0$, is situated in the upper half-plane, and three remaining poles lie in the lower half-plane. We close the integration contour in the upper half-plane and take a residue in α at

$$\alpha = \gamma - \frac{m^2 - k_\perp^2}{(1 - \beta)s} \simeq \frac{k_\perp^2}{s}.$$

The integral over $k_\perp$ converges at $|k_\perp^2| \sim m^2$, and the integral over β is dominated by the region $m^2/s \ll \beta \ll 1$. Bearing this in mind we can simplify the remaining propagators in (8.28) as

$$\left.\begin{aligned} m^2 - k^2 &\Longrightarrow m^2 - k_\perp^2, \\ m^2 - (q - k)^2 &\Longrightarrow m^2 - (q_\perp - k_\perp)^2, \\ m^2 - (p_2 + k)^2 &\Longrightarrow -\beta s. \end{aligned}\right\} \tag{8.30}$$

As a result we get

$$f(s,t) \simeq g^4 \int \frac{d^2 k_\perp}{(2\pi)^3} \int_{m^2/s}^1 \frac{d\beta\, s/2}{(m^2 - k_\perp^2)[m^2 - (q_\perp - k_\perp)^2]\, s\,(-\beta s)}$$
$$\simeq -\frac{g^4}{4\pi s} \ln(-s) \left(\right), \tag{8.31}$$

where the symbol $\left(\right)$ denotes the Feynman loop in the two-dimensional space resulting from the graph in Fig. 8.5 when one contracts the two lines whith large momenta $\mathcal{O}(p_1)$ and $\mathcal{O}(p_2)$ into a point:

$$k_\perp \quad q_\perp - k_\perp \equiv \int \frac{\mathrm{d}^2 k_\perp}{(2\pi)^2} \frac{1}{(m^2 - k_\perp^2)[m^2 - (q_\perp - k_\perp)^2]}. \tag{8.32}$$

In (8.31) we have replaced the lower limit of integration over β roughly by $1/s$, taking no care of the precise value of the numerator of the ratio (of the order of m^2). This is legitimate in our logarithmic approximation. Indeed, rescaling $1/s \to c/s$ produces a finite correction to the large logarithm, $\ln s \to \ln s - \ln c$, which gives rise to a negligible correction $\mathcal{O}(g^4)/s$ to the amplitude.

Nevertheless, we have kept the factor -1 in the argument of the logarithm, $\ln(-s)$, since the graph under consideration has an imaginary part for $s > 0$, corresponding to a two-particle intermediate state in the s-channel. This is the largest contribution to the imaginary term of order g^4 for $s \to \infty$, so we do not exceed our accuracy by taking it into account.

The asymptotics of the graph in Fig. 8.4b is calculated similarly. This graph differs from that of Fig. 8.4a by the replacement $s \to u$, hence its asymptotics is of the following form:

$$f(u, t) \simeq -\frac{g^4}{4\pi u} \ln(-u) \left(\right). \tag{8.33}$$

Consider now the graph in Fig. 8.4c. It differs from the previous ones in that large momenta p_1 and p_2 enter not in two but in three internal lines:

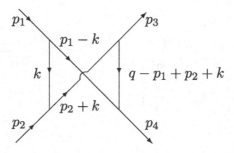

Therefore, at $s \to \infty$ this graph falls as $1/s^2$ (and not as $1/s$) and does not have any relation to Regge poles.

8.2.4 Order $g^6 \ln^2 s$

We pass now to the next order of perturbation theory and consider the sixth order diagram of Fig. 8.6.

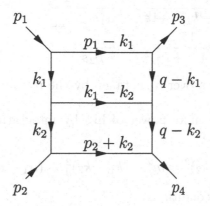

Fig. 8.6. Ladder diagram contributing $g^6 \ln^2 s$

The terms of interest to us in the asymptotics of this graph have the form $\ln^2 s/s$. To analyse the asymptotics of this graph we will again apply the Sudakov method.

We extract from the momenta k_1 and k_2 the longitudinal parts (along p_1 and p_2) and the transverse parts, and integrate first over α_1 and α_2 and then over β_1 and β_2. Let us find out what regions of α_i and β_i contribute mainly to the asymptotics.

We recall that in the previous example such a region was $\alpha = \mathcal{O}(m^2/s)$, $m^2/s \ll \beta \ll 1$, so that

$$\beta s \gg m^2, \quad \alpha\beta s \ll m^2.$$

These strong inequalities allowed us to simplify the propagators according to (8.30). In particular, in the propagator of the horizontal line with momentum $p_2 + k$ we have kept the only large term,

$$\frac{1}{m^2 - (p_2 + k)^2} \simeq \frac{1}{-\beta s},$$

which gave rise to $\ln s$ upon integration over β. The integral over α was calculated by closing the contour around the pole of the propagator of the other horizontal line:

$$\frac{1}{m^2 - (p_1 - k)^2 - \mathrm{i}\varepsilon} \simeq \frac{1}{\alpha s - k_\perp^2 - \mathrm{i}\varepsilon}.$$

Applying similar considerations to the diagram of Fig. 8.6, we find that we get a large contribution to the asymptotics if we approximate the propagators of the upper and lower lines as

$$\frac{1}{m^2 - (p_1 - k_1)^2 - \mathrm{i}\varepsilon} \simeq \frac{1}{\alpha_1 s - k_{11\perp}^2 - \mathrm{i}\varepsilon} \qquad (\beta_1 \ll 1),$$

$$\frac{1}{m^2 - (p_2 + k_2)^2} \simeq \frac{1}{-\beta_2 s} \qquad (|\alpha_2| \ll 1).$$

Then the integration over α_1 is taken by the residue, and the integration over β_2 gives $\ln s$.

One can get one more power of $\ln s$ by integrating the propagator of the middle line,

$$m^2 - (k_1 - k_2)^2 = m^2 - (k_1 - k_2)_\perp^2 - (\alpha_1 - \alpha_2)(\beta_1 - \beta_2)s.$$

Let us impose the conditions

$$\beta_1 \gg \beta_2, \qquad |\alpha_2| \gg |\alpha_1| \sim \frac{1}{s}.$$

Then the propagator of the middle line becomes

$$\frac{1}{m^2 - (k_1 - k_2)^2 - \mathrm{i}\varepsilon} \simeq \frac{1}{m^2 - (k_1 - k_2)_\perp^2 + \alpha_2 \beta_1 s - \mathrm{i}\varepsilon}.$$

If the conditions $|\alpha_2| \gg |\alpha_1|$ and $\beta_1 \gg \beta_2$ are fulfilled, then the original integral over α_i and β_i reduces to the following symbolic expression:

$$\int \frac{s^2 \, \mathrm{d}\alpha_1 \, \mathrm{d}\alpha_2 \, \mathrm{d}\beta_1 \, \mathrm{d}\beta_2}{(\alpha_1 s + 1 - \mathrm{i}\varepsilon)(\alpha_2 \beta_1 s + 1 - \mathrm{i}\varepsilon)(\beta_2 s + 1)}. \qquad (8.34)$$

The integrals over α_1 and α_2 are calculated by residues, and the remaining expression

$$\int_{1/s}^1 \frac{\mathrm{d}\beta_1}{\beta_1} \int_{1/s}^{\beta_1} \frac{\mathrm{d}\beta_2}{\beta_2}$$

gives $\frac{1}{2} \ln^2 s$, that is the power of $\ln s$ we have been looking for.

So for the diagram of the sixth order the region that gives the main contribution to the asymptotics is defined by the conditions

$$\left. \begin{array}{lll} 1 \gg |\alpha_2| \gg |\alpha_1| \gg 1/s, & 1 \gg \beta_1 \gg \beta_2 \gg 1/s, \\ |\alpha_1|\beta_1 s \ll 1, & |\alpha_2|\beta_2 s \ll 1, & |\alpha_2|\beta_1 s \sim 1. \end{array} \right\} \qquad (8.35)$$

Under these conditions only the transverse components $\mathbf{k}_{1\perp}$, $\mathbf{k}_{2\perp}$ and $\mathbf{q}_\perp$ should be kept in the propagators of the vertical lines, containing k_1, k_2 and q, so that the remaining transverse momentum integral has the

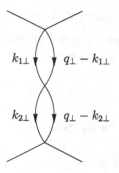

Fig. 8.7. Reduced two-dimensional graph corresponding to the Feynman diagram of Fig. 8.6

structure of the two-dimensional Feynman integral for the reduced graph of the type of Fig. 8.7. The integrations over $\mathbf{k}_{1\perp}$ and $\mathbf{k}_{2\perp}$ are performed independently in each loop, therefore the integral is equal to the product of the two single loops (8.32).

The conditions (8.35) which we have derived have a clear physical meaning. They correspond to the fact that the particles in the intermediate states of the s-channel are almost real. In addition, in the laboratory frame, where the second particle is at rest, and the first one has a large momentum $p_1 \sim s$, the first emitted particle $k_1 \sim \beta p_1$ carries away a small momentum fraction, and the second one carries still smaller momentum, $1 \gg \beta_1 \gg \beta_2$, in order for the virtualities k_1^2, k_2^2 of the momenta transferred at each stage of the process to be small.

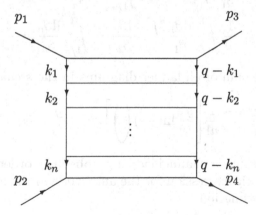

Fig. 8.8. Ladder diagrams in $g\phi^3$ theory contributing to the leading logarithmic approximation

8.2.5 Ladder diagrams in all orders

These conditions can be easily generalized to the case of the ladder diagram in Fig. 8.8 of an arbitrary order:

$$
\left.
\begin{aligned}
&\frac{1}{s} \ll \alpha_1 \ll \alpha_2 \ll \cdots \ll \alpha_n \ll 1; \\[2mm]
&1 \gg \beta_1 \gg \beta_2 \gg \cdots \gg \beta_n \gg \frac{1}{s}.
\end{aligned}
\right\}
\tag{8.36}
$$

Under these conditions the propagators of the longitudinal lines give, from top to bottom,

$$
\frac{1}{\alpha_1 s + 1}, \quad \frac{1}{\alpha_2 \beta_1 s + 1}, \quad \frac{1}{\alpha_3 \beta_2 s + 1}, \ldots, \frac{1}{-\beta_n s}.
$$

In the propagators of the vertical lines there remain only the transverse components $k_{i\perp}$ and $q_\perp$, and the integrals over $k_{i\perp}$ factorize.

Therefore, the asymptotic expression for the diagram of Fig. 8.8 acquires the form

$$
-\frac{g^2}{s} \left(\frac{g^2}{4\pi} \right)^n \int_{1/s}^1 \frac{\mathrm{d}\beta_1}{\beta_1} \int_{1/s}^{\beta_1} \frac{\mathrm{d}\beta_2}{\beta_2} \cdots \int_{1/s}^{\beta_{n-1}} \frac{\mathrm{d}\beta_n}{\beta_n} \times \left[\big(\!\big) \right]^n.
\tag{8.37}
$$

The ordered n-fold integral over $\beta_1, \beta_2, \ldots, \beta_n$ can be transformed according to the formula

$$
\begin{aligned}
&\int_{1/s}^1 \frac{\mathrm{d}\beta_1}{\beta_1} \int_{1/s}^{\beta_1} \frac{\mathrm{d}\beta_2}{\beta_2} \cdots \int_{1/s}^{\beta_{n-1}} \frac{\mathrm{d}\beta_n}{\beta_n} \\[2mm]
&= \frac{1}{n!} \int_{1/s}^1 \frac{\mathrm{d}\beta_1}{\beta_1} \int_{1/s}^1 \frac{\mathrm{d}\beta_2}{\beta_2} \cdots \int_{1/s}^1 \frac{\mathrm{d}\beta_n}{\beta_n} = \frac{1}{n!} [\ln(-s)]^n.
\end{aligned}
\tag{8.38}
$$

As a result the sum of all ladder diagrams in the s-channel equals

$$
-\frac{g^2}{s} \sum \frac{1}{n!} \left[\frac{g^2}{4\pi} \ln(-s) \big(\!\big) \right]^n = -\frac{g^2}{s} (-s)^{\frac{g^2}{4\pi} \big(\!\big)}.
\tag{8.39}
$$

Taking into account u-channel ladder graphs in all orders, we finally arrive at the asymptotic expression for the amplitude in the leading logarithmic (ladder) approximation:

$$
A(s, q^2) = g^2 \left[(-s)^{\alpha(t)} + (-u)^{\alpha(t)} \right],
\tag{8.40}
$$

with the Regge pole tragectory

$$\alpha(t) = -1 + \frac{g^2}{4\pi} \left(\right)$$

$$= -1 + \frac{g^2}{4\pi} \int \frac{\mathrm{d}^2 k_\perp}{(2\pi)^2} \frac{1}{(m^2 - k_\perp^2)[m^2 - (q-k)_\perp^2]}. \qquad (8.41)$$

One can verify that the expression (8.41) for $\alpha(t)$ coincides with the one previously derived, (8.22). The formula (8.41) describes a motion of the rightmost Regge pole, which at $q^2 = -\infty$ resides at the point $\ell = -1$.

As is seen from (8.12), when the coupling constant g^2 is small, there are fixed poles not only at $\ell = -1$, but also at all negative integer points – the poles of the Legendre function Q_ℓ. In higher orders in g^2 these poles become reggeized as well. Trajectories of these poles are similar to the trajectory of the rightmost pole:

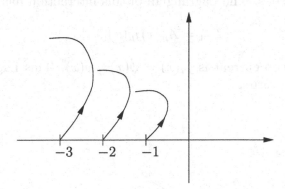

8.2.6 *Non-ladder diagrams*

Thus, taking account of the ladder diagrams in the t-channel leads to Regge pole singularities of partial wave amplitudes. The remaining non-ladder graphs can be divided into two groups.

In the first group there enter those diagrams which fall for $s \to \infty$ faster than a ladder graph of the same order in g^2. For instance, the diagram of Fig. 8.9a gives a contribution which is $\ln^2 s$ times less than the one shown in Fig. 8.9b. Therefore, these diagrams can be neglected compared with the ladder graphs.

In the second group there enter diagrams of the type of Fig. 8.3b. This diagram does not depend on s at all and is, therefore, much larger than the ladder graphs which fall as $1/s$, modulo logarithms. There is an infinite number of such diagrams, all of them having singularities in the ℓ plane of a non-Regge-type (e.g. δ_{l0}) corresponding to non-reggeized particles.

When the coupling constant g is small, the motion of the Regge poles happens near the points $\ell = -1, -2, \ldots$, therefore it is non-Regge singularities that govern the asymptotics of the whole amplitude, as they are situated to the right of all Regge poles. However, it is possible that, if one takes a larger coupling constant g, the pole $\alpha(t)$ will pass through the point $\alpha(m^2) = 0$, i.e. in the $g\phi^3$ theory reggeization of the scalar particle could take place. The latter possibility cannot be checked within the framework of perturbation theory. This makes the $g\phi^3$ theory less interesting from a pedagogical point of view.

8.3 Interaction with vector mesons

There is one more relativistic model in which one can succeed in studying the behaviour of Regge poles with the help of perturbation theory. It is the model of the interaction of spinor particles (let us call them nucleons) with vector mesons. The Lagrangian of this interaction has the form

$$\mathcal{L}_{\text{int}} = A_\mu(x) j_\mu(x), \tag{8.42}$$

where the nucleon current is $j_\mu(x) = \bar{\psi}(x)\gamma_\mu\psi(x)$. This Lagrangian corresponds to the vertex

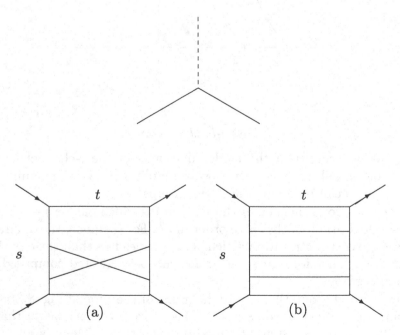

Fig. 8.9.   On comparison of ladder and non-ladder diagrams

In order to avoid infrared divergences we shall assume a non-zero mass λ of the vector meson (in the case of a vanishing mass this theory coincides with conventional electrodynamics).

Let us consider the diagrams of Fig. 8.10 describing Compton scattering. We introduce, as usual, the Mandelstam variables: $s = (p_1 + k_1)^2$,

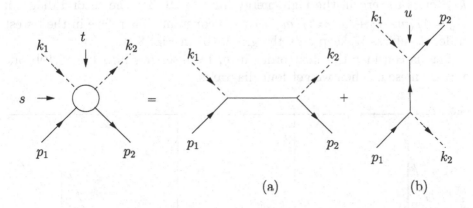

(a)   (b)

Fig. 8.10.   Lowest order Compton scattering diagrams

$t = (p_1 - p_2)^2$ and $u = (p_1 - k_2)^2 \equiv q^2$, and study the asymptotics of the process for $s \to \infty$, $u = $ const (rather than $t = $ const). The region $s \to \infty$, $t = $ const is not interesting for us since here there are only diagrams without single nucleon intermediate states, for instance those of Fig. 8.11, which contribute to the asymptotics of the amplitude at fixed t.

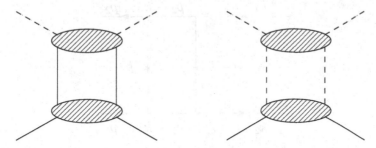

Fig. 8.11.   Diagrams contributing to Compton scattering at fixed t and $s \to \infty$

Therefore, the asymptotics of the amplitude at $s \to \infty$, $t = $ const has no relation to the issue of nucleon reggeization.

At the same time, our investigation of the asymptotics of Compton backward scattering $s \to \infty$, $u = $ const will show that at higher orders of perturbation theory the fixed pole that corresponds to a nucleon is reggeized. This proves the fallacy of the opinion popular in the past that if one introduces into the theory an elementary particle (fixed pole), then it remains elementary to all orders of perturbation theory.

Let us consider the lowest order of perturbation theory, i.e. the two diagrams of Fig. 8.10a,b. These diagrams give an analytic expression of the type $\bar{u}(p_2)\hat{e}_{2(1)}A_{a(b)}\hat{e}_{1(2)}u(p_1)$. For the first diagram $A_a \sim 1/(m-\hat{p}_1-\hat{k}_1)$ tends to zero in the high energy limit, while for the second diagram $A_b \sim 1/(m-\hat{p}_2+\hat{k}_1) = 1/(m-\hat{q})$ stays constant. Therefore in the lowest order it suffices to keep only the graph (b) in Fig. 8.10.

Let us consider the next order in g, by inserting into Fig. 8.10b one virtual meson. Then we get four diagrams

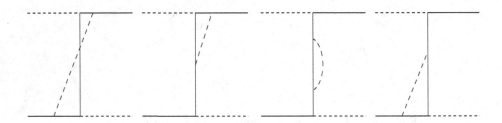

The latter three diagrams do not depend on s at all, whereas the first one is of the order of $\ln s$, as will be shown below. Therefore, to the relevant order it is sufficient to calculate only the first diagram. Redraw that in the form

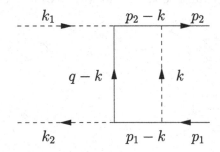

The first graph corresponds to the Feynman integral

$$g^4 \int \frac{\mathrm{d}^4 k}{(2\pi)^4} \frac{\bar{u}(p_2)\gamma_\alpha(m+\hat{p}_2-\hat{k})\hat{e}_1(m+\hat{q}-\hat{k})\hat{e}_2(m+\hat{p}_1-\hat{k})\gamma_\alpha u(p_1)}{[m^2-(p_2-k)^2][m^2-(q-k)^2][m^2-(p_1-k)^2][\lambda^2-k^2]}.$$
$$(8.43)$$

This expression differs from the one previously considered, (8.24), mainly in the numerator. In the numerator of (8.43) we should keep only the largest terms, which give the main contribution to the asymptotics. For example, accounting for the fact that in (8.43), as in (8.24), the main contribution comes from relatively small k, one can neglect the quantities m and $\hat{k}$ compared with the large external momenta, $\hat{p}_1$ and $\hat{p}_2$.

Indeed, consider the expression in the right part of the numerator,

$$\hat{p}_1 \gamma_\alpha u(p_1) = (-\gamma_\alpha \hat{p}_1 + 2p_{1\alpha}) u(p_1)$$
$$= (-\gamma_\alpha m + 2p_{1\alpha}) u(p_1) \simeq 2p_{1\alpha} u(p_1), \qquad (8.44)$$

then contract $p_{1\alpha}$ with the second γ_α factor. Then the left part of (8.43) gets transformed as follows:

$$\bar{u}(p_2) \hat{p}_1 \hat{p}_2 = \bar{u}(p_2)\{-\hat{p}_2 \hat{p}_1 + 2p_1 p_2\} = \bar{u}(p_2)\{-m\hat{p}_1 + s\} \simeq \bar{u}(p_2) s. \quad (8.45)$$

Thus the terms kept lead to the appearance in the numerator of the large quantity s. Now one can compute the asymptotics of the integral (8.43) quite in the same way as in the case of the $g\phi^3$ theory. As a result we obtain

$$\bar{u}(p_2) \hat{e}_1 \left\{ \frac{g^2}{4\pi} \ln(-s) \left(\begin{array}{c} \\ \end{array} \right) \right\} \hat{e}_2 u(p_1), \qquad (8.46)$$

where $\left(\begin{array}{c} \\ \end{array} \right)$ denotes the two-dimensional Feynman integral of the nucleon self-energy type:

$$\left(\begin{array}{c} \\ \end{array} \right) \equiv \int \frac{d^2 k_\perp}{(2\pi)^2} \frac{1}{m - \hat{q}_\perp + \hat{k}_\perp} \frac{1}{\lambda^2 - k_\perp^2}. \qquad (8.47)$$

Taking this expression together with that of the lowest order in g^2, we get the amplitude in the form of $\bar{u}(p_2) \hat{e}_1 A \hat{e}_2 u(p_1)$, where

$$A = \frac{g^2}{m - \hat{q}} \left\{ 1 + \frac{g^2}{4\pi} (m - \hat{q}) \left(\begin{array}{c} \\ \end{array} \right) \ln(-s) \right\}. \qquad (8.48)$$

So the coefficient of $\ln(-s)$ appears to depend on q, which corresponds to a moving pole in the j plane. However, in order to assert that this moving pole is the reggeized nucleon, we must convince ourselves that on the mass shell this pole comes to the point $j = \frac{1}{2}$, i.e. that the coefficient of $\ln(-s)$ tends to zero at $\hat{q} = m$.

In our case such a vanishing does really take place. However, the reason for this is trivial, namely just the fact that we have taken out $1/(m - \hat{q})$ in (8.48) as an overall factor. Therefore the answer to the question about the reggeization of the nucleon pole can come only from the next order of perturbation theory, i.e. from the diagram

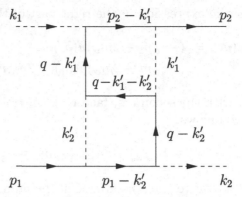

Consider the numerator of this diagram:

$$\bar{u}(p_2)\gamma_\mu[m+\hat{p}_2-\hat{k}_1']\hat{e}_1[m+\hat{q}-\hat{k}_1']\gamma_\nu[m+\hat{q}-\hat{k}_1'-\hat{k}_2']$$
$$\times\gamma_\mu[m+\hat{q}-\hat{k}_2']\hat{e}_2[m+\hat{p}_1-\hat{k}_2']\gamma_\nu u(p_1). \qquad (8.49)$$

In order to obtain s in the numerator we will proceed as in the previous case, i.e. move $\hat{p}_1$ to the right from γ_ν, and $\hat{p}_2$ to the left from γ_μ, and keep only the terms $p_{1\nu}$ and $p_{1\mu}$. Then the numerator is transformed to the form

$$\bar{u}(p_2)\hat{e}_1[m+\hat{q}-\hat{k}_1']\hat{p}_1[m+\hat{q}-\hat{k}_1'-\hat{k}_2']\hat{p}_2[m+\hat{q}-\hat{k}_2']\hat{e}_2u(p_1). \qquad (8.50)$$

Now our task is to commute $\hat{p}_1$ and $\hat{p}_2$, as a result of which the large factor $2p_1p_2 \simeq s$ will appear in the numerator. To do so we first neglect the longitudinal components proportional to $\hat{p}_1$ ($\hat{p}_2$) in the expression $m+\hat{q}-\hat{k}_1'-\hat{k}_2'$, because these would give only small terms of order m^2 when sandwiched between $\hat{p}_1$ and $\hat{p}_2$. Thus, keeping only the transverse components we get

$$\hat{p}_1[m+\hat{q}_\perp-\hat{k}_{2\perp}'-\hat{k}_{1\perp}']\hat{p}_2 \simeq s[m-\hat{q}_\perp+\hat{k}_{2\perp}'+\hat{k}_{1\perp}']$$

and the numerator reduces to

$$s\cdot\bar{u}(p_2)\hat{e}_1[m+\hat{q}-\hat{k}_1'][m-\hat{q}+\hat{k}_1'+\hat{k}_2'][m+\hat{q}-\hat{k}_2']\hat{e}_2u(p_1).$$

Let us represent $m-\hat{q}+\hat{k}_1'+\hat{k}_2'$ as the sum of three inverse fermion propagators:

$$m-\hat{q}+\hat{k}_1'+\hat{k}_2' = -(m-\hat{q})+(m-\hat{q}+\hat{k}_1')+(m-\hat{q}+\hat{k}_2'). \qquad (8.51)$$

Then, keeping in the numerator only the first term $m-\hat{q}$ and calculating the asymptotics of the expression obtained, we find it to be equal to

$$\bar{u}(p_2)\hat{e}_1\left\{\left(\frac{g^2}{4\pi}\right)\left(\right)(m-\hat{q})\frac{1}{2}\ln^2(-s)\right\}\hat{e}_2u(p_1). \qquad (8.52)$$

Taking into account terms of the lower orders, we find that

$$A = \frac{g^2}{m - \hat{q}} \left\{ 1 + \frac{g^2}{4\pi} \left(\right) (m - \hat{q}) \ln(-s) \right.$$

$$\left. + \frac{1}{2!} \left(\frac{g^2}{4\pi} \right)^2 (m - \hat{q}) \left(\right) (m - \hat{q}) \left(\right) \ln^2(-s) \right\}. \quad (8.53)$$

Such an asymptotics of the amplitude, in which for each power of $\ln s$ there is a factor of $m - \hat{q}$, shows that the nucleon is indeed reggeized.

As for the remaining terms $(m - \hat{q} + \hat{k}_1')$ and $(m - \hat{q} + \hat{k}_2')$ in (8.51), they cancel the propagators of the adjoining propagator lines, leading to the graphs

The asymptotics of these graphs for $s \to \infty$ is actually large: it contains not only $\ln^2 s$, but even $\ln^3 s$. However, their contributions are cancelled completely by those of other graphs not considered by us, such as

The reason for this cancellation is the conservation of the vector current $j_\mu(x)$.

9
Reggeization of an electron

In this lecture we shall continue to consider the perturbation theory of quantum electrodynamics with a photon with a non-zero mass λ. The conventional QED with $\lambda = 0$ has its own specific features, which will be addressed later.

The reason for considering this theory and for accounting for all complications related to spin, is that in the field theory with vector mesons one may expect the appearance of Regge particles at small values of the coupling constant g.* We note that although in the $g\phi^3$ theory there is a Regge pole, it is situated not near $j = 0$, which is required for reggeization of a scalar particle, but near $j = -1$. From the point of view of Feynman diagrams this is related to the fact that in the t-channel with a fixed momentum transfer we had two particles instead of one in the Born pole diagram. Each additional line in the t-channel reduces the asymptotics of a diagram by one power of the large variable s. A compensation of this suppression is possible owing to the *Azimov spin shifting*, which is made manifest by the appearance of the factor s^σ in the numerator of the Feynman amplitude, with σ the spin of the added particle. One can see from this that to make reggeization of a 'nucleon' possible in perturbation theory we should have a particle with $\sigma = 1$. Such a situation is realized in QED (see Fig. 9.1).

The solid lines here correspond to electrons, and the dashed ones denote photons. The regions interesting from the point of view of reggeization are

1. $s \to \infty$ with fixed t,

2. $s \to \infty$ with fixed u.

---

* as was observed by Gell-Mann, Goldberger and Low [ed]

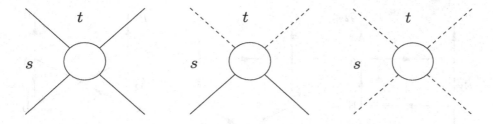

Fig. 9.1.   Two-particle scattering processes in QED

In the first region one can study the reggeization of a photon, in the second one we look for electron reggeization. Besides this, in the first region apart from the photon, which has a negative signature $P_j = -1$, the exchange of a 'particle' with a positive signature $P_j = +1$ and the quantum numbers of vacuum is possible.

9.1 Electron exchange in $\mathcal{O}(g^6)$ Compton scattering amplitude

Let us return now to the question of the electron reggeization in perturbation theory. In the previous lecture we have calculated the asymptotics of the Compton scattering amplitude for $s \to \infty$ and u fixed in g^4 and g^6 orders of perturbation theory. We have obtained

$$F = F_0(1 + \alpha(\hat{q})\ln s), \qquad (9.1)$$

where F_0 is the Born term, and

$$\alpha(\hat{q}) = (m - \hat{q})\frac{g^2}{4\pi} \int \frac{\mathrm{d}^2 k_\perp}{(\lambda^2 - k_\perp^2)(m - \hat{q} + \hat{k}_\perp)}. \qquad (9.2)$$

For $\hat{q} = m$ we have $\alpha = 0$, so $j \equiv \alpha + \frac{1}{2} = \frac{1}{2}$, and the electron lies on the Regge trajectory.

It was also shown that the contribution of the diagram of Fig. 9.2a can be split into three pieces: the one giving $F_0 \cdot \frac{1}{2}\alpha^2(\hat{q})\ln^2 s$, and two more terms that correspond to cancelling the line ab or cd. These two terms have the numerators

$$-4s\hat{e}_1(m - \hat{q} - \hat{k}_1)\hat{e}_2, \qquad -4s\hat{e}_1(m - \hat{q} - \hat{k}_2)\hat{e}_2, \qquad (9.3)$$

and the propagators of the line ab or cd, correspondingly, are absent.

Let us show now that the first term is cancelled by the contribution of the diagram shown in Fig. 9.2c.

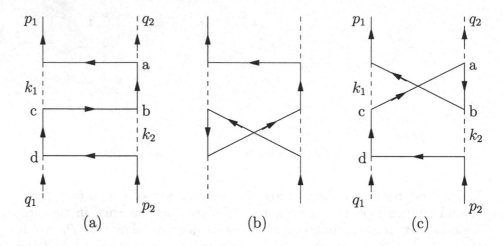

Fig. 9.2.   Compton scattering diagrams $\mathcal{O}(g^6)$

Using the Dirac equation for the wave functions of external spinors and keeping only the leading terms in s we get

$$\gamma_\nu \hat{p}_2 \hat{e}_1 (m - \hat{q} - \hat{k}_2)\gamma_\mu(m - \hat{q} - \hat{k}_1 - \hat{k}_2)\hat{e}_2(m - \hat{p}_1 - \hat{k}_1 - \hat{k}_2)\gamma_\nu \hat{p}_1 \gamma_\mu$$

$$\simeq 4\hat{e}_1(m - \hat{q} - \hat{k}_2)\hat{p}_1(m - \hat{q} - \hat{k}_1 - \hat{k}_2)\hat{e}_2(m - \hat{p}_1 - \hat{k}_1 - \hat{k}_2)\hat{p}_2$$

$$\simeq -4s\hat{e}_1(m - \hat{q} - \hat{k}_2)\hat{p}_1(m - \hat{q} - \hat{k}_1 - \hat{k}_2)\hat{e}_2$$

$$\simeq (\alpha_1 + \alpha_2)s \cdot 4s\hat{e}_1(m - \hat{q} - \hat{k}_2)\hat{e}_2.$$

The propagator corresponding to the line ab has the form

$$m^2 - (p_1 - k_1 - k_2)^2 \simeq (\alpha_1 + \alpha_2)s.$$

After the cancellation we get in the numerator

$$4s\hat{e}_1(m - \hat{q} - \hat{k}_2)\hat{e}_2. \tag{9.4}$$

Thus we see that the contribution (9.4) differs from that of (9.3) only by a sign. Similarly one can show that the second *extra* term from Fig. 9.2a cancels the contribution of the diagram in Fig. 9.2b. The reason for these cancellations is the gauge invariance of the theory.

Note now that t and s alternate in the ladder graphs of different orders: the graphs with an even power of g^2 have imaginary parts with respect to t, and the graphs with an odd power have imaginary parts with respect to s. Since we consider the region $s \to \infty$, and u is fixed, $t \simeq -s$. So in the perturbative expansion there are contributions proportional to $\ln s$ or $\ln(-s)$. In the leading approximation that we have been following, the difference between s and $-s$ is not seen. Taking into account the above remark, however, we may write down the sum of the first three orders of perturbation theory in the following form:

$$F = F_0 \left(1 + \alpha(\hat{q}) \ln s + \frac{1}{2!} \alpha(\hat{q}) \ln^2(-s) \right). \qquad (9.5)$$

To the chosen accuracy the factor of F_0 in (9.5) is just an expansion of the exponent.

9.2 Electron Regge poles

It seems likely that higher orders of perturbation theory give further terms of the exponent expansion, and we shall assume this in the following. An additional argument in favour of this assumption is provided by the investigation of the reggeization problem from the point of view of the t-channel partial waves, which will be performed later on.

By summing up separately even and odd terms in g^2, we get

$$
\begin{aligned}
F &= F_0 \left[\frac{1}{2} \left((-s)^{\alpha(\hat{q})} + (-s)^{-\alpha(\hat{q})} \right) + \frac{1}{2} \left(s^{\alpha(\hat{q})} - s^{-\alpha(\hat{q})} \right) \right] \\
&= F_0 \left[\frac{1}{2} \left((-s)^{\alpha(\hat{q})} + s^{\alpha(\hat{q})} \right) + \frac{1}{2} \left((-s)^{-\alpha(\hat{q})} - s^{-\alpha(\hat{q})} \right) \right] \\
&= F^+(s) + F^-(s).
\end{aligned}
\qquad (9.6)
$$

This expression corresponds to the contribution of two Regge poles with $P_j = \pm 1$ which differ from each other by the substitution $\alpha(\hat{q}) \leftrightarrow -\alpha(\hat{q})$. Besides the fact that the powers of $\ln s$ have 'exponentiated', another remarkable feature that emerged is the *degeneracy in signature*.

In the Born approximation there is no symmetry between s and t, that is there is no symmetry in signature. On the other hand, the signature is a characteristic feature of a Regge pole, which follows from the general theory. The only way to reconcile the contradictory requirements within perturbation theory is the degeneracy in signature near $\alpha = 0$.

For large s, only one contribution with the largest Re α survives, and a new asymptotic symmetry arises, which is absent in the exact amplitude.

9.2.1 Conspiracy in perturbation theory

There is one more important consequence of the general theory, which was discussed in the seventh lecture: for negative u and a given signature there should be two fermion poles of opposite parity, which differ from one another by a Hermitian conjugation. Let us show that this requirement is also fulfilled in perturbation theory.

The most general expression for $\alpha(\hat{q})$ reads

$$\alpha(\hat{q}) = a(q^2) + \hat{q}\, b(q^2), \qquad u = q^2, \tag{9.7}$$

as there are no other vectors besides q_μ at our disposal. We collect separately even and odd terms from the exponent and get

$$\left.\begin{aligned}
s^{\alpha(\hat{q})} &= s^a e^{b\hat{q}\xi}, \qquad \xi = \ln s, \\
e^{b\hat{q}\xi} &= \left[\cosh(b\xi\sqrt{u}) + \frac{\hat{q}}{\sqrt{u}}\sinh(b\xi\sqrt{u})\right] \\
&= s^{b\sqrt{u}} \cdot \frac{1}{2}\left(1 + \frac{\hat{q}}{\sqrt{u}}\right) + s^{-b\sqrt{u}} \cdot \frac{1}{2}\left(1 - \frac{\hat{q}}{\sqrt{u}}\right).
\end{aligned}\right\} \tag{9.8}$$

Multiplying by the Born term and taking into account that

$$\frac{1}{m - \hat{q}} \cdot (\sqrt{u} \pm \hat{q}) = \frac{1}{m \mp \sqrt{u}} \cdot (\sqrt{u} \pm \hat{q}), \tag{9.9}$$

we get for the positive signature amplitude

$$F^+ = [(-s)^{\alpha_+} + s^{\alpha_+}]\frac{g^2}{m - \sqrt{u}}\frac{\sqrt{u} + \hat{q}}{2\sqrt{u}} + \{\sqrt{u} \to (-\sqrt{u})\}, \tag{9.10a}$$

$$\alpha_\pm \equiv a(u) \pm b(u)\sqrt{u}. \tag{9.10b}$$

We see that, indeed, two Regge poles (9.10b) arise that conspire at $u = 0$ and in the physical region of the s-channel ($u < 0$) become complex conjugate.

These two poles have opposite parity because in the cms of the u-channel, where $\boldsymbol{q} = 0$, we have

$$\sqrt{u} \pm \hat{q} = q_0(1 \pm \gamma_0), \tag{9.11}$$

and $(1 \pm \gamma_0)$, as we already know, are the projection operators onto the state with parity $P = \pm 1$.

It is possible to transform (9.10) in such a way as to extract the signature factor. Indeed, for small g^2 we have

$$\left.\begin{aligned}
\alpha^+ &= (m - \sqrt{u})\frac{g^2}{\pi}f(q^2), \qquad f = \mathcal{O}(1), \\
\sin\pi\alpha^+ &\simeq \pi(m - \sqrt{u})\frac{g^2}{\pi}f, \\
\frac{g^2}{m - \sqrt{u}} &\simeq \frac{g^4 f}{\sin\pi\alpha_+},
\end{aligned}\right\} \tag{9.12}$$

i.e. the residue in the pole turns out to be $\mathcal{O}(g^4)$.

Note that from the expression (9.2) for the trajectory we have Re $\alpha > 0$, so the asymptotics of the scattering amplitude in (9.6) is determined by the pole with $P_j = 1$, i.e. by the physical electron.

Up to this point we have considered charged particles with spin 1/2. If we took a scalar particle instead, then the corresponding trajectory would not cross the point $\alpha = 0$ at $q^2 = m^2$, i.e. a scalar particle would not reggeize.

9.2.2 Reggeization in QED (with massless photon)

A few words about the situation in conventional electrodynamics.

If the photon mass λ is equal to zero, then the region of small $k_\perp$ in the integral (9.2) is significant. Therefore it seems that one can omit $k_\perp$ in the denominator of the fermion propagator in (9.2). Then the factor $(m - \hat{q})$ cancels out, and the trajectory does not pass through $\alpha = 0$ at $\hat{q} = m$. This approximation is, however, not correct since the integral (9.2) is logarithmically divergent for $\lambda = 0$.

Physically this divergence is related to the fact that in electrodynamics any process is accompanied by emission of infinitely many bremsstrahlung photons, and the probability of a purely elastic process is equal to zero. In order for the amplitude of the elastic scattering, and the trajectory $\alpha(\hat{q})$, to be meaningful one has to cut off the integral in (9.2) at small $k_\perp$.

We could do that by treating the electron not as a real particle but as a virtual one: $p^2 - m^2 = \Delta m^2$. Then the asymptotics of the amplitude, first of all, depends on the introduced quantity Δm^2 and, secondly, there appear large double logarithmic terms of the order of $g^2 \ln^2 s \propto \alpha_{\text{e.m.}} \ln^2 s$ ($\alpha_{\text{e.m.}} \simeq 1/137$). In such circumstances it is not sufficient to calculate only the leading terms in asymptotics in order to determine the trajectory.

Another way around is to calculate the probabilities of the elastic and inelastic processes and sum them up. In the resulting differential cross section the double logarithmic terms cancel out, and formally it has the form of a Regge pole contribution:

$$d\sigma = d\sigma_0 \exp(2\beta \ln s). \qquad (9.13)$$

The expression for β tends to zero at $\hat{q} = m$. But this expression depends on experimental conditions, i.e. on the value of the experimental resolution in $k_\perp$ for registered bremsstrahlung photons.

To conclude, one could say that the electron in QED is reggeized, but its trajectory is not universal, rather it depends on experimental conditions. The situation for scalar particles in conventional electrodynamics is the same as for spinor particles, contrary to the case when $\lambda \neq 0$.

9.3 Electron reggeization from the cross-channel point of view: nonsense states

Let us consider now the reggeization of particles from the point of view of the cross-channel partial waves.

There are several ways of defining partial waves for particles with spin. It is convenient to characterize them by the quantity which is called helicity. The helicity of a particle is the projection of its spin on the direction of its motion: $\mu = \boldsymbol{\sigma} \cdot \boldsymbol{p}/|\boldsymbol{p}|$. Because the projection of the orbital momentum on the direction of motion is zero, the helicity coincides with the projection of the full momentum j. One can define a partial wave amplitude by three numbers: j, m, and m', where m (m') is the full helicity of the initial (final) state, which is equal to the difference between the helicities of the incident (produced) particles. To simplify our qualitative discourse we have dropped the dependence of the partial wave amplitudes on the helicities of the individual particles.

The u-channel unitarity condition for particles with spin reads, symbolically,

$$\Delta\phi^j_{mm'} \simeq \frac{k^{2j+1}}{\omega} \sum_{m''} \phi^j_{mm''}\phi^{*j}_{m''m'} + \cdots . \tag{9.14}$$

In our case m takes on values from $-\frac{3}{2}$ to $+\frac{3}{2}$, since we have a spinor and a vector particle in the u-channel. For the value of the full momentum $j = \frac{1}{2}$ under interest, the states with $m = \pm 3/2$ cannot really occur. They can, however, contribute away from the physical point $j = \frac{1}{2}$.

Such states are called *nonsense* states.

For large j in the sum (9.14) there can be states with any $|m| \leq j$. Let us now take j tending to $\frac{1}{2}$. Then, because in the unitarity condition for physical amplitudes there are only real intermediate states, the contribution of nonsense states should disappear. In the expression for a partial wave amplitude of the transition sense–nonsense there enter not simply Legendre polynomials but more complicated functions. In our case these functions contain explicitly the multiplier $(j - \frac{1}{2})^{1/2}$. Thus, the natural quantity for the transition sense–nonsense is not the conventional partial wave amplitude ϕ^j_{sn}, but rather

$$f^j_{\mathrm{sn}} = \frac{\phi^j_{\mathrm{sn}}}{\sqrt{j - \frac{1}{2}}} .$$

Here subscripts s and n denote sense and nonsense.

The unitarity condition (9.14), being rewritten via f^j_{sn}, contains the factor $j - \frac{1}{2}$, which ensures the disappearance at $j = \frac{1}{2}$ of contributions from nonsense states to $\mathrm{Im}\,\phi_{\mathrm{ss}}$.

Let a Regge pole be at $j = \alpha + \frac{1}{2}$. Then the partial wave amplitudes should have the form

$$\phi^j_{ss} = \frac{r_{ss}}{j - \frac{1}{2} - \alpha(u)}, \quad \frac{\phi^j_{sn}}{\sqrt{j - \frac{1}{2}}} = \frac{r_{sn}}{j - \frac{1}{2} - \alpha(u)}, \quad \phi^j_{nn} = \frac{r_{nn}}{j - \frac{1}{2} - \alpha(u)}. \quad (9.15)$$

The requirement of the residue factorization imposess an extra restriction, which in the region close to the pole reads

$$r^2_{sn}\alpha(u) = r_{ss}r_{nn}. \quad (9.16)$$

Let us return now to the perturbation theory. We consider first the Born approximation:

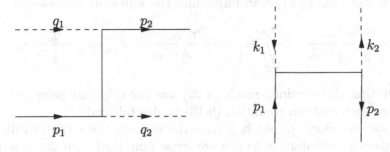

The direct calculation of partial wave amplitudes results in

$$\phi^j_{ss} = -\eta^2_s \delta_{j, \frac{1}{2}}, \quad \phi^j_{sn} = \frac{\eta_s \eta_n}{\sqrt{j - \frac{1}{2}}}, \quad \phi^j_{nn} = \frac{\eta^2_n}{j - \frac{1}{2}}, \quad (9.17)$$

where

$$\left.\begin{array}{c} \eta_n = \left(\dfrac{g^2(E+m)(\sqrt{u}-m)}{8\pi \sqrt{u}\, k^2} \right)^{\frac{1}{2}}, \\[3mm] \eta_{s0} = \left(\dfrac{g^2(E+m)\lambda^2}{8\pi \sqrt{u}\, k^2 (\sqrt{u}-m)} \right)^{\frac{1}{2}}, \\[3mm] \eta_{s1} = \left(\dfrac{g^2(E+m)(E-m-\omega)}{16\pi \sqrt{u}\, k^2 (\sqrt{u}-m)} \right)^{\frac{1}{2}}. \end{array}\right\} \quad (9.18)$$

Here the indices 0 and 1 mark helicities of the vector meson (massive photon). The helicity of the electron is equal to $\frac{1}{2}$. The energies of the electron and photon in the cms of the u-channel are denoted by E and ω respectively; k is the momentum.

From (9.17) one can envisage the fundamental rôle of nonsense states for reggeization.

Iteration of the sense–sense amplitude $\phi_{ss} \propto \delta_{j,\frac{1}{2}}$ using the two-particle unitarity condition reproduces again $\delta_{j,1/2}$. At the same time, iteration of the nonsense–nonsense partial wave amplitude ϕ_{nn} can yield a series of the type

$$\frac{1}{j-\frac{1}{2}} + \frac{\alpha}{(j-\frac{1}{2})^2} + \frac{\alpha^2}{(j-\frac{1}{2})^3} + \cdots = \frac{1}{j-\frac{1}{2}-\alpha}, \qquad (9.19)$$

Note also that in (9.17) the coefficients in front of $(-\delta_{j,\frac{1}{2}})$, $1/\sqrt{j-\frac{1}{2}}$ and $1/(j-\frac{1}{2})$ are factorized. Only due to this factorization in the Born approximation does the complete perturbation theory amplitude turn out to be factorized as well.

After iteration (9.17) transforms into the following expression:

$$\phi_{ss}^j = \frac{\eta_s^2 \alpha}{j-\frac{1}{2}-\alpha}, \qquad \frac{\phi_{sn}^j}{\sqrt{j-\frac{1}{2}}} = \frac{\eta_s \eta_n}{j-\frac{1}{2}-\alpha}, \qquad \phi_{nn}^j = \frac{\eta_n^2}{j-\frac{1}{2}-\alpha}. \qquad (9.20)$$

We see that the obtained result (9.20) has the required form (9.15), and that the factorization condition (9.16) is also fulfilled.

Of the two Born graphs it is only the second, the s-channel diagram, that gives a contribution to the nonsense amplitude. In other words, it becomes a kernel that produces the reggeized electron upon the u-channel iteration. That is, of course, seen directly from the ladder structure of the relevant Feynman graphs in high orders in perturbation theory.

The first diagram, which contains the factor s as compared with the second one, does not contribute to the nonsense amplitude and therefore does not participate in reggeization, since for this amplitude $u(p_1)\hat{e}_2 = 0$. This reflects the fact that a particle with spin $j = \frac{1}{2}$ cannot have the projection of the spin $m = \frac{3}{2}$.

10
Vector field theory

In the previous lecture we have demonstrated how the reggeization of a fermion (nucleon) occurs within the framework of QED with a massive photon. In the Born approximation the asymptotics of backward Compton scattering is determined by the Feynman graph shown in Fig. 10.1a, which describes the u-channel exchange of a fermion. The matrix element corresponding to the graph Fig. 10.1b falls with energy as $\sqrt{s}/(s - m^2)$.

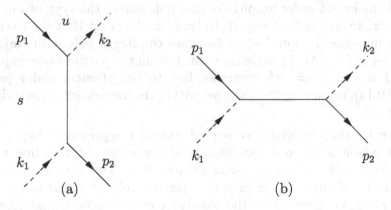

Fig. 10.1. u- And s-channel Born graphs for Compton scattering

Considering the diagrams of higher orders which are responsible for the reggeization of the nucleon, we have convinced ourselves that the main asymptotic contribution arises from the ladder-type graphs shown in Fig. 10.2.

They appear as an iteration of the simplest s-channel diagram of Fig. 10.1b in spite of the fact that its contribution is small. Thus, curiously, the reggeization of the u-channel nucleon pole of Fig. 10.1a occurs due to the presence of the s-channel Born graph (b) whose own contribution to the asymptotics is negligible.

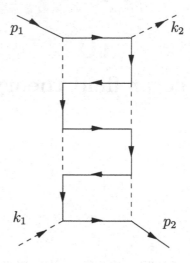

Fig. 10.2.   Ladder graphs responsible for reggeization of a fermion

It is interesting to compare the situation in QED with that in the $g\phi^3$ model, which was considered in one of the previous lectures. In this model the lowest order graphs are also pole-like of the type of Fig. 10.1. However, an iteration of Fig. 10.1b leads to the effect that the fixed pole at the unphysical point $\ell = -1$ becomes the Regge pole with trajectory $\ell = -1 + g^2 f(t)$. At the same time, the pole in the partial wave amplitude with $j = 0$ at $u = m^2$, corresponding to the physical scalar particle (Fig. 10.1a), is not reggeized at least within the framework of perturbation theory.

In vector theories with a conserved current the presence of spin of the vector meson leads to a contribution of the same order as that of diagram Fig. 10.1a for any cross-channel iteration. This corresponds to the appearance of singularities in partial wave amplitudes near the physical point $j = \frac{1}{2}$, i.e. there arises the possibility of reggeizing the nucleon with spin $\frac{1}{2}$.

10.1 Rôle of spin effects in reggeization

In this lecture we consider in more detail the essential rôle of spin effects, which in particular cause the reggeization of spinor particles.

It is convenient to specify every particle by its momentum p_j and helicity m_j which is the projection of its spin on the direction of motion.

The amplitude can be expanded into t-channel partial wave amplitudes:

$$A_{m_1 m_2 m_3 m_4} = \sum_{j=0}^{\infty} (2j+1) f^j_{m_1 m_2 m_3 m_4}(t) d^j_{m_1-m_2, m_3-m_4}(\Theta). \quad (10.1)$$

Here the function $d^j_{m,m'}(v)$ is a natural generalization of the Legendre polynomial, which takes into account the spin of the participating particles. The continuation of the helicity amplitudes $f^j_{\{m\}}(t)$ to complex j permits one to write (10.1) in the form of a Sommerfeld–Watson integral.

The unitarity condition for the amplitudes $f^j_m(t)$ has a form similar to the spinless case:

$$\operatorname{Im} f^j_{m_1 m_2 m_3 m_4}(t) = \frac{k}{16\pi\omega} \sum_{m'_1 m'_2} f^j_{m_1 m_2 m'_1 m'_2}(t) \left(f^j_{m'_1 m'_2 m_3 m_4}(t) \right)^*. \quad (10.2)$$

10.1.1 Nonsense states in the unitarity condition

In the previous lecture in the perturbation theory framework we have discovered the fundamental rôle for high energy scattering of the 'nonsense' state with the maximal spin projection onto the direction of motion in the cms and a sufficiently small total angular momentum j.

Let σ_1, σ_2 denote the spins of colliding particles. The maximal spin projection $\sigma_1 + \sigma_2$ is realized when the helicities of colliding particles are maximal and differ in sign. It is clear that the projection of the total angular momentum j onto the direction of motion is equal to the difference of the particle helicities, since the projection of the *orbital* angular momentum is zero. So the states with spin projection equal to $\sigma_1 + \sigma_2$ and total angular momentum $j \leq \sigma_1 + \sigma_2 - 1$ cannot be realized physically. They are the nonsense states discussed above. For example, the state formed by two particles with $\sigma_1 = \sigma_2 = 1$ having total angular momentum $j = 1$ and spin projection onto the direction of motion $m_1 - m_2 = 2$ is a nonsense state.

At fixed j there is a set of sense states $|i\rangle$, $|k\rangle$, ... The two-particle unitarity condition can be separated into three groups of relations for sense–sense, sense–nonsense and nonsense–nonsense transition amplitudes:

$$\text{Im } f_{ik} = \frac{k}{16\pi\omega} \sum_{k'} f_{ik'} f^*_{k'k} + \frac{k}{16\pi\omega} \sum_{n'} f_{in'} f^*_{n'k}, \qquad (10.3a)$$

$$\text{Im } f_{in} = \frac{k}{16\pi\omega} \sum_{k'} f_{ik'} f^*_{k'n} + \frac{k}{16\pi\omega} \sum_{n'} f_{in'} f^*_{n'n}, \qquad (10.3b)$$

$$\text{Im } f_{nm} = \frac{k}{16\pi\omega} \sum_{k'} f_{nk'} f^*_{k'm} + \frac{k}{16\pi\omega} \sum_{n'} f_{nn'} f^*_{n'm}, \qquad (10.3c)$$

where n, m, n' mark nonsense states. It is clear that any transitions to the unphysical states in (10.3a) are absent for integer j. So the condition $f^j_{in} \to 0$ at $j \to \sigma_1 + \sigma_2 - 1$ has to be satisfied.

It is interesting to see how the perturbative partial wave amplitudes f^j of Compton scattering, considered in the previous lecture, behave in this respect. If one calculates the helicity amplitudes $A_{m_1 m_2 m_3 m_4}$ for the two lowest order graphs of Fig. 10.1a,b then, using the orthogonality condition for $d^j_{m,m'}(\Theta)$, one can find from (10.1) all partial wave amplitudes $f^j_{m_1 m_2 m_3 m_4}$. Near the point $j = \sigma_1 + \sigma_2 - 1$ (in our case $\sigma_1 = 1$ and $\sigma_2 = \frac{1}{2}$) we can obtain expressions of the form (see the previous lecture)

$$f^j_{ik} \sim g^2 \delta_{j,\frac{1}{2}}, \qquad (10.4a)$$

$$f^j_{in} \sim \frac{g^2}{j+1-\sigma_1-\sigma_2} \cdot \sqrt{j+1-\sigma_1-\sigma_2}, \qquad (10.4b)$$

$$f^j_{nm} \sim \frac{g^2}{j+1-\sigma_1-\sigma_2}. \qquad (10.4c)$$

As was mentioned above, the vanishing of the sense–nonsense transition amplitude at $j = -1 + \sigma_1 + \sigma_2$ follows from physical reasons. From the formal point of view, the appearance of the multiplier $\sqrt{j+1-\sigma_1-\sigma_2}$ in (10.4b) has a general reason, related to the properties of the spherical functions $d^j_{m,m'}(\Theta)$.

As for the pole factors in (10.4b) and (10.4c), they originate from the use of the perturbation theory.* For instance, the amplitude, corresponding to a nonsense–nonsense transition, calculated from the Born graph Fig. 10.1b, has the form

$$f^j_{nn} \sim \frac{g^2}{j+1-\sigma_1-\sigma_2}, \qquad (10.5)$$

since the absorptive part $A_{1,nn}$ of this perturbative graph is proportional to $\delta(s - m^2)$. Note that in the $g\phi^3$ theory, where $\sigma_1 = \sigma_2 = 0$, the state

---

* cf. (8.14)–(8.18) [ed]

with $\ell = -1$ is in some respects a nonsense state for the projection of spin $\sigma_1 + \sigma_2 = 0$, so in the lowest order of perturbation theory the amplitude f^ℓ has a pole at the point $\ell = -1 + \sigma_1 + \sigma_2 = -1$.

10.1.2 Iteration of the unitarity condition

Let us investigate now higher order corrections to f^j, using for the unitarity condition an iteration procedure, applied to the Born expressions (10.4). It is evident that in the relation (10.3c) the maximal contribution $g^4/(j + 1 - \sigma_1 - \sigma_2)^2$ comes from the second term, as compared with the first term $\sim g^4/(j+1-\sigma_1-\sigma_2)$. So at $j \simeq \sigma_1+\sigma_2-1$ the unitarity condition (10.3c) relates to only nonsense–nonsense transitions:

$$\text{Im } f_{nm} = \frac{k}{16\pi\omega} \sum_{n'} f_{nn'} f_{n'm}^*. \tag{10.6}$$

This equation together with the dispersion relation for f_{nm} allows us to find the trajectory of the Regge pole, following the procedure that we have developed in Lecture 8 considering the $g\phi^3$ theory. Perturbation theory, used in the previous lectures, is only required to prove both an applicability of a subtraction-free dispersion relation for f_{nm}^j and the possibility of using only the two-particle unitarity condition (10.6).

After finding the nonsense–nonsense amplitudes we easily calculate the sense–nonsense amplitudes from the linear relation (10.3b), and by inserting them into (10.3a), we derive the sense–sense amplitudes. So in fact only the nonsense states work in all intermediate states, at least in the leading logarithmic approximation.

10.1.3 Nonsense states from the s- and t-channel points of view

What are the physical arguments supporting the dominance of the nonsense state from the standpoint of the s-channel?

Let us consider, for example, the case when two particles with spins 1 are present in the intermediate state of the t-channel (they are represented by the dashed lines in Fig. 10.3a). We choose the z axis along the 3-vector k_1 in the cms of the t-channel, and the x axis is chosen in the reaction plane (see Fig. 10.3b).

Then the state of a photon with momentum k_1, polarized along the x- or y-axis, will be described by vectors $(e_x)_\mu$ or $(e_y)_\mu$, respectively. The states of the photon with helicities ± 1 are described by the polarization vectors

$$e^\pm = \frac{e_x \pm i\, e_y}{\sqrt{2}}. \tag{10.7}$$

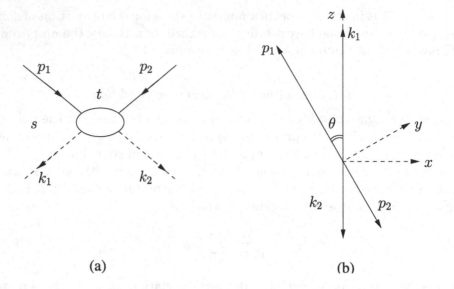

Fig. 10.3.   Kinematics of Compton scattering

The vectors e^+, e^- have the following properties:

$$e^+ = (e^-)^*; \quad (e^+ e^-) = -1; \tag{10.8a}$$
$$(e^+ e^+) = (e^- e^-) = (e^\pm k_1) = (e^\pm k_2) = 0. \tag{10.8b}$$

In the same way the state of the photon with momentum k_2 and helicity $+1$ (-1) is described by the vector e^- (e^+) since the direction of its momentum is opposite to that of the momentum k_1 in the cms of t-channel. Therefore, the two-photon state with spin projection 2 onto the z axis will be described by the tensor $e^+_{\mu_1} e^+_{\mu_2}$.

The fact that the nonsense state does not exist at $j = 1$, $\sigma_1 + \sigma_2 = 2$ readily follows from the observation that, due to (10.8b), it is impossible to construct a 4-vector from the tensor $e^+_\mu e^+_\nu$ and momenta $k_{1\sigma}$, $k_{2\rho}$.

The most important feature of t-channel nonsense states is that it is these intermediate states that give the main asymptotic contribution to the s-channel scattering amplitudes.

Nonsense states in two-photon exchange. Let us demonstrate this on a particular example. Consider the diagrams with two vector mesons (massive photons) in the t-channel intermediate state (see Fig. 10.4).

Then the spin structure for propagators of each vector meson can be written in the following form:

$$-g_{\mu\nu} + \frac{k_\mu k_\nu}{k^2} = e^+_\mu e^-_\nu + e^-_\mu e^+_\nu + e^0_\mu e^0_\nu, \tag{10.9}$$

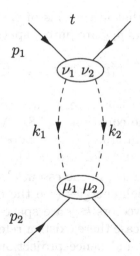

Fig. 10.4. Two-photon t-channel exchange

where e^0 is the longitudinal polarization vector (for $k_\mu = (k_x, k_y, k_z, k_0) = (0, 0, |\boldsymbol{k}|, k_0)$, $e^0_\mu = (0, 0, k_0, |\boldsymbol{k}|)/\sqrt{k^2}$). Its contribution depends on the gauge choice but, as we shall see below, it is inessential for the description of high energy processes.

To build up explicit expressions for $e^\pm$ we find first two vectors $\tilde{p}_{1,2}$ orthogonal to $k_{1,2}$. We shall search for them in the form

$$\tilde{p}_{1,2} = p_{1,2} + a_{1,2}k_1 + b_{1,2}k_2, \qquad (10.10a)$$

where the coefficients a_i, b_i can easily be found from the four conditions

$$(\tilde{p}_1 k_1) = (\tilde{p}_1 k_2) = (\tilde{p}_2 k_1) = (\tilde{p}_2 k_2) = 0. \qquad (10.10b)$$

Given the two vectors $\tilde{p}_1$, $\tilde{p}_2$ it is straightforward to construct their linear combinations $e^\pm$,

$$e^\pm = \gamma_1^\pm \tilde{p}_1 + \gamma_2^\pm \tilde{p}_2, \qquad (10.11)$$

that satisfy the orthonormality conditions

$$(e^+ e^+) = (e^- e^-) = 0, \quad (e^+ e^-) = -1.$$

The explicit expressions are

$$e^\pm = \frac{1}{\sqrt{2}} \left\{ \pm \frac{\tilde{p}_2}{\sqrt{\tilde{p}_2^2}} + \frac{\tilde{p}_1 \tilde{p}_2^2 - \tilde{p}_2 (\tilde{p}_1 \tilde{p}_2)}{\sqrt{[(\tilde{p}_1 \tilde{p}_2)^2 - \tilde{p}_1^2 \tilde{p}_2^2] \tilde{p}_2^2}} \right\}. \qquad (10.12)$$

From (10.10) we observe that in the cms of the t-channel (where k_1 and k_2 are time-like) the vectors $\tilde{p}_{1,2}$ are purely spatial and obey the relations

$$\tilde{p}_{1,2}^2 = -\tilde{\boldsymbol{p}}_{1,2}^2 < 0, \quad \tilde{p}_1^2 \tilde{p}_2^2 > (\tilde{p}_1 \tilde{p}_2)^2, \quad (\tilde{\boldsymbol{p}}_{1,2} \cdot \boldsymbol{k}) = 0, \qquad (10.13)$$

where $\boldsymbol{k} = \boldsymbol{k}_1 = -\boldsymbol{k}_2$. In this case $e^{\pm}$ are complex, and it is easy to check that they satisfy the equations (10.8). Vectors $e^{\pm}$ so constructed will have the standard form (10.7) if one chooses the z axis along $\boldsymbol{k}$, the y axis along $\tilde{\boldsymbol{p}}_2$ and the x axis along $[\tilde{\boldsymbol{p}}_2 \times \boldsymbol{k}]$.

Conversely, in the s-channel, in the essential integration region for the diagram of Fig. 10.4 at high energies (see the calculation of the box diagram in Lecture 8), the vectors $k_{1,2}$ are space-like, so that $\tilde{p}_{1,2}$ become time-like. (Indeed, in this case there exists a reference frame in which $k_{1,2}$ are purely spatial vectors and hence projection of the time-like vectors $p_{1,2}$ onto the subspace orthogonal to $k_{1,2}$ will give the time-like vectors $\tilde{p}_{1,2}$.) So in the s-channel we have

$$\tilde{p}_{1,2}^2 > 0, \qquad \tilde{p}_1^2 \tilde{p}_2^2 < (\tilde{p}_1 \tilde{p}_2)^2, \qquad (10.14)$$

and the vectors $e^{\pm}$ in (10.12) are real.

In particular, at large energies $s \gg m^2$, where we can approximate

$$\tilde{p}_{1,2} \simeq p_{1,2}, \qquad (10.15)$$

the following simple expressions for $e^{\pm}$ follow from (10.12):

$$e_{\mu}^+ \simeq p_{1\mu} \frac{\sqrt{2\tilde{p}_2^2}}{s}, \quad e_{\mu}^- \simeq -p_{2\mu} \sqrt{\frac{2}{\tilde{p}_2^2}}; \qquad s = 2p_1 p_2. \qquad (10.16)$$

It follows from (10.9) and (10.16) that the main contribution to the asymptotics in the s-channel really arises from a nonsense intermediate state of two photons in the t-channel which is described by the tensor $e_{\mu_1}^+ e_{\mu_2}^+$:

$$\left[g_{\mu_1 \nu_1} - \frac{k_{1\mu_1} k_{1\nu_1}}{k_1^2} \right] \left[g_{\mu_2 \nu_2} - \frac{k_{2\mu_2} k_{2\nu_2}}{k_2^2} \right] \simeq e_{\mu_1}^+ e_{\mu_2}^+ \, e_{\nu_1}^- e_{\nu_2}^- + \cdots . \quad (10.17)$$

This follows from the observation that the lower (upper) blob in Fig. 10.4 contains a single vector with large components $p_{2\mu}$ ($p_{1\nu}$) so that its main contribution should be proportional to $p_{2\mu_1} p_{2\mu_2}$ ($p_{1\nu_1} p_{1\nu_2}$). The convolution of these tensors with the propagators of intermediate photons then singles out the nonsense–nonsense component (10.17).

10.1.4 The $j = \frac{1}{2}$ pole in the perturbative nonsense–nonsense amplitude

Let us explain now why in the case of Compton backward scattering in the amplitude corresponding to the t-channel nonsense–nonsense transition there appears a pole at $j = \frac{1}{2}$.

To this end we have to build up an expression corresponding to the total projection $\frac{3}{2}$ from the spinor $u(p_1)$ and the polarization vector $e_\mu(k_2)$ (see Fig. 10.1). Since for the total momentum $j = \frac{1}{2}$ it is a nonsense state, the relation

$$\hat{e}_2 u(p_1) = 0 \qquad (10.18a)$$

should be satisfied. Here $e_{2\mu}$ and $u(p_1)$ describe a vector meson and a nucleon, both polarized in the direction of the nucleon momentum $\boldsymbol{p}_1$ in cms of the u-channel. If the u-channel *final* state is also nonsense, then we have a similar condition for the wave functions of the final particles:

$$\bar{u}(p_2)\hat{e}_1 = 0, \qquad (10.18b)$$

where $e_{1\mu}$ and $\bar{u}(p_2)$ describe the particles polarized both in the direction $\boldsymbol{p}_2$.

The conditions (10.18) show that the diagram of Fig. 10.1a does not contribute to the nonsense–nonsense transition amplitude in u-channel.

For the diagram of Fig. 10.1b we have

$$A_{\mathrm{nn}} = \bar{u}(p_2)\,\hat{e}_2\frac{m + \hat{p}_1 + \hat{k}_1}{m^2 - s}\hat{e}_1\,u(p_1).$$

The conditions (10.18) allow us to simplify this expression significantly:

$$A_{\mathrm{nn}} = 2(e_1 e_2)\,\bar{u}(p_2)\frac{-\hat{k}_1}{m^2 - s}u(p_1) = 2(e_1 e_2)\,\bar{u}(p_2)\frac{\hat{q} - m}{m^2 - s}u(p_1). \quad (10.19)$$

The last relation follows from the Dirac equation $\bar{u}(p_2)(\hat{p}_2 - m) = 0$. From (10.18b) we observe that the nonsense–nonsense amplitude vanishes at $q^2 = m^2$, as expected.

Now we are in a position to find an order of magnitude of this amplitude at high energies. The first step is to use (10.7) for the polarization vectors of the incident photon $e_1 = e^\pm$. To obtain then the polarization vector e_2 of the second photon, we have to rotate $e^\pm$ by the scattering angle Θ_u around the axis y:

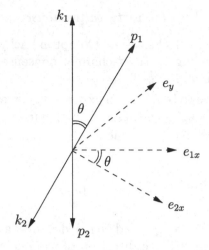

It is easy to verify that

$$(e_1 e_2) = \frac{\pm 1 + z}{2} \propto z,$$

$$\bar{u}(p_2)\,(\hat{q} - m)\,u(p_1) \;\propto\; \sin\frac{\Theta_u}{2} \propto \sqrt{z},$$

with z the cosine of the scattering angle in the u-channel, $z \propto s$. As a result, the magnitude of the nonsense–nonsense transition amplitude (10.19) at large energies is

$$A_{\mathrm{nn}} \sim s \cdot s^{1/2} \Big/ s \sim s^{1/2}.$$

This implies that the corresponding u-channel nonsense–nonsense partial wave amplitude has a pole at $j = \frac{1}{2}$.

In the spinless case of $\lambda\phi^3$ theory we had a pole at $\ell = -1$. In the presence of spin this pole is shifted, as we have shown in this lecture, to the point $j = \sigma_1 + \sigma_2 - 1$. Thus for nonsense states there is a trivial additive relation between j, σ and ℓ. Another important lesson is that the spin and coordinate parts of the wave function have a factorized form.

For instance if we have two external particles of spin 1, then in order to construct a tensor of rank j we can use (due to the orthogonality conditions (10.8)) only functions of the scalar product $(k_1 k_2)$, multiplied by arbitrary tensors built from the vectors k_1, k_2 and $e_{1,2}^{\pm}$. The number of vectors $e_{1,2}^{\pm}$ used in this procedure will give the spin of the state while the number of vectors $k_{1\mu}$, $k_{2\nu}$ used will give the orbital angular momentum of this state in cms of the cross-channel.

With this remark we conclude the discussion of nonsense states.

10.2 QED processes with photons in the t-channel

We have considered the QED process of backward Compton scattering whose physical region on the Mandelstam plane is marked as a in Fig. 10.5. The region b corresponds to the related process of two-photon annihilation of a e^+e^- pair.

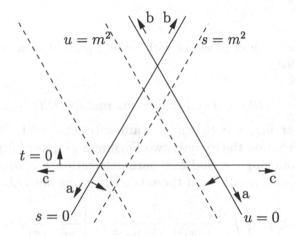

Fig. 10.5. Physical regions of two-photon annihilation (b) and of Compton scattering (a,c) on the Mandelstam plane

Let us consider now the high energy processes with finite t and photons in the t-channel. It could be forward Compton scattering (the region c in Fig. 10.5) or small angle electron–electron scattering; see Fig. 10.6.

It is natural to separate these processes into two types having essentially different asymptotic behaviour, namely those with an *odd* (Fig. 10.6a) and *even* number of photons (see Fig. 10.6b,c) in any t-channel section. Such a separation is possible due to the charge conjugation invariance of the theory.

Processes of the type of Fig. 10.6a correspond to the t-channel transition between states with negative charge parity. In the case of spinless charged particles, processes (a) and (b,c) differ by signature $P_j = (-1)^j$. That is, the processes with an odd number of t-channel photons have $P_j = -1$ and those with an even number of photons have $P_j = +1$. In the presence of spin in both cases a small admixture of states with another signature will be present.

The processes with $P_j = -1$ are related to the reggeization problem of the vector meson in QED, whereas the processes with $P_j = +1$ allow us to investigate an interesting possibility of existence of a vacuum Regge pole in a simple QFT model.

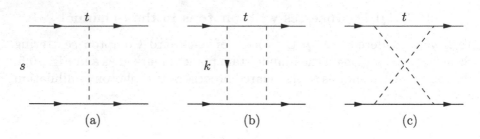

Fig. 10.6. Separation of processes with an odd (a) and even (b,c) number of exchange photons

10.2.1 The vacuum channel in QED

Let us consider first the t-channel state with $P_j = +1$. In the lowest order of perturbation theory only two Feynman graphs of Fig. 10.6b,c are relevant. Introducing the Sudakov parametrization $k = \alpha p_2 + \beta p_1 + k_\perp$, for the sum of contributions of these two diagrams one obtains

$$M_{b,c}^{(0)} = e^4 (2p_1 2p_2)^2 \frac{s}{2} \int \frac{d\alpha \, d\beta \, d^2 k_\perp}{(2\pi)^4 i} \frac{1}{-\alpha s + i\varepsilon} \left[\frac{1}{\beta s + i\varepsilon} + \frac{1}{-\beta s + i\varepsilon} \right]$$
$$\times \frac{1}{[k_\perp^2 + s\alpha\beta - \lambda^2 + i\varepsilon][(q-k)_\perp^2 + s\alpha\beta - \lambda^2 + i\varepsilon]}$$
$$= 2s \, e^2 \cdot i\, \alpha_{\text{e.m.}} \int \frac{d^2 k_\perp}{\pi} \frac{1}{[k_\perp^2 - \lambda^2][(q-k)_\perp^2 - \lambda^2]}, \quad \alpha_{\text{e.m.}} = \frac{e^2}{4\pi}.$$

So in this approximation the amplitude is purely imaginary, $M^{(0)} \sim i s$. This corresponds to constant total cross section of $e^- e^-$ scattering which in the same approximation is equal to the total cross section of $e^- e^+$ scattering. The situation is reminiscent of that for the vacuum pole in strong interactions.

From (10.19) we see, however, that the position of this pole in the j plane does not depend on t. The fixed pole in the t-channel partial wave amplitude $f_j \sim 1/(j-1)$, which corresponds to the asymptotics (10.19), has the general form of (10.4c): $e^4/(j - \sigma_1 - \sigma_2 + 1)$. It is related, as we have shown above, to the dynamical rôle of nonsense states of two photons in the t-channel.

One can ask the question: can the higher order approximations lead to the reggeization of this pole?

In the next order of perturbative expansion the main contribution to the asymptotics with $P_j = +1$ will arise from the three diagrams of Fig. 10.7:

$$M^{(1)} \sim i \, s e^4 f_1(t) \cdot e^4 \ln s. \tag{10.20}$$

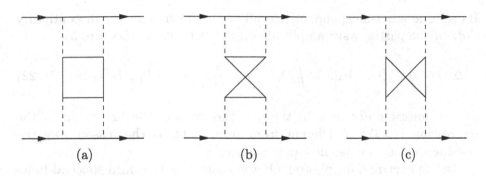

Fig. 10.7.   Higher order contributions to $P_j = +1$ t-channel exchange

In the general case when $e^4 \ln s \sim 1$ one can restrict oneself to summing the sequence of ladder-type diagrams shown in Fig. 10.8, where the blob of the light–light scattering includes a sum of six diagrams[†] of the lowest order (Fig. 10.7).

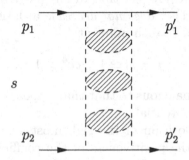

Fig. 10.8.   Iteration of light–light scattering blobs

Thus, in the approximation $e^4 \ln s \mathcal{O}(1)$, $e^2 \ll 1$, the expansion of the scattering amplitude looks as follows:

$$M = i e^2 s \sum_{n=0}^{\infty} f_n(t)(e^4 \ln s)^n. \qquad (10.21)$$

The appearance of an additional factor $\ln s$ in each successive order of perturbative series is related to the fact that the light–light scattering amplitude corresponding to the nonsense–nonsense transitions has a pole with its residue depending both on t and on the virtual masses of external photons.

In the $g\phi^3$ theory the pole $1/(\ell + 1)$ appeared from the divergence of the integral on the left cut $\int_{-\infty}^{0} (dt' \Delta \phi_\ell(t')/(t' - t))$ at $\ell \to 1$. Therefore

---

[†] three topologies; two directions of the electron arrow

its residue was t-independent. Recall the expression for the discontinuity $\Delta\phi_\ell$ of the partial wave amplitude on the left cut (see Lecture 3):

$$\Delta\phi_\ell \sim \int P_\ell(-z)A_1\mathrm{d}s + \int Q_\ell\left(\frac{2s}{4m^2 - t + \mathrm{i}\varepsilon} - 1\right)\rho_{su}(s,t)\,\mathrm{d}s. \quad (10.22)$$

In the *lowest order* in g in the $g\phi^3$ theory only the first term in the expression (10.21) is different from zero. This is the reason why the residue in this case has no t-dependence.

On the contrary, for photon–photon scattering the third spectral function $\rho_{su}(s,t)$ is non-zero (see Fig. 10.7c). This means that the discontinuity on the left cut, as follows from the generalization of (10.22) to the case of particles with spin, acquires a pole at $j = 1$ with a t-dependent residue.

In the next lecture it will be shown that in such a situation the iteration of this pole produces, in general, an infinite series of poles condensing near $j = 1$. It turns out however that in our particular QED problem the coefficient before $(j - 1)^{-1}$ *does not decrease* at large t. As a result there appears instead a fixed branch point at

$$j = j_0 \simeq 1 + ce^4 > 1.$$

This branch point is analogous to that that appears in the non-relativistic theory with the $1/r^2$ potential.

The fact that the position of the rightmost singularity exceeds unity follows directly from the positiveness of the coefficients f_n in (10.21) at $t = 0$. The latter property reflects the positiveness of the production cross section of several e^+e^- pairs.

As a result the total cross section, which by the optical theorem is proportional to the imaginary part of the forward scattering amplitude, grows at large s as

$$\sigma_{\mathrm{tot}} \sim s^{j_0-1} = s^{ce^4}. \quad (10.23)$$

As we know, such a behaviour of the total cross section contradicts the Froissart theorem. This contradiction is explained by the fact that our leading approximation violates s-channel unitarity. Using the methods which will be developed in Lectures 12 and 13 one can show that the correction terms to (10.20), of the order of $se^8 f(e^4 \ln s)$, will exceed the main terms for $e^4 \ln s \gg 1$.

10.2.2 The problem of the photon reggeization

Let us consider now the diagrams of type Fig. 10.6a with an odd number of photons in the t-channel. They have negative signature, $P_j = -1$, and are interesting in view of the problem of reggeization of vector mesons (massive photons) within a perturbative QFT framework. The contribution of diagram Fig. 10.6a has the form

$$F \simeq e^2 \frac{2s}{\lambda^2 - t}. \qquad (10.24)$$

In the next order of perturbation theory we have to consider the six diagrams shown in Fig. 10.9. We proceed by a close analogy with (10.19) and expand particle momenta in term of the Sudakov variables α_i, β_i and $k_{i\perp}$. Performing the α_i integrations by taking residues in the upper virtual electron lines, we arrive at the following integral over $\beta_{1,2}$:

$$\int_{m^2/s}^{1} d\beta_1 d\beta_2 \left[\frac{1}{(\beta_1 + i\varepsilon)(\beta_1 + \beta_2 + i\varepsilon)} + \frac{1}{(\beta_2 + i\varepsilon)(\beta_1 + \beta_2 + i\varepsilon)} \right.$$
$$+ \frac{1}{(\beta_2 + i\varepsilon)(-\beta_1 + i\varepsilon)} + \frac{1}{(\beta_1 + i\varepsilon)(-\beta_2 + i\varepsilon)}$$
$$\left. + \frac{1}{(-\beta_1 + i\varepsilon)(-\beta_1 - \beta_2 + i\varepsilon)} + \frac{1}{(-\beta_2 + i\varepsilon)(-\beta_1 - \beta_2 + i\varepsilon)} \right],$$
$$(10.25)$$

where we kept in the denominators only potentially large terms $\beta_i s \gg m^2 \sim |t|$.

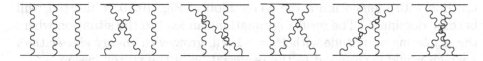

Fig. 10.9.   Three-photon exchange graphs generating (10.25)

Large contributions containing $\ln s$ are indeed present in separate graphs but, as can be seen from (10.25), they cancel out in the full sum. The remaining integral over the transverse momenta $k_{1\perp}$ and $k_{2\perp}$ does not give logarithmic contributions either. This exercise shows that if the reggeization of vector mesons takes place in this theory, it may occur only due to diagrams containing electron loops.

The simplest diagram in which the logarithmic terms do appear is shown in Fig. 10.10.

It behaves asymptotically as $e^{10} s \ln s$. One can easily find the diagrams of higher orders in powers of $e^4 \ln s$ if the blob in this diagram is replaced by a set of ladder-type diagrams of Fig. 10.8. Summing up the contributions of all such diagrams results in a quantity of the order of $e^6 s f(e^4 \ln s)$. Therefore in the approximation $e^4 \ln s = \mathcal{O}(1)$, $e^2 \ll 1$ the contribution (10.22) from diagram Fig. 10.6a is much larger, which means that the vector meson in this approximation is not reggeized.

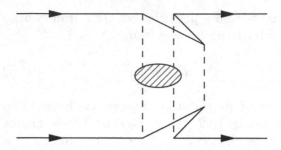

Fig. 10.10.   Structure of diagrams that may contribute to photon reggeization

Will the reggeization of vector mesons take place at $e^4 \ln s \gg 1$?

From s-channel unitarity (positivity of multiple e^+e^- pair production cross sections) it follows that the asymptotics of the amplitude with $P_j = +1$ calculated above (see (10.23)) is

$$A_{P_j=+1}(s,t) \sim \mathrm{i}\, s\, e^2\, s^{e^4 \beta(t)}, \qquad \beta(t) > 0.$$

So there exists a region in energy where the total contribution of diagrams of type Fig. 10.10, $\mathcal{O}\big(e^6 s f(e^4 \ln s)\big)$, becomes comparable with the contribution of the diagram Fig. 10.6a, $\mathcal{O}\big(c^2 s\big)$. This means that at high enough energies the set of diagrams Fig. 10.10, where instead of the photon line with momentum k the whole blob of Fig. 10.10 is inserted, will become dominant. The resulting equation can lead to the situation where the scattering amplitude with $P_j = -1$ will grow with energy *slower* than s which would correspond to the reggeization of the vector meson.

If such a possibility is realized, then in the diagrams of type Fig. 10.8 we have to substitute scattering amplitudes with $P_j = -1$ instead of photon lines. This will suppress the contribution of these diagrams at high energies. Hence the scattering amplitude with $P_j = +1$ will not grow faster than s like a power of s and the discrepancy with the Froissart theorem would disappear.

We conclude by remarking that the problem of vector meson reggeization, as well as the problem of the existence of a vacuum pole in a theory with conserved vector current, [‡] inevitably leads to the necessity of considering the strong interaction regime even in the case of the small coupling, $e^2 \ll 1$.

[‡] With the advent of non-Abelian gauge field theories (Glashow–Weinberg–Salam $SU(2)$ theory of weak interactions; QCD) Gribov's approaches and calculation techniques were applied in 1976 to demonstrate that vector mesons (gluons; intermediate bosons W, Z) *reggeize* in perturbation theory (L. Lipatov; L. Frankfurt and V. Sherman), and so do fermions (quarks; V. Fadin and V. Sherman). These fields reggeize in a similar manner, and in the same sense, as an electron reggeizes in QED [ed].

11

Inconsistency of the Regge pole picture

So far we have assumed that the only singularities of partial wave amplitudes are the poles. But a theory having only poles in the ℓ plane appears to be not self-consistent. To understand the problem we consider first a simple phenomenon.

Let us recall the way complex angular momenta were introduced. To begin with, for integer positive ℓ we had

$$
\begin{aligned}
f_\ell &= \frac{1}{2} \int_{-1}^{1} P_\ell(z) A(z) \, dz \\
&= \frac{1}{\pi} \int_{z_0}^{\infty} Q_\ell(z_s) A_1(s) \, dz_s + \frac{(-1)^\ell}{\pi} \int_{z_0}^{\infty} Q_\ell(z_u) A_2(u) \, dz_u. \quad (11.1)
\end{aligned}
$$

Both integrals can be easily continued to complex ℓ. They give $f_\ell^{(1)}$ and $f_\ell^{(2)}$, respectively. However, the multiplier $(-1)^\ell$ cannot be continued unambiguously. Therefore we have introduced separately

$$
f_\ell^+ = f_\ell^{(1)} + f_\ell^{(2)}, \qquad f_\ell^- = f_\ell^{(1)} - f_\ell^{(2)}, \quad (11.2)
$$

that is we have been forced to present the amplitude as a sum of separate contributions of left and right cuts. Now we are going to pay for that.

11.1 The pole $\ell = -1$ and restriction on the amplitude fall-off

When $t > 0$, nothing unusual happens. But we are interested in the region $t < 0$. There the third spectral function ρ_3 is present (see Fig. 11.1). What is its effect? Consider the dispersion relation at fixed t:

$$
A(s,t) = \frac{1}{\pi} \int_{4\mu^2}^{\infty} A_1(s',t) \frac{ds'}{s'-s} + \frac{1}{\pi} \int_{4\mu^2}^{\infty} A_2(u',t) \frac{du'}{u'-u}. \quad (11.3)
$$

137

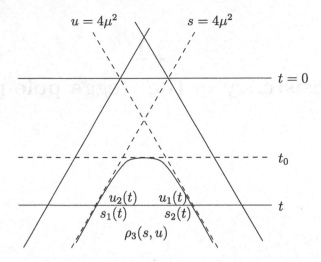

Fig. 11.1.   The third spectral function on the Mandelstam plane

It turns out that due to the third spectral function each term behaves at large energies differently as compared with their sum.

Let us take for instance the imaginary part of the amplitude at $s > 4\mu^2$ and $u < 0$, $t < t_0$ (this is in the physical region of the s-channel). From (11.3) we obtain

$$\operatorname{Im}_s A(s,t) = A_1(s,t) + \frac{1}{\pi} \int_{s_1(t)}^{s_2(t)} \rho_3(s',t)\, \frac{\mathrm{d}s'}{s'-s}$$

$$+ \frac{1}{\pi} \int_{u_1(t)}^{u_2(t)} \rho_3(u',t)\, \frac{\mathrm{d}u'}{u'-u}. \tag{11.4}$$

Here the first term appeared due to the denominator in (11.3) being complex valued and the remaining ones are the contributions from the numerators A_1, A_2 being complex valued as well.

At first sight (11.4) contradicts the statement that the imaginary part of the amplitude in the physical region is equal *by definition* to A_1. There is no contradiction, however, since the sum of the two integral terms is zero. Indeed, we have $s + t + u = 4\mu^2$ and $s' + t + u' = 4\mu^2$ so that $u' - u = -(s' - s)$ and we obtain

$$\int \rho_3 \left[\frac{1}{s'-s} - \frac{1}{s'-s} \right]\, \mathrm{d}s' = 0.$$

If we were to examine separately the contribution of one of these cuts, say the right one,

$$A^{\text{right}} = \frac{1}{\pi} \int_{4\mu^2}^{\infty} A_1(s',t) \frac{ds'}{s'-s}, \qquad (11.5)$$

then its imaginary part would be a sum of two terms:

$$\text{Im}_s A^{\text{right}}(s,t) = A_1 + \frac{1}{\pi} \int_{s_1(t)}^{s_2(t)} \rho_3(s',t) \frac{ds'}{s'-s}. \qquad (11.6)$$

Suppose now that A and A_1 decrease rapidly with $s \to \infty$:

$$|A_1| < s^{-n}, \qquad |A| < s^{-n}, \qquad n > 1. \qquad (11.7)$$

And how does A^{right} in (11.5) behave? In (11.6), A_1 falls rapidly, whereas the second term is of the order of

$$\sim \frac{1}{s} \cdot \int_{s_1}^{s_2} \rho_3(s',t) \frac{ds'}{\pi}.$$

That is,

$$\text{Im}\, A^{\text{right}} \sim \frac{1}{s}, \qquad \text{if} \quad \int_{s_1(t)}^{s_2(t)} \rho_3(s',t)\, ds' \neq 0.$$

For arbitrary $t < t_0$ one cannot prove that this integral is non-zero. But there exists an interval of t where ρ_3 can be calculated explicitly and where $\rho_3 > 0$.

This is related to the explicit form of the two-particle unitarity condition in the s-channel. Indeed in the region below the first inelastic threshold we have

$$A_1(s, z_s) = \int A(s, z_1) A^*(s, z_2) d\Omega \qquad (11.8)$$

(we drop the coefficients that are not essential for us). To obtain $\rho_3 \equiv \rho_{su}$ it is necessary to continue (11.8) analytically to the region $u > 4\mu^2$, i.e. to large enough negative t. Then the singularity of the integral arises when $A(s, t_1)$ and $A^*(s, t_2)$ in the integrand simultaneously hit their proper singularities in t_1, t_2 (see Fig. 11.2a). Therefore in some region in t (corresponding to u_1, u_2 *below* the next singularities) the third spectral function is given by a single Feynman diagram Fig. 11.2b and is positive definite. The same result can be obtained starting from the u-channel.

So $\text{Im}\, A^{\text{right}} \sim 1/s$. The same conclusion is valid for $\text{Im}\, A^{\text{left}}$.

As $f_\ell^{(1)}$ and $f_\ell^{(2)}$ are the partial wave amplitudes of A^{right} and A^{left}, respectively, close to $\ell = -1$ we have

$$\text{Im}\, f_\ell^{(1)} \simeq \frac{\ell_1(t)}{\ell+1}, \quad \text{Im}\, f_\ell^{(2)} \simeq \frac{\ell_2(t)}{\ell+1}, \quad \ell_1 = \ell_2 \propto \int_{s_1}^{s_2} \rho_3(s',t) ds'. \quad (11.9)$$

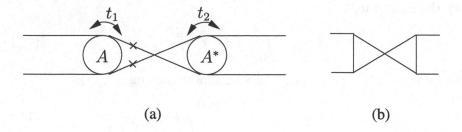

(a)   (b)

Fig. 11.2.   Diagrams giving rise to the third spectral function ρ_{su}

Constructing partial wave amplitudes with definite signature we observe that the pole in f_ℓ^+ remains whereas in f_ℓ^- it is cancelled. This corresponds to the fact that an *odd* ℓ is a 'proper' point for the negative signature amplitude f_ℓ^- whereas for f_ℓ^+ it is 'foreign' – unphysical.

We have assumed from the start that the amplitude is fast falling with energy. For the assumption (11.7) to be valid, the pole at $\ell = -1$ must not give any contribution to the asymptotics of the full amplitude. And this is what actually happens. Indeed, f_ℓ^+ contributes to the amplitude according to the Sommerfeld–Watson formula

$$A^+(s,t) = \frac{i}{4} \int (2\ell + 1) f_\ell^+(t) \frac{[P_\ell(z) + P_\ell(-z)]}{\sin \pi \ell} \, d\ell. \qquad (11.10)$$

At $z \equiv z_s \to \infty$ we have

$$P_\ell(z) \simeq \frac{2^\ell \Gamma(\ell + \frac{1}{2})}{\sqrt{\pi}\, \Gamma(\ell + 1)} z^\ell,$$

so that

$$A_{z \to \infty}^+ \simeq \frac{i}{2\pi} \int d\ell\, \Gamma(-\ell) \frac{2^\ell\, \Gamma(\ell + \frac{3}{2})}{\sqrt{\pi}} f_\ell^+(t) \left[z^\ell + (-z)^\ell \right]. \qquad (11.11)$$

Note that the zeros of $\sin \pi \ell$ at $\ell \le -1$ have been compensated in the denominator by the poles of $\Gamma(\ell + 1)$.

Now we see that the pole of f_ℓ^+ at $\ell = -1$ is compensated by the vanishing of the square bracket $[z^\ell + (-z)^\ell]$. Due to this fact the contour of integration in (11.11) may be shifted to the left, Re $\ell < -1$, in agreement with the assumption (11.7).

So the pole f_ℓ^+ seems to be inessential. But f_ℓ^+ enters the unitarity condition and this imposes definite restrictions. What do they lead to?

11.2 Contradiction with unitarity

Let us consider the partial wave $\phi_\ell = k^{-2\ell} f_\ell^+$ as a function of t. On the right cut in the t plane $(4\mu^2 \le t \le 16\mu^2)$ we have

$$\Delta^{\text{right}} \phi_\ell(t) = \frac{k_t^{2\ell+1}}{16\pi\omega_t} \phi_\ell(t + \mathrm{i}\,\varepsilon) \phi_\ell(t - \mathrm{i}\,\varepsilon). \qquad (11.12)$$

As we have stressed repeatedly in previous lectures, a pole $\phi_\ell \simeq c/(\ell+1)$ contradicts the unitarity condition (11.12) (a simple pole on the l.h.s., but a second order pole on the r.h.s.). Note however that we have seen the pole of ϕ_ℓ at $\ell = -1$ only at negative $t < t_0$, so the continuation to positive t is needed.

To this end we will use analytic properties of ϕ_ℓ as a function of t. In the t plane, ϕ_ℓ has right and left cuts. The discontinuity on the right cut is given by (11.12). On the left cut at $t < t_0$ the discontinuity is given by (see (3.8))

$$\Delta^{\text{left}} \phi_\ell(t) \sim -\frac{1}{2} \int P_\ell(-z_t)(A_1 + A_2) \frac{\mathrm{d}s}{(-k_t^2)^{\ell+1}}$$
$$+ \frac{1}{\pi} \int Q_\ell(-z_t) \rho_3(s,t) \frac{\mathrm{d}s}{(-k_t^2)^{\ell+1}}. \qquad (11.13)$$

(For $t_0 < t < 0$ the second term is absent.) It is essential that the integration in (11.13) is performed over finite intervals. Q_ℓ has a pole at $\ell = -1$ with unit residue. So $\Delta^{\text{left}} \phi_\ell(t)$ when $t < t_0$ has a pole at $\ell = -1$ with a residue depending on t and proportional to $\int \rho_3(s,t)\mathrm{d}s$.

Write down now the dispersion relation for ϕ_ℓ:

$$\phi_\ell = \frac{1}{\pi} \int_{4\mu^2}^\infty \Delta^{\text{right}} \phi_\ell(t') \frac{\mathrm{d}t'}{t' - t} + \frac{1}{\pi} \int_{-\infty}^0 \Delta^{\text{left}} \phi_\ell(t') \frac{\mathrm{d}t'}{t' - t}. \qquad (11.14)$$

The contribution of the left cut contains a term proportional to $1/(\ell+1)$. On the right cut the amplitude ϕ_ℓ itself and also its discontinuity are restricted by the unitarity condition,

$$\phi_\ell = \frac{16\pi\omega}{k^{2\ell+1}} \sin\delta_\ell\, \mathrm{e}^{\mathrm{i}\,\delta_\ell}.$$

Therefore the contribution of the right cut cannot compensate the unbounded contribution of the left cut. (Even if at some t these contributions accidentally compensate each other this cannot be true for other values of t: these contributions have different analytic properties and cannot compensate each other identically, i.e. at any t.) Thus at $t > 4\mu^2$ a contradiction with unitarity inevitably arises. Where is the way out?

11.3 Poles condensing at $\ell = -1$

Let us assume for the sake of argument that a contribution to the pole at $\ell = -1$ comes from a single point $t = \tilde{t} < t_0$, i.e. the residue of the pole in $\Delta^{\text{left}}\phi(t)$ is equal to $a\pi\delta(t - \tilde{t})$. Then the pole term in ϕ_ℓ is equal to $a/(\ell + 1)(t - \tilde{t})$.

Now we recall that the dispersion relation (11.14) is written for large positive ℓ. When one decreases ℓ, other *moving* poles can appear at $t < 4\mu^2$ from the right cut whose contributions should be included explicitly in (11.14). The contribution of one such pole has the form $r(\ell)/(t(\ell) - t)$. If this moving pole were such that $t(\ell = -1) = \tilde{t}$ and $r(\ell) \simeq (-a)/(\ell + 1)$, then ϕ_ℓ would not have a pole at $\ell = -1$.

Thus the problem with unitarity would disappear if a moving pole in $\phi_\ell(t)$ existed whose position depended on ℓ, and which at $\ell = -1$ came to the negative point $t = \tilde{t}$ that yielded a pole $\sim 1/(\ell+1)$ in the discontinuity of the amplitude.

In our toy model there was one such point. In reality, the pole at $\ell = -1$ of the left-cut discontinuity of ϕ_ℓ arises from the whole interval of negative t-values. Therefore, to compensate for it an infinite sequence of moving poles is needed. With ℓ decreasing down to $\ell = -1$, these poles will come onto the physical sheet and move to the left cut. At $\ell = -1$ they should completely fill the interval of the left cut that provides a pole at $\ell = -1$.

How does this phenomenon look in the ℓ plane?

Imagine that the poles in the t plane emerge through the tip of the two-particle threshold at $t = 4\mu^2$ and move along the real axis to the left cut. Let us fix some value of t and push $\ell \to -1$. Then in the t plane we will have an infinite number of poles passing through our point. Hence in the ℓ plane we shall see an infinite number of Regge poles, accumulating towards the point $\ell = -1$. (If the poles in the t plane appear on the physical sheet through infinity, then it is convenient to choose a large negative t to detect their accumulation in the ℓ plane.) In general not only the position of separate poles but also the pattern of their accumulation (in particular, from which side the poles approach the point $\ell = -1$) depends on t.

The picture given here is the only possibility of rescuing a theory which contains in the ℓ plane nothing but poles. This statement can be proved rigorously by solving the dispersion relation for the amplitude whose discontinuity on the left cut contains a pole in ℓ and that on the right cut is determined by a two-particle unitarity condition.

11.3.1 Amplitude cannot fall faster than $1/s$

The poles accumulating in the ℓ plane give a non-zero contribution to the asymptotics of the total amplitude, since the signature factor in (11.11) vanishes only at the point $\ell = -1$. Therefore the amplitude at arbitrary t and $s \to \infty$ cannot decrease faster than $1/s$. Together with the Froissart bound we obtain for the amplitudes in the physical region of the s-channel

$$\frac{1}{s^{1+\varepsilon}} < A(s,t) < s\ln^2 s, \qquad s \to \infty. \tag{11.15}$$

Here the infinitesimal positive quantity ε takes account of the possibility of the poles accumulating on the left of $\ell = -1$.

Thus the scattering amplitude in a theory containing only Regge poles must obey the rather rigid constraint (11.15).

11.4 Particles with spin: failure of the Regge pole picture

In reality the situation is worse, since we have assumed the particles to be spinless. For the interaction of particles with spins σ_1 and σ_2, the point $\ell = -1$ is equivalent to the point $j = \sigma_1 + \sigma_2 - 1$ as we have shown in the previous lecture.

Let us consider as an example the case of two vector particles, $\sigma_1 = \sigma_2 = 1$:

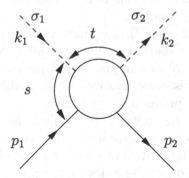

Let us write $P = p_1 + p_2$. Then the matrix element of t-channel annihilation of particles with polarizations λ_1 and λ_2 is $e_\mu^{\lambda_1} A_{\mu\nu} e_\nu^{\lambda_2}$, where the tensor amplitude $A_{\mu\nu}$ is

$$A_{\mu\nu} = P_\mu P_\nu A_1 + P_\mu k_{1\nu} A_2 + P_\nu k_{2\mu} A_3 + k_{1\mu} k_{2\nu} A_4 + g_{\mu\nu} A_5. \tag{11.16}$$

To avoid kinematical singularities we must write the Mandelstam representation for A_i. By separating the contributions of the cuts we find, as above, $A_i^{\text{right}} \sim 1/s$.

The invariant amplitude contains in addition the kinematical factors. The largest one at $s \to \infty$ is the factor $(P_\mu e_\mu^+)(P_\nu e_\nu^+)$, which corresponds to vector particle helicities $\lambda_1 = -\lambda_2 = 1$. Indeed in the case of identical particles we have in the t-channel cms $P_\mu = (0, 2\boldsymbol{p}_t)$, and we get $(\boldsymbol{p} \cdot \boldsymbol{e}^+)^2 \simeq p_t^2 \sin^2 \Theta_t \simeq P_t^2 \cos^2 \theta_t \sim s^2$, so the total amplitude is $A^{\text{right}} \sim s$. This means that there is a singularity at $j = 1$, which by virtue of unitarity will be passed over to all channels that are connected to the pair of vector particles.

In the last example we have considered hypothetical particles. And what can one say about real particles? In principle it could be possible that for real hadrons the integral of ρ_3 is zero. But generally this is not so.

For nucleons, in particular, there certainly exists a singularity at $j = 0$, since in this case there is a region where ρ_3 is determined by the single Feynman diagram

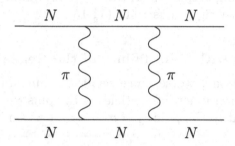

The Ω^- hyperon (a stable hadron with $j = \frac{3}{2}$) gives a singularity already at $j = 2$, which violently contradicts the Froissart theorem, derived from the s-channel unitarity condition.

It is this very contradiction that we have mentioned at the beginning of this lecture. It becomes sharper if one takes into account the higher spin resonances. Although we cannot succeed in proving it, nevertheless it is natural to expect the resonances as well as the stable particles to produce singularities at integer points in the j plane.

How can one resolve this contradiction? The answer lies in the reggeization of all particles. This automatically removes the difficulties but necessarily introduces singularities of another type – branchings – in the j plane. The accumulation of the poles at the point $j = \sigma_1 + \sigma_2 - 1$ does not arise; there appears only a simple pole which does not show up in the asymptotics. In the presence of branchings such a pole does not contradict the unitarity condition, since it is located not on the real axis but on a side of the cut in the j plane.

If there exists at least one non-reggeized particle then the corresponding accumulation of the Regge poles is inevitable.

12

Two-reggeon exchange and branch point singularities in the ℓ plane

12.1 Normalization of partial waves and the unitarity condition

Let us now discuss the question of normalization. We are about to redefine partial wave amplitudes in order to simplify, and generalize, the unitarity condition.

Earlier we have used the representation of the amplitude in the form of the Sommerfeld–Watson integral:

$$A^\pm = \frac{i}{4} \int_{a-i\infty}^{a+i\infty} d\ell (2\ell+1) \frac{P_\ell(-z_t) \pm P_\ell(z_t)}{\sin \pi \ell} f_\ell^\pm(t).$$

At large z the Legendre function can be approximated as

$$P_\ell(z_t) \simeq \frac{\Gamma(\ell+\frac{1}{2})}{\sqrt{\pi}\Gamma(\ell+1)} \cdot (2z_t)^\ell \simeq \frac{\Gamma(\ell+\frac{1}{2})}{\sqrt{\pi}k_t^{2\ell}\Gamma(\ell+1)} \cdot s^\ell, \quad z_t \simeq \frac{s}{2k_t^2}.$$

12.1.1 Redefinition of partial wave amplitudes

Inserting this expression we shall write the Sommerfeld–Watson representation in the form

$$A^+(s,t) = \frac{i}{4} \int d\ell (2\ell+1) \frac{s^\ell + (-s)^\ell}{\sin \pi \ell} \varphi_\ell^+(t), \qquad (12.1a)$$

in terms of the redefined partial wave amplitudes

$$\varphi_\ell^+ \equiv \frac{\Gamma(\ell+\frac{1}{2})}{\sqrt{\pi}k_t^{2\ell}\Gamma(\ell+1)} f_\ell^+ = \frac{\Gamma(\ell+\frac{1}{2})}{\sqrt{\pi}\,\Gamma(\ell+1)} \phi_\ell^+. \qquad (12.1b)$$

The unitarity condition for φ_ℓ now takes the form

$$\delta_{\text{right}}\varphi_\ell = \frac{k_t^{2\ell+1}}{8\pi\sqrt{t}} \frac{\sqrt{\pi}\Gamma(\ell+1)}{\Gamma(\ell+\frac{1}{2})} \varphi_\ell \varphi_\ell^* = C_\ell \frac{\Gamma(\ell+1)}{\Gamma(\ell+\frac{1}{2})} \varphi_\ell \varphi_\ell^*, \qquad (12.2a)$$

145

where

$$C_\ell = \frac{1}{8\sqrt{\pi}} \frac{k_t^{2\ell+1}}{\sqrt{t}}. \tag{12.2b}$$

In the previous lecture it was shown that if the third spectral function ρ_{su} exists, then $f_\ell^+ \sim \phi_\ell^+ \sim C/(\ell+1)$ near $\ell = -1$. From (12.1b) we see that φ_ℓ^+ is finite at this point. The singularity at $\ell = -1$ manifests itself in the fact that the l.h.s. in (12.2) is finite near $\ell = -1$ whereas the r.h.s. tends to infinity due to the factor $\Gamma(\ell+1)$, which may be looked upon as being a part of the phase volume continued to non-integer ℓ.

12.1.2 Particles with spin in the unitarity condition

An analogous redefinition of partial wave amplitudes and the unitarity condition one can carry out for particles with non-zero spin. Consider the amplitude with two spinless particles in the initial state and two particles with spins σ_1 and σ_2 and helicities m_1 and m_2 in the final state:

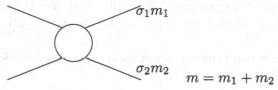

$$m = m_1 + m_2$$

$$A_{m_1,m_2} \equiv A_m = \sum (2j+1) f_{jm}(t) Y_{jm}(z); \tag{12.3a}$$

$$Y_{jm}(z) = \sqrt{\frac{\Gamma(j-m+1)}{\Gamma(j+m+1)}} P_{jm}(z) \quad (\phi = 0); \tag{12.3b}$$

$$f_{jm} = \frac{1}{2}\sqrt{\frac{\Gamma(j-m+1)}{\Gamma(j+m+1)}} \int_{-1}^{1} P_{jm} A_m \, dz_s. \tag{12.3c}$$

The formula (12.3c) for f_{jm} is valid only for integer (or half-integer) j.

The analytical continuation to arbitrary j can be performed in the usual way, using the dispersion relation for A and the transition from the function P_{jm} to Q_{jm}.

Substituting the expression for A_m from (12.3a,b),

$$A_m = \sum (2j+1) f_{jm}(t) \sqrt{\frac{\Gamma(j-m+1)}{\Gamma(j+m+1)}} P_{jm}(z),$$

into the Sommerfeld–Watson integral, we obtain

$$A_m^\pm = \frac{i}{4} \int_{a-i\infty}^{a+i\infty} \frac{dj\,(2j+1)}{\sin \pi j} \sqrt{\frac{\Gamma(j-m+1)}{\Gamma(j+m+1)}} [P_{jm}(-z_t) \pm P_{jm}(z_t)] f_{jm}^\pm(t).$$

Now, using the asymptotic expression for $P_{jm}(z)$ at large z,

$$P_{jm}(z_t) \simeq \frac{\Gamma(j + \frac{1}{2})}{\Gamma(j - m + 1)} \frac{(2z_t)^j}{\sqrt{\pi}}, \qquad z \simeq \frac{s}{2k_t^2},$$

we arrive at

$$A_m^\pm = \frac{i}{4} \int_{a - i\infty}^{a + i\infty} dj (2j + 1) \frac{[(-s)^j \pm s^j]}{\sin \pi j} \varphi_{jm}^\pm,$$

where

$$\varphi_{jm} = \frac{\Gamma(j + \frac{1}{2})}{\sqrt{\pi \Gamma(j - m + 1)\Gamma(j + m + 1)}} k^{-2j} f_{jm}^\pm.$$

The unitarity condition in the case when only internal particles possess non-zero spins,

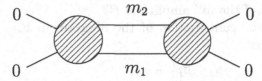

takes the following form:

$$\left.\begin{aligned}
\delta_{\text{right}} \varphi_j &= C_j \sum_{m = -(\sigma_1 + \sigma_2)}^{\sigma_1 + \sigma_2} \varphi_{jm} \frac{\Gamma(j - m + 1)\Gamma(j + m + 1)}{\Gamma(j + \frac{1}{2})\Gamma(j + 1)} \varphi_{jm}^*, \\
C_j &= \frac{1}{8\sqrt{\pi}} \frac{k_t^{2j+1}}{\sqrt{t}}.
\end{aligned}\right\} \quad (12.4)$$

Once again, the factor $\Gamma(j - m + 1)\Gamma(j + m + 1)/\Gamma(j + \frac{1}{2})\Gamma(j + 1)$ on the r.h.s. of (12.4) can be considered as a phase volume which contains poles, the rightmost being at $j = \sigma_1 + \sigma_2 - 1$. It is the familiar $j = -1$ pole shifted to the right due to the presence of spin. If the third spectral function were zero then φ_{jm} would vanish at $j = \sigma_1 + \sigma_2 - 1$.

We note that in the spinless case, $m \equiv 0$, the unitarity condition (12.4) coincides with (12.2) previously derived.

The unitarity condition

$$\delta_{\text{right}} f_j = C \sum_m f_{jm} f_{jm}^*$$

contains the wave function normalization factors of the initial and of the final state only once, on both the left- and the right-hand side of the equation. Therefore if the spins of the initial and final particles are non-zero, then the corresponding factors related to these spins cancel out on both sides and (12.4) remains valid.

12.2 Particle scattering via a two-particle intermediate state

We consider now a specific diagram for two-particle collision shown in Fig. 12.1.* Let us assume that the amplitude $A(s_i, t)$ is a rapidly falling

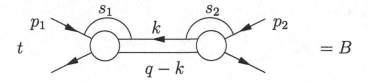

Fig. 12.1.   Scattering via a two-particle intermediate state

function of energy (faster than $1/s_i$ at $s_i \to \infty$). But, as we already know, the contributions of its left and right cuts separately decrease with energy only as $1/s_i$. How will such a behaviour of the block amplitude A affect the asymptotics of the full amplitude B?

To calculate the contribution of the diagram in Fig. 12.1 we use the Sudakov variables:

$$k = \alpha p_2' + \beta p_1' + k_\perp;$$

$$t = q^2 = q_\perp^2 + s\alpha_q\beta_q \simeq q_\perp^2, \quad \alpha_q \sim \beta_q \sim \frac{\mu^2}{s}; \qquad (12.5a)$$

$$s_1 = (p_1 + k)^2 \simeq \alpha s, \quad s_2 = (p_2 - k)^2 \simeq -s\beta. \qquad (12.5b)$$

Since $A(s_i)$ falls rapidly with increasing s_i, the essential values of $s_{1,2}$ in the integral are of order μ^2, so that $\alpha \sim |\beta| \sim \mu^2/s$. Therefore the propagators of the intermediate lines in Fig. 12.1 can be approximated as

$$\frac{1}{(\mu^2 - k_\perp^2 - s\alpha\beta)} \cdot \frac{1}{\mu^2 - (q - k_\perp)^2 - s(\alpha - \alpha_q)(\beta - \beta_q)}$$

$$\simeq \frac{1}{\mu^2 - k_\perp^2} \frac{1}{\mu^2 - (q_\perp - k_\perp)^2}.$$

Inserting these propagators, for B we obtain

$$B = \frac{s}{2} \int \frac{d^2k_\perp d\alpha \, d\beta}{2! \, (2\pi)^4 i} \frac{A(s_1, k_\perp, q_\perp) \, A(s_2, k_\perp, q_\perp)}{[\mu^2 - k_\perp^2][\mu^2 - (q - k)_\perp^2]},$$

with $\frac{1}{2!}$ the combinatorial symmetry factor. Using (12.5b) we represent B in terms of the integrals of the subamplitudes A over their proper energies s_i:

$$B = \frac{i}{4s} \int \frac{d^2k_\perp}{(2\pi)^2} \frac{1}{[\mu^2 - k_\perp^2][\mu^2 - (q - k)_\perp^2]} \int_\Gamma \frac{ds_1}{2\pi i} A(s_1) \int_\Gamma \frac{ds_2}{2\pi i} A(s_2).$$

---

* It is implied that the block A does not contain a two-particle cut in the t-channel.

The contour of integration in the s_1 (s_2) plane has the form

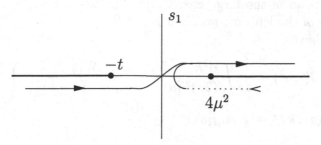

in accordance with the Feynman rules for bypassing singularities of the amplitude. Note that the contour runs along those edges of the s- and u-cuts where $A(s_1, t)$ coincides with the physical amplitude in the s- and u-channel, correspondingly.

Since $A(s_1)$ rapidly falls as $s_1 \to \infty$, the integration contour can be deformed to go around the right cut:

$$\int_\Gamma \frac{ds_1}{2\pi i} A = \int_{4\mu^2}^\infty A_1 \frac{ds_1}{\pi} \equiv N_1(q_\perp, k_\perp).$$

One important remark is needed. This integral would actually be zero if the amplitude A had only the right cut. Indeed in this case the integration contour can be shifted to the left and one immediately gets zero. Recall that the absence of the left cut in the s_1 plane of the function $A(s_1, t)$ means that the third spectral function $\rho(s, u)$ is equal to zero.

Let us consider the contribution of the right cut alone:

$$A_{\text{right}}(s) = \frac{1}{\pi} \int_{4\mu^2}^\infty \frac{A_1(s')}{s' - s} ds'.$$

Then it is evident that N_1 can be written as

$$N_1 = \int_{4\mu^2}^\infty A_1(s') \frac{ds'}{\pi} = \int_C A_{\text{right}}(s) \frac{ds}{2\pi i}.$$

The function A_{right} does not have the left cut by definition, so why then is the integral non-zero? In the previous lecture we have shown that the amplitude A_{right} cannot fall faster than $1/s$:

$$\text{Im } A_{\text{right}} = A_1 + \frac{1}{\pi} \int \frac{\rho_{su}(s', t)}{s' - s} ds' \sim -\frac{1}{s} \cdot \frac{1}{\pi} \int \rho_{su}(s', t) ds'.$$

In these circumstances the contour cannot be closed on the left due to the contribution of the large circle, and N_1 does not vanish, in spite of the absence of the left cut, *provided $\rho_{su} \neq 0$*.

Thus we have

$$\left. \begin{aligned}
B(s, q^2) &= \frac{i}{4s} \int \frac{d^2 k_\perp}{(2\pi)^2} \frac{N_1(q, k_\perp)\, N_2(q, k_\perp)}{[\mu^2 - k_\perp^2][\mu^2 - (q - k)_\perp^2]}, \\
N_i(q_\perp, k_\perp) &= \int A_{1,i}(s', q_\perp, k_\perp) \frac{ds'}{\pi}.
\end{aligned} \right\} \tag{12.6}$$

Hence despite the fact that A decreases faster than $1/s$, the full amplitude $B \sim 1/s$.

This is the same result that we have obtained using the t-channel unitarity condition for partial waves. The structure of B is consistent with the unitarity condition, and $1/s$ here is in one-to-one correspondence with $\Gamma(\ell + 1)$ in the previous analysis.

12.3 Two-reggeon exchange and production vertices

So we learned that the amplitude A describing the t-channel scattering blocks in Fig. 12.1 must have a non-zero third spectral function ρ_{su}. Let us examine the structure of the process described by the amplitude B in more detail. To this end we invoke the field-theoretical $g\phi^3$ model to replace A by the first perturbative graph with $\rho_{su} \neq 0$. It is the crossed box diagram, as shown in Fig. 12.2.

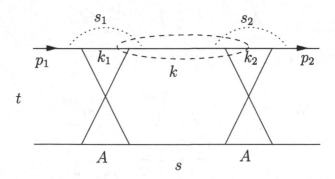

Fig. 12.2.   The simplest double exchange diagram having $\rho_{su} \neq 0$

It is easy to verify that at large s_1 the box diagram behaves as $A(s_1) \sim 1/s_1^2$, so that the s_i-integrals that define N_1 and N_2 in (12.6) converge. Moreover, the transverse momentum loop integral over $k_{1\perp}^2$ converges as well. This means that the essential values of $s_1, k_{1\perp}^2$ are of the order of

μ^2, and for the left block we have

$$\left.\begin{aligned} (p_1 - k_1)^2 &\simeq -\alpha_1 s \sim \mu^2, \quad k_1^2 = k_{1\perp}^2 + s\alpha_1\beta_1 \sim \mu^2, \\ \beta_1 &\sim 1, \quad |\alpha_1| \sim \frac{\mu^2}{s}, \quad k_{1\perp}^2 \sim \mu^2. \end{aligned}\right\} \qquad (12.7a)$$

In the same way we have for the right block in Fig. 12.2 we obtain the estimates

$$\alpha_2 \sim 1, \quad |\beta_2| \sim \frac{\mu^2}{s}, \quad k_{2\perp}^2 \sim \mu^2. \qquad (12.7b)$$

Have a look now at the scattering subprocess $k_1, k_2 \to (k_1 - k), (k_2 + k)$ shown by a dashed blob in Fig. 12.2. Under the conditions (12.7) the energy invariant in this four-point function is large, of the order of total s:

$$s_{12} = (k_1 + k_2)^2 = s(\alpha_1 + \alpha_2)(\beta_1 + \beta_2) + (k_1 + k_2)_\perp^2 \simeq s\alpha_2\beta_1 = \mathcal{O}(s).$$

(The regions $\beta_1, \alpha_2 \sim \mu^2/s$ give negligible contributions due to small phase volume.) Meanwhile the square of the momentum transfer is of the order of μ^2 due to the convergence of the integral for B with respect to $k_\perp$. Consequently the four-point function enters into the the diagram

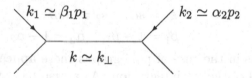

Fig. 12.3. One-particle approximation for the blob in Fig. 12.2

Fig. 12.2 for B in the *asymptotic regime*. Hence there is no reason to believe that one can restrict oneself to a pole graph only, as we know that at high energies the behaviour of the amplitude is quite different from that of a pole graph.

We assume now that the amplitude of Fig. 12.3 behaves like a reggeon and write for it the Regge pole expression:

$$-g_1(k_1, k)g_2(k_2, k)\frac{(-s_{12})^{\gamma(k_\perp^2)} \pm s_{12}^{\gamma(k_\perp^2)}}{\sin \pi\gamma(k_\perp^2)}, \qquad s_{12} \simeq \beta_1\alpha_2 s. \qquad (12.8)$$

As we have shown above, a reggeon in QFT emerges due to the diagrams of the type (a),

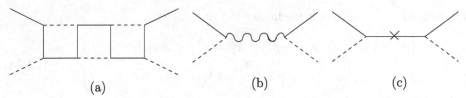

(a)   (b)   (c)

and can be represented by a single Regge pole line (b) ('particle with varying spin'). Near the mass shell, $k^2 = m^2$ where $\gamma(m^2) = 0$, the reggeon propagator turns into the propagator of an ordinary scalar particle (c).

Pulling out the factors[†] of α_2, β_1,

$$\frac{(-s_{12})^\gamma \pm s_{12}^\gamma}{\sin \pi \gamma} = (\beta_1)^\gamma (\alpha_2)^\gamma \cdot \frac{(-s)^\gamma \pm s^\gamma}{\sin \pi \gamma}.$$

and substituting for the particle propagator in the expression for B we obtain

$$B = \frac{1}{4s} \int \frac{d^2 k_\perp}{(2\pi)^4 i} \frac{(-s)^\gamma \pm s^\gamma}{\sin \pi \gamma} \frac{(-s)^{\gamma'} \pm s^{\gamma'}}{\sin \pi \gamma'}$$

$$\times \int ds_1 \frac{d^4 k_1 \, g_1 g_1' (\beta_1)^\gamma (\beta_1')^{\gamma'}}{(2\pi)^4 i \, (\ )(\ )(\ )(\ )} \int ds_2 \frac{d^4 k_2 \, g_2 g_2' (\alpha_2)^\gamma (\alpha_2')^{\gamma'}}{(2\pi)^4 i \, (\ )(\ )(\ )(\ )}. \quad (12.9)$$

Here

$$\gamma = \gamma(k_\perp^2), \quad \gamma' = \gamma((q-k)_\perp^2),$$

$$s_1 = s\alpha, \quad s_2 = -s\beta,$$

$$\beta_1' = 1 - \beta_1, \quad \alpha_2' = 1 - \alpha_2.$$

As well as in the case of pole graphs there appears a factorization in the integrand of the $k_\perp$ integration. As a result (12.9) can be written in the following form:

$$\left. \begin{array}{l} B = \dfrac{i}{4} \displaystyle\int \dfrac{d^2 k_\perp}{(2\pi)^2} \, N_{\gamma\gamma'}^2 (k_\perp, q_\perp) \, \xi_\gamma \xi_{\gamma'} s^{\gamma + \gamma' - 1}, \\[12pt] \xi_\gamma = -\dfrac{e^{-i\pi\gamma} \pm 1}{\sin \pi \gamma}, \end{array} \right\} \quad (12.10a)$$

$$N_{\gamma\gamma'} = \int \frac{ds_1}{2\pi i} \int \frac{d^4 k_1 \, g_1 g_1' \, \beta_1^\gamma (1-\beta_1)^{\gamma'}}{(2\pi)^4 i \, (\ ) \cdot (\ ) \cdot (\ ) \cdot (\ )}. \quad (12.10b)$$

One can present (12.10) in the form of the graph where the wavy lines correspond to the reggeons and the blobs describe the amplitudes of reggeon production N given in (12.10b); see Fig. 12.4.

So we have met two types of reggeon production graphs shown in Fig. 12.5.

[†] The integration in (12.9) is actually performed over the region $0 < \beta_1, \alpha_2 < 1$.

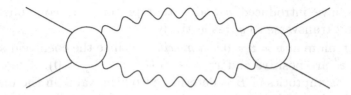

Fig. 12.4.   Two-reggeon exchange diagram

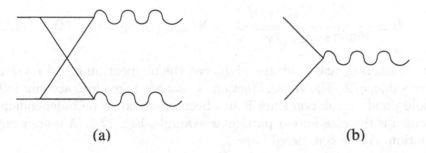

(a)   (b)

Fig. 12.5.   Two-particle–two-reggeon (a) and two-particle–one-reggeon (b) production amplitudes

The factors of the type $\beta_1^\gamma \simeq (2k_1p_2/s)^\gamma$ in (12.10b) have a simple meaning. They are the asymptotics of the vertex for the transition of two scalar particles into a particle with complex spin γ in Fig. 12.5b. The vertex of the transition to a spin-1 particle, considered in one of the previous lectures, can be invoked as an example.

12.4 Asymptotics of two-reggeon exchange amplitude

The $s \to \infty$ asymptotics of the integral in (12.10a) can be evaluated by the saddle point method. To this end we change the transverse momentum integration from $k_\perp$ to the two-dimensional variable x,

$$k_\perp = \frac{q_\perp}{2} + x, \quad (q-k)_\perp = \frac{q_\perp}{2} - x,$$

$$k^2 \simeq \frac{q^2}{4} + x^2 + (q_\perp x), \quad (q-k)^2 \simeq \frac{q^2}{4} + x^2 - (q_\perp x),$$

and expand the sum of the trajectories ($t = q^2 \simeq q_\perp^2 < 0$),

$$\gamma(k^2) \simeq \gamma(q^2/4) + \gamma'(x^2 + (qx)) + \gamma'' \frac{(qx)^2}{2},$$

$$\gamma(k^2) + \gamma((q-k)^2) \simeq 2\gamma(q^2/4) + 2\left(\gamma' + \frac{\gamma''}{2}q^2\right)y^2 + 2\gamma'z^2,$$

where we have introduced y and z to represent the components of x parallel and transverse to q, respectively.

The extremum at $y = z = 0$ is a *maximum* since the coefficients in front of y^2 and z^2 are positive and $x^2, y^2 < 0$ ($k_\perp^2 = -\boldsymbol{k}_\perp^2 \leq 0$). The exponent of s in the asymptotics of B is determined by the value in the maximum; the Gaussian integral provides a pre-exponential factor depending on $\ln s$. Evaluating the y- and z-integrals we arrive at

$$B \simeq \frac{\mathrm{i}}{16\pi \ln s} \frac{s^{2\gamma(q^2/4)-1}}{\sqrt{2\gamma'(2\gamma' + \gamma'' q^2)}} N_{\gamma_0 \gamma_0}^2 \xi_{\gamma_0}^2, \quad \gamma_0 = \gamma(q^2/4). \quad (12.11)$$

In the factors N and ξ_γ we have replaced the momentum $k_\perp$ by its saddle point value, $q/2$. The vertex function N was not taken into account in the saddle point calculation since it has been assumed to be independent of s, which is the case for our particular example, Fig. 12.2. A more general situation will be considered later.

By examining the expression (12.11) we conclude that

1. The presence of $1/\ln s$ means that the singularity of the t-channel partial wave amplitude in the j plane is a branch point, not a pole,

2. This branch point is located at $j = 2\gamma(t/4) - 1$, because its position is determined by the exponent of s.

12.5 Two-reggeon branching and $\ell = -1$

It is important that the position of the singularity we have obtained does not coincide with the initial one.

First, we had a particle ($1/s$, $j = -1$). Then we have introduced a reggeon $\gamma(t)$ ($\gamma(\mu^2) = 0$ at $t = \mu^2 > 0$). Finally we have found that the asymptotics of B corresponds to the singularity of φ_j at a new point $j = 2\gamma(t/4) - 1$.

In terms of partial wave amplitudes, the asymptotics of the amplitude, as we already know, is determined by the equations

$$B^\pm = -\frac{1}{4\mathrm{i}} \int \varphi_j^\pm(t) \frac{(-s)^j \pm s^j}{\sin \pi j} \mathrm{d}j, \quad B_1 = \frac{1}{4\mathrm{i}} \int \varphi_j s^j \mathrm{d}j.$$

Let φ_j have its rightmost singularity at some $j = j_0$. If this is a pole, then its residue gives a power asymptotics s^{j_0}. If it is a branch point, then, closing the contour around the cut,

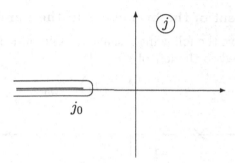

we obtain

$$B_1(s) = -\frac{1}{2}\int_{-\infty}^{j_0}\Delta\varphi_j s^j\mathrm{d}j = -\frac{1}{2}s^{j_0}\int_{-\infty}^{j_0}\Delta\varphi_j e^{(j-j_0)\xi}\mathrm{d}j$$

$$= -\frac{1}{2}s^{j_0}\int_{-\infty}^{0}e^{x\xi}\Delta\varphi(x)\,\mathrm{d}x.$$

The latter integral falls obviously like $1/\xi^n$ at large $\xi = \ln s$. Indeed, if it fell faster, like s^{-k}, then

$$\varphi_j \propto \int s^{-j-1}B_1(s)\mathrm{d}s$$

would have no singularities at $j > j_0 - k$, while the integral converges.

As is known from quantum mechanics, the trajectory $\gamma(t)$ grows with t:

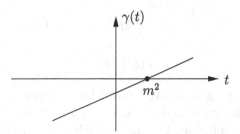

For $t < 0$ the amplitude B is small, in fact it falls faster than $1/s$. This means that a compensation has occurred of the contribution of the initial pole at $j = -1$ (a) by other graphs (b):

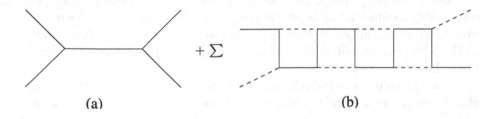

(a)   (b)

12.6 Movement of the branching in the t and j planes

In the j plane we have the following situation. The branch point $j = j_0(t)$
for $t < 4\mu^2$ is situated to the left of $j = -1$:

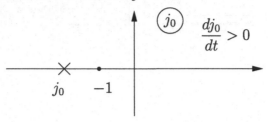

The singularity $j_0(t)$ comes to the point $j = -1$ at $t = 4\mu^2$.

In the t plane the corresponding picture looks as in Fig. 12.6, where
the position of the branch point is shown as a function of j_0.

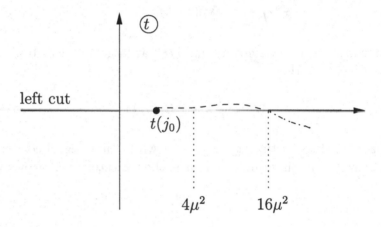

Fig. 12.6.

When we increase j_0 starting from a large negative value, the branch
point moves to the right and at $j_0 = -1$ coincides with the tip of the
right cut of the amplitude, $t = 4\mu^2$. With j_0 increasing, at some point
$j_0 = j^*$ it hits the four-particle threshold at $t = 16\mu^2$ (see the dashed
line in Fig. 12.6 showing the movement of the branch point). For $j_0 > j_*$
the position of the branching $t(j_0)$ becomes complex, since the trajectory
$\gamma(t/4)$ is complex valued when $t > 16\mu^2$. Since no singularity at complex
t is allowed on the physical sheet by virtue of causality (and crossing), this
complexity means that at large enough $j > j^*$ the branch point disappears
from the physical sheet: it dives onto the (second) unphysical sheet related
to the four-particle cut in the t plane as shown by the dot–dashed line in
Fig. 12.6.

We know that for large enough Re j the partial wave $\varphi_j(t)$ is a regular function in the entire complex t plane. We also know that with j

decreasing moving *Regge poles* emerge from beneath the two-particle cut at $t = 4\mu^2$. Now we see that also the moving *branch point* singularity appears on the physical sheet with decrease of j; it comes through the tip of the cut related to the four-particle intermediate state, $t = 16\mu^2$. This is due to the fact that instead of the usual state of two *particles*,

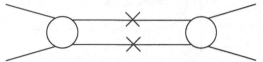

we have now a system of two *reggeons*. As we have seen before, each reggeon can be represented by a two-particle ladder. Therefore the reggeon–reggeon branching corresponds (at least) to the four-particle threshold (see Fig. 12.7).

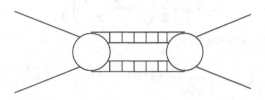

Fig. 12.7.   Branching of reggeons as two-particle ladders.

In the previous lecture we showed, in great detail, that the unitarity condition for φ_j in the form of (12.2) leads to the presence of an essential singularity of the partial wave amplitude φ_j at $j = -1$.

Now we have been convinced that at $j = -1$ a new singularity – the reggeon–reggeon branching $t(j)$ – arrives at the point $t = 4\mu^2$. In such circumstances the r.h.s. of the unitarity condition (12.2) accounts only for a *part* of the discontinuity of $\varphi_j(t)$ (curve d in Fig. 12.8). The total discontinuity (curve c in Fig. 12.8) on the sum of the two cuts does not contain in general a pole at $j = -1$ and the singularity in φ_j at that point does not appear.

12.7 Signature of the two-reggeon branching

Let us find now the signature of the two-reggeon branching (12.10) in the case when the signatures of the reggeons entering B in Fig. 12.7 are different.

The signature factor ξ_γ can be rewritten as

$$\xi_\gamma = -\frac{e^{-i\frac{\pi}{2}\gamma} \pm e^{i\frac{\pi}{2}\gamma}}{\sin \pi\gamma} e^{-i\frac{\pi}{2}\gamma} = -\frac{1}{\zeta_\gamma} e^{-i\frac{\pi}{2}(\gamma + \frac{P-1}{2})},$$

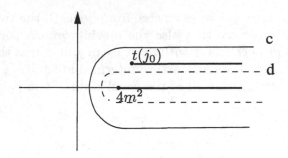

Fig. 12.8.   Overlapping cuts in the t plane due to the two-particle threshold and the moving two-reggeon branching

with

$$\zeta_\gamma = \sin \frac{\pi}{2}\left(\gamma + \frac{P-1}{2}\right).$$

Adopting this notation we have

$$\xi_{\gamma_1}\xi_{\gamma_2} = \frac{-i}{\zeta_{\gamma_1}\zeta_{\gamma_2}} e^{-i\frac{\pi}{2}(\gamma_1+\gamma_2-1+\frac{P_1-1}{2}+\frac{P_2-1}{2})}. \qquad (12.12)$$

Thus the expression for B can be reduced to the form[‡]

$$B = (\pm)\int \frac{N^2 d^2 k}{(2\pi)^2} \frac{\zeta_{\gamma_1+\gamma_2-1}}{\zeta_{\gamma_1}\zeta_{\gamma_2}} \frac{(-s)^{\gamma_1+\gamma_2-1} + P \cdot s^{\gamma_1+\gamma_2-1}}{\sin\frac{\pi}{2}(\gamma_1+\gamma_2-1)}, \qquad (12.13a)$$

where

$$P = P_{j_1} P_{j_2}. \qquad (12.13b)$$

Thus the signature of the branching P is equal to the product of the signatures of the poles.

[‡] The overall phase factor $(\pm)$ equals (-1) only for $P_1 = P_2 = -1$.

13
Properties of Mandelstam branch singularities

In the previous lecture we have calculated the contribution of the graph of Fig. 13.1 to the asymptotics of the total amplitude. Here we shall find

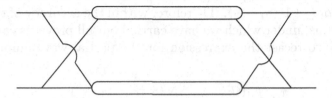

Fig. 13.1. Rescattering graph generating Mandelstam branching

its contribution to the singular part of the partial wave amplitude.

13.1 Branchings as a generalization of the $\ell = -1$ singularity

13.1.1 Branchings in the j plane

To do this we use the inverse Mellin transform

$$\phi_j = \frac{2}{\pi} \int s^{-j-1} A_1(s,t)\, \mathrm{d}s. \qquad (13.1)$$

Taking into account results of Lecture 12, we cast this equation in the following form (changing the order of integration):

$$\phi_j = \int \frac{\mathrm{d}^2 k_\perp}{(2\pi)^2} N^2(k_\perp, q_\perp) \operatorname{Re}(\xi_1\xi_2) \cdot \frac{2}{\pi} \int\limits_{1}^{\infty} s^{-j-1} s^{\gamma_1+\gamma_2-1} \mathrm{d}s$$

$$= \frac{2}{\pi} \int N^2(k_\perp, q_\perp) \frac{\mathrm{d}^2 k_\perp}{(2\pi)^2} \operatorname{Re}(\xi_1\xi_2) \frac{1}{j+1-\gamma_1-\gamma_2}. \qquad (13.2)$$

159

The expression for the product of the signature factors $\xi_1\xi_2$ has been calculated in the previous lecture (see (13.12)):

$$\mathrm{Re}\,(\xi_1\xi_2) = \frac{\sin\frac{\pi}{2}(\gamma_1 + \gamma_2 - 1 + \frac{P_1 + P_2}{2} - 1)}{\sin\frac{\pi}{2}(\gamma_1 + \frac{P_1 - 1}{2})\sin\frac{\pi}{2}(\gamma_2 + \frac{P_2 - 1}{2})}, \qquad (13.3)$$

where P_1 and P_2 are the signatures of the poles ($P_i = +1$ for positive and $P_i = -1$ for negative signature). We have shown there that the signature of the two-reggeon contribution to the amplitude is equal to the product of the signatures of the poles.

The expression (13.3) may be rewritten as

$$\mathrm{Re}\,(\xi_1\xi_2) = (\pm)\frac{\zeta_{\gamma_1 + \gamma_2 - 1}^{\mathrm{cut}}}{\zeta_{\gamma_1}\zeta_{\gamma_2}}, \qquad \zeta_\gamma^{\mathrm{cut}} = \sin\frac{\pi}{2}\left(\gamma + \frac{P_1 P_2 - 1}{2}\right). \qquad (13.4)$$

One can see from (13.2) some important features of the amplitude.

First, the j plane singularity of the partial wave is determined by the denominator $(j + 1 - \gamma_1 - \gamma_2)$. Therefore, within the accuracy of the leading singular terms, under which we have carried out all previous calculations, we are able to recast the expression for the partial wave amplitude into the form

$$\phi_j = \sin\frac{\pi}{2}j \int \frac{\mathrm{d}^2 k_\perp}{(2\pi)^2} \frac{N^2}{\sin\frac{\pi}{2}\gamma_1 \sin\frac{\pi}{2}\gamma_2} \frac{1}{j + 1 - \gamma_1 - \gamma_2}, \qquad (13.5)$$

where we have restricted ourselves to poles with positive signature (and thus the positive signature amplitude). This expression vanishes at physical points $j = 2k$.

13.1.2 Branch singularity in the unitarity condition

How are these formulæ related to the unitarity condition?

Let us consider two particles with spins σ_1 and σ_2. As we know, the most dangerous intermediate state in the unitarity condition is the one that possesses the maximal spin projections $m_1 = \sigma_1$ and $m_2 = \sigma_2$. As we have shown in the previous lecture, the unitarity condition for partial wave amplitudes (12.4) looks as follows:

$$\delta\phi_j = c\phi_j \frac{1}{j + 1 - \sigma_1 - \sigma_2}\phi_j^+. \qquad (13.6)$$

In (13.5) the denominator is the same as in (13.6), but we have to integrate over the 'masses' of particles and, besides, the spin of these particles is varied as a function of mass: $\sigma_1 \to \gamma_1(k_\perp^2)$, $\sigma_2 \to \gamma_2((q - k)_\perp^2)$.

We can rewrite the expression for partial wave amplitude (13.5) in the form of an integral over the reggeon masses $t_1 = k_\perp^2$ and $t_2 = (q - k)_\perp^2$ ($t_1, t_2 < 0$):

$$\left. \begin{aligned} \phi_j(t) &= -\frac{\sin \frac{\pi j}{2}}{(8\pi)^2} \int \frac{dt_1\,dt_2}{\sqrt{\mathcal{L}(t, t_1, t_2)} \sin \frac{\pi}{2}\gamma_1(t_1) \sin \frac{\pi}{2}\gamma_2(t_2)} \frac{N^2}{j+1-\gamma_1-\gamma_2}, \\ \mathcal{L} &= [t - (\sqrt{t_1} + \sqrt{t_2})^2][t - (\sqrt{t_1} - \sqrt{t_2})^2]. \end{aligned} \right\}$$

(13.7)

Where do the threshold singularities in t arise from?

The integral becomes singular when the singularities of the integrand do not allow the deformation of the integration contours necessary to avoid the root singularity of the Jacobian factor. Such singularities in t_1, t_2 are connected to zeros of $\sin \frac{\pi}{2}\gamma(t_j)$. In particular, at $t_1 = t_2 = m^2$ the singularity in t emerges when $\sqrt{t} = \sqrt{t_1} + \sqrt{t_2}$, i.e. $t = 4m^2$.

So we have established that the Mandelstam branch point is a generalization of the singularity at $\ell = -1$, which appears for spinless particles, to the case of a non-integer spin.

13.2 Branchings in the vacuum channel

We have conjectured that the four-point function is determined asymptotically by the Regge pole. Then we have investigated the self-consistency of this assumption by considering the more complicate diagram of Fig. 13.1 and substituting the Regge poles $j = \alpha(t)$ into the internal four-point functions. As a result we have found that the high energy behaviour of such amplitude is described by a branching at $j_2 = 2\alpha(t/4) - 1$.

Next if we suppose further that the internal four-point functions are determined by the pole $\alpha(t)$ as well as by the branching $j_2 = 2\alpha(t/4) - 1$, then we will get new branchings at the points

$$j_3 = 3\alpha\left(\frac{t}{9}\right) - 2, \quad j_4 = 4\alpha\left(\frac{t}{16}\right) - 3.$$

Iterating this procedure we will arrive at a set of branchings:

$$j = n\alpha\left(\frac{t}{n^2}\right) - n + 1.$$

Thus the conjecture of the existence of a pole will lead to an infinite sequence of branchings.

Fixed poles located at σ_1 and σ_2 in the j plane generated a sequence of poles accumulating to the point $j = \sigma_1 + \sigma_2 - 1$. That picture was not satisfactory because for large σ_i the poles would have moved arbitrarily far to the right, thus violating the Froissart theorem.

13.2.1 The pattern of branch points in the j plane

In the case of a moving pole $\alpha(t)$, an infinite sequence of branchings appears,

$$j_n = n\left[\alpha(0) + \alpha'\frac{t}{n^2}\right] - n + 1 = n(\alpha(0) - 1) + \alpha'\frac{t}{n} + 1,$$

the structure of which at large n depends crucially on the value of $\alpha(0)$. Indeed, if $\alpha(0) < 1$, then the branch points move to the left,

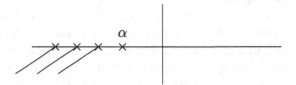

and at small t stay to the left of the pole,

$$j_n = \alpha(0) + (n-1)(\alpha(0) - 1) + \frac{\alpha't}{n} < \alpha(0).$$

If $\alpha(0) > 1$ (which is forbidden, by the way, by the Froissart theorem), then the branchings are located arbitrarily far to the right and the amplitude A increases faster than any power. Such a scenario contradicts all our initial assumptions.

The case $\alpha(0) = 1$ is of special interest. Then $j_n = 1 + \alpha't/n$, and for positive t the branchings are situated to the right of unity whereas for negative t they are shifted to the left:

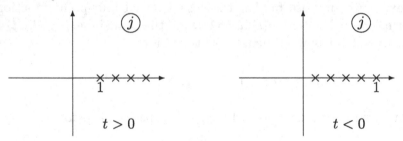

Thus, if the Pomeranchuk pole (which corresponds to a constant cross section) exists, then one has an interesting physical picture.

13.2.2 The Mandelstam representation in the presence of branchings

Let us discuss the question of existence of the Mandelstam representation in the presence of branch points. This question includes two aspects to be mentioned here, namely the analytic properties of the scattering amplitudes and the explicit form of the representation in terms of the double integral

$$\int \rho_{st}(s', t') \frac{\mathrm{d}s'\mathrm{d}t'}{(s'-s)(t'-t)}$$

with a finite number of subtractions. It turns out that generally the analytic properties are preserved whereas the explicit form does not exist.

Indeed, if we consider very large s, $s \gg -t$, then the contribution of the nth reggeon branching to the asymptotics has the form $s^{n\alpha(t/n^2)-n+1}$.

Provided that the trajectory $\alpha(t)$ at large t continues to grow, it is evident that an infinite number of subtractions is required.

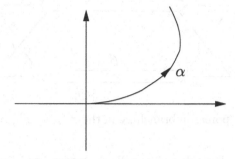

Suppose $\alpha(t)$ does turn over and the rightmost point is $\alpha(t_1)$ with Re $\alpha(t_1) > 1$. We can choose n such that $t/n^2 = t_1$, i.e. $n = \sqrt{t/t_1}$. At given t the rightmost branching corresponds to $n = \sqrt{t/t_1}$ and its asymptotic contribution is

$$s^{1+[\alpha(t_1)-1]\sqrt{t/t_1}}.$$

This shows that the growh of the exponent of s with increasing t is unbounded, so the single dispersion relations exist whereas the double dispersion representation is not valid.

13.3 Vacuum–non-vacuum pole branchings

Next we consider also other reactions, for instance the pion–nucleon charge exchange process shown in Fig. 13.2.

Here a pole with the quantum numbers of the ρ meson and trajectory $\beta(t)$ is exchanged. How will branchings contribute to such processes?

The contribution of the graph of Fig. 13.3 can be expressed, as before, in terms of an integral over the loop momentum k, and the integrand will contain the factor

$$\frac{1}{j+1-\alpha(k^2)-\beta((q-k)^2)}.$$

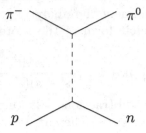

Fig. 13.2. $\pi^- p \to \pi^0 n$ Scattering with t-channel exchange of ρ-trajectory

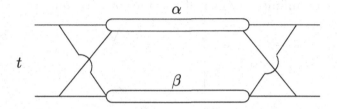

Fig. 13.3.   Reggeon–pomeron branching in the $\pi^- p \to \pi^0 n$ scattering amplitude

The position of the branch point will be determined by the extremum of $\alpha + \beta$. To find this extremum is a slightly more complicated exercise than in the case of identical poles.

At small q and k we have

$$\alpha(k^2) + \beta((q-k)^2) = \alpha(0) + \beta(0) + \alpha' k^2 + \beta'(q-k)^2.$$

The position of the branching turns out to be

$$\alpha(0) + \beta(0) + \frac{\alpha' \beta'}{\alpha' + \beta'} t - 1.$$

When $t = 0$, the branching is situated at $j = \beta(0) + \alpha(0) - 1$. If we consider the branch points, corresponding to the exchange of a single ρ meson along with many vacuum poles (pomerons P) with $\alpha_P(0) = 1$, then at small t they are situated at the points

$$j_n(t) = \beta(0) + \frac{\alpha' \beta'}{\alpha' + n\beta'} t.$$

So the pole $j_0 = \beta(t)$ is transformed to the series of cuts accumulating at the point $\beta(0)$:

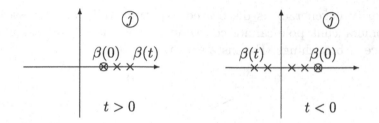

If previously the asymptotics of the scattering amplitude in the physical region had the form $s^{\beta(t)}$, then now we have $s^{\beta(0)}f(\xi,t)$ with f logarithmically depending on s. We conclude that the shape of the t-distribution is no longer determined by the pole.

In some sense we have a sort of a phase transition here. Indeed, in the t-channel we had a stable pole and thresholds at large positive t. When we approach the physical region of the s-channel ($t \leq 0$) the pole becomes unstable. It would be strange if the pole remained stable after it had collided with the branchings: we would acquire a bound state in the continuous spectrum. All this means that the Regge pole must move to another sheet.

Thus, if both the vacuum pole and the related branchings exist, then we actually do not know the angular dependence of the scattering amplitude.

Nevertheless there exist specific features which permit one to establish experimentally the existence of branch points without knowing the details (such as the angular distributions and so on).

13.4 Experimental verification of branching singularities

A Regge pole has definite quantum numbers: P_j, P, G, T, S, etc.

For instance a ρ meson pole ($G = +1$) can be exchanged in the reaction $\pi^- p \to \pi^0 n$. Other reactions, such as $\pi^- p \to \eta^0 n$ ($G = -1$) cannot have the ρ pole, but can be controlled by some other pole, β^*.

There also exist reactions without definite G-parity in the t-channel, for instance the process $K^- + p \to K^0 + n$, in which case different non-vacuum poles contribute and the asymptotics of the amplitude has the form

$$s^{\beta(0)} + cs^{\beta^*(0)}.$$

It is important that the poles have definite spatial parity. On the other hand an exchange by *two* vacuum poles has all quantum numbers of the vacuum except parity (due to the presence of orbital angular momentum). Therefore the state in the t-channel becomes parity degenerate when we take into account vacuum branchings.

If parity degeneracy is discovered experimentally in processes where the Pomeranchuk pole cannot contribute, then this will be a signal of the presence of branchings. For instance the reaction

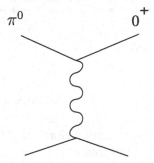

cannot occur by vacuum pole exchange, only by the exchange of other poles. Therefore the cross section, in the pole approximation, has to fall rapidly with energy. But an exchange of two vacuum poles is possible and therefore the amplitude should be proportional to s, modulo logarithm. That is, the cross sections of some reactions should decrease more slowly with energy, as compared with the expectation of the Regge pole picture, due to the contribution of branchings.

13.4.1 Branchings and conspiracy

We address now the question of branchings and conspiracy.

Let us recall that in nucleon–nucleon scattering, exchange of a pole with pion quantum numbers leads to the appearance of a term of the type $(\boldsymbol{\sigma}_1 \cdot \boldsymbol{q})(\boldsymbol{\sigma}_2 \cdot \boldsymbol{q})$ in the amplitude, which vanishes for $q \to 0$:

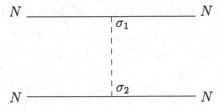

The contribution of this term can be made non-zero, provided it is divided by q^2, but then we get a singular, rotationally non-invariant expression in the $q \to 0$ limit,

$$\frac{(\boldsymbol{\sigma}_1 \cdot \boldsymbol{q})(\boldsymbol{\sigma}_2 \cdot \boldsymbol{q})}{q^2} \quad \to \quad \sigma_{1x}\sigma_{2x}.$$

On the other hand, the exchange of a pole of the same quantum numbers but with positive internal parity ($P_r = +1$) will contribute

$$\frac{[\boldsymbol{\sigma}_1 \times \boldsymbol{q}][\boldsymbol{\sigma}_2 \times \boldsymbol{q}]}{q^2} \quad \to \quad \sigma_{1y}\sigma_{2y}.$$

Their sum is an invariant expression

$$\sigma_{1x}\sigma_{2x} + \sigma_{1y}\sigma_{2yx} = (\boldsymbol{\sigma}_{1\perp} \cdot \boldsymbol{\sigma}_{2\perp}).$$

If there is a branching, however, the contribution of the graph corresponding to the two-reggeon branching, which appears when we add the vacuum pole,

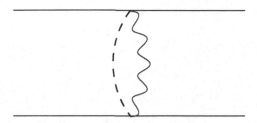

may also give a contribution proportional to $(\boldsymbol{\sigma}_{1\perp} \cdot \boldsymbol{\sigma}_{2\perp})$, which will be suppressed only by $\ln s$ compared with the contribution of the leading pole. This makes it difficult to establish the fact of conspiracy, to distinguish the contribution of two conspiring poles from that of branchings.

14

Reggeon diagrams

I have already said how the branch points arise. Now I shall show how the scattering process is described in terms of these branchings.

There are two ways to do that:

1. by considering the multi-particle t-channel unitarity conditions, generalized to the case of complex j;

2. by analysing the structure of Feynman diagrams.

The more convenient approach turns out to be the one based on diagrams. We adopt as a basic hypothesis the following postulates:

- there exist Regge poles;

- they generate branch points;

- these singularities together determine the structure of the j plane;

- there are no other singularities.

So let us assume that the elastic scattering amplitude has a Regge pole:

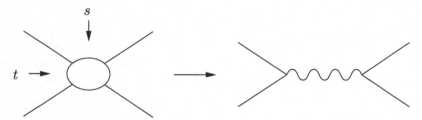

Following the path we have chosen let us study the diagram shown in Fig. 14.1, where the solid lines correspond to exchanges by ordinary particles, whereas the wavy lines describe reggeons. I will justify the

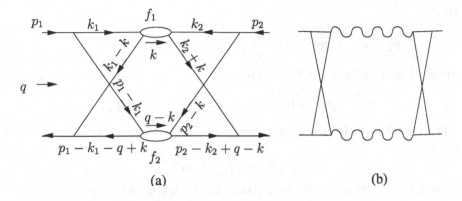

Fig. 14.1. Kinematics of double scattering (a) and the corresponding two-reggeon exchange graph (b)

possibility of this replacement which corresponds to inserting Regge poles into the blobs f_1 and f_2.

The graph of Fig. 14.1a contains two four-point functions whose amplitudes are denoted by f_1 and f_2. They are functions of momenta k_1, k_2 and k and should be integrated according to the Feynman rules. We consider now the various regions of integration over k.

In the region where $f_i(k_1, k_2, k)$ enter a non-asymptotic regime, this diagram does not give any essential contribution and cannot be distinguished from many other arbitrary diagrams (this region gives an additive contribution to a Regge pole in the full amplitude, if there is such a pole).

It is clear that this diagram can give something special only if we have Regge kinematics for $f_{1,2}$. If, inserting the poles into f_1 and f_2, I obtain an amplitude that is not power-suppressed at large s and is expressed in terms of exact trajectories, then this scattering topology is worth studying.

The amplitude f_1 is a function of virtual masses k_1^2, k_2^2, $(k_1 - k)^2$ and $(k_2 + k)^2$, of the momentum transfer k^2 and the squared energy $(k_1 + k_2)^2$. We will be able to calculate the asymptotic contribution of the graph of Fig. 14.1 easily if in the asymptotic regime the regions of large momentum transfers and large masses are suppressed. This is what we shall assume henceforth.

We express all internal momenta in Fig. 14.1 in terms of Sudakov parameters:

$$\left. \begin{array}{l} k = \alpha p_2' + \beta p_1' + k_\perp, \\ k_1 = \alpha_1 p_2' + \beta_1 p_1' + k_{1\perp}, \\ k_2 = \alpha_2 p_2' + \beta_2 p_1' + k_{2\perp}, \end{array} \right\} \tag{14.1}$$

where

$$p'_1 = p_1 - \kappa p_2, \quad \kappa \simeq \frac{p_1^2}{s}, \quad p'_2 = p_2 - \kappa' p_1, \quad \kappa' \simeq \frac{p_2^2}{s},$$

are by now familiar light-like vectors,

$$p'^2_1 = p'^2_2 = 0, \quad 2p'_1 p'_2 \simeq s, \quad s = (p_1 + p_2)^2.$$

For any internal momentum I require the condition $|k_i^2| \lesssim m^2$ to be satisfied, i.e.

$$|\alpha_i \beta_i s + k_{i\perp}^2| \lesssim m^2.$$

Consider the virtuality of the particle in the left-hand loop:

$$|k_1^2| = |\alpha_1 \beta_1 s + k_{1\perp}^2| \lesssim m^2. \tag{14.2a}$$

The condition $|(p_1 - k_1)^2| \lesssim m^2$, taken together with (14.2a), imposes an additional constraint on α_1 and β_1:

$$|\alpha_1 s + \beta_1 m^2| \lesssim m^2. \tag{14.2b}$$

This shows clearly that there cannot be any compensation between $\alpha_1 \beta_1 s$ and $k_{1\perp}^2$, i.e. the following relations hold:

$$|\alpha_1 \beta_1| s \lesssim m^2, \quad |k_{1\perp}^2| \lesssim m^2; \quad |\alpha_1| s \lesssim m^2, \quad |\beta_1| m^2 \lesssim m^2. \tag{14.3}$$

Similarly we have for the right-hand loop

$$|\alpha_2 \beta_2| s \lesssim m^2, \quad |k_{2\perp}^2| \lesssim m^2, \quad |\beta_2| s \lesssim m^2, \quad |\alpha_2| m^2 \lesssim m^2. \tag{14.4}$$

For the Sudakov parameters of k, it follows from the conditions

$$\left.\begin{aligned}
k^2 &= s\alpha\beta + k_\perp^2 \lesssim m^2, \\
(k - k_1)^2 &= (\alpha_1 - \alpha)(\beta_1 - \beta)s + (k_1 - k)_\perp^2 \lesssim m^2, \\
(k + k_2)^2 &= (\alpha_2 + \alpha)(\beta_2 + \beta)s + (k_2 + k)_\perp^2 \lesssim m^2,
\end{aligned}\right\} \tag{14.5}$$

that

$$|k_\perp^2| \lesssim m^2, \qquad |\alpha| \sim |\beta| \lesssim \frac{m^2}{s}. \tag{14.6}$$

Thus our region of integration is defined by the following inequalities:

$$\left.\begin{aligned}
|\alpha_1| &\lesssim \frac{m^2}{s}, \quad \beta_1 \sim 1; \quad -k_{1\perp}^2 \lesssim m^2; \\
\alpha_2 &\sim 1, \quad |\beta_2| \lesssim \frac{m^2}{s}; \quad -k_{2\perp}^2 \lesssim m^2; \\
|\alpha| &\sim |\beta| \sim \frac{m^2}{s}; \quad -k_\perp^2 \lesssim m^2.
\end{aligned}\right\} \tag{14.7}$$

The physical meaning of these conditions is clear:

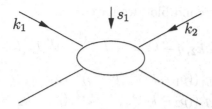

in order to have here a large energy invariant $s_1 \simeq \alpha_2 \beta_1 s \sim s$, while keeping the virtual masses of all internal lines small, it suffices to produce two particles k_1 and $p_1 - k_1$ which carry finite fractions β_1, $(1 - \beta_1)$ of the incident particle momentum p_1. Each of these offspring then scatters with a small momentum transfer $k \simeq k_\perp$ $(q - k \simeq q_\perp - k_\perp)$ on its counterpart with momentum $\alpha_2 p_2$ $((1 - \alpha_2)p_2)$ from the splitting of p_2, and the scattered particles finally merge into outgoing momenta $p_3 = p_1 - q$ and $p_4 = p_2 + q$.

In such kinematical conditions the invariant energies of the scattering blobs f_1 and f_2 are of the order of the total s:

$$s_1 = (k_1 + k_2)^2 = (\alpha_1 + \alpha_2)(\beta_1 + \beta_2)s + (k_{1\perp} + k_{2\perp})^2 \simeq \beta_1 \alpha_2 s, \quad (14.8a)$$

$$s_2 = (p_1 - k_1 + p_2 - k_2)^2 \simeq (1 - \beta_1)(1 - \alpha_2)s, \quad (14.8b)$$

$$s_1 \sim s_2 = \mathcal{O}(s).$$

Let us now write down the amplitude f_1 *as if* in the form of a Sommerfeld–Watson integral:

$$f_1(k_1, k_2, k) = -\int_C \frac{\mathrm{d}\ell}{4\mathrm{i}} \, \xi_\ell s_1^\ell \, J_\ell(k^2)$$
$$\times g_1(k_1^2, (k - k_1)^2, k^2) \, g_2(k_2^2, (k + k_2)^2, k^2). \quad (14.9a)$$

This amplitude corresponds to the Regge pole graph

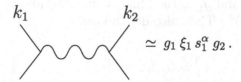

$$\simeq g_1 \, \xi_1 \, s_1^\alpha \, g_2.$$

I slightly generalized the latter expression by writing (14.9a) in the form of an integral over $\mathrm{d}\ell$ of the simple pole $J_\ell = 1/(\ell - \alpha(t))$ with the contour of integration located to the right of $\ell_0 = \alpha(t)$. This generalization has been made for future convenience in taking account of any number of reggeons.

Similarly for the second blob we have

$$
\begin{aligned}
f_2(p_1-k_1, p_2-k_2, q-k) = -\int \frac{\mathrm{d}\ell'}{4\mathrm{i}} \xi_{\ell'} s_1^{\ell'} J_{\ell'}((q-k)^2) \\
\times g_1'((p_1-k_1)^2, (k-q)^2, (p-k_1+k-q)^2) \\
\times g_2'((p_2-k_2)^2, (q-k)^2, (p_1-k_2+q-k)^2).
\end{aligned}
\tag{14.2}
$$

If I substitute (14.9) into the corresponding Feynman integral and take account of the restrictions (14.7), then the integration variables in the left-hand and right-hand loops get separated and a factorization occurs.

The amplitude of the graph of Fig. 14.1 will then acquire the following form:

$$
\begin{aligned}
F(s,t) = \frac{\mathrm{i}}{2\cdot 2!} \iint \frac{\mathrm{d}\ell\,\mathrm{d}\ell'}{4\mathrm{i}\,4\mathrm{i}} \xi_\ell \xi_{\ell'} \int \frac{\mathrm{d}^2 k_\perp}{(2\pi)^2} J_\ell(k_\perp^2) J_{\ell'}((q_\perp-k_\perp)^2) s^{\ell+\ell'-1} \\
\times \int \frac{s\,\mathrm{d}\alpha}{2\pi\mathrm{i}} \int \frac{\mathrm{d}^4 k_1}{(2\pi)^4\mathrm{i}} \frac{g_1 g_1' \beta_1^\ell (1-\beta_1)^{\ell'}}{[1][2][3][4]} \\
\times \int \frac{s\,\mathrm{d}\beta}{2\pi\mathrm{i}} \int \frac{\mathrm{d}^4 k_2}{(2\pi)^4\mathrm{i}} \frac{g_2 g_2' \alpha_2^\ell (1-\alpha_2)^{\ell'}}{[1][2][3][4]}.
\end{aligned}
\tag{14.10}
$$

Here we have taken the functions J_ℓ and $J_{\ell'}$ out of the integrals over α, β since, as was mentioned above, in the region of interest they may be treated as functions of two-dimentional space-like vectors: we have $\alpha \sim \beta \sim m^2/s$ so that $k^2 \simeq k_\perp^2$ and $(q-k)^2 \sim (q-k)_\perp^2$.

Counting powers of s we should remember the factor s arising from the Jacobian due to Sudakov variables:

$$
\mathrm{d}^4 k = \frac{s}{2}\,\mathrm{d}\alpha\,\mathrm{d}\beta\,\mathrm{d}^2 k_\perp.
$$

14.1 Two-particle–two-reggeon transition amplitude

14.1.1 Structure of the vertex

The functions g_1 and g_1' in the l.h.s. of the diagram depend, respectively, on $k_\perp^2$ and $(q-k)_\perp^2$. They also depend on
g_1 –

$$
\left.
\begin{aligned}
k_1^2 &= \alpha_1 \beta_1 s + k_{1\perp}^2, \\
(k-k_1)^2 &\simeq (\alpha_1 - \alpha)\beta_1 s + (k-k_1)_\perp^2
\end{aligned}
\right\}
\tag{14.11a}
$$

$-\,g_1'$ –

$$
\left.
\begin{aligned}
(p_1-k_1)^2 &\simeq -\alpha_1(1-\beta_1)s + k_{1\perp}^2, \\
(p_1-k_1+k-q)^2 &\simeq (\alpha-\alpha_1-\alpha_q)(1-\beta_1)s + (k_1-k+q)_\perp^2.
\end{aligned}
\right\}
\tag{14.11b}
$$

We see that the integral over k_1 is a function only of α and $k_\perp$. Similarly the integral over k_2 is a function of β and $k_\perp$ only.

The expressions [1], [2], [3], [4] on the second line of (14.10) are the usual Feynman denominators which contain the squared particle momenta (14.11).

The analysis of the location of singularities in the complex plane α_1 (zeros of $-m^2 + \mathrm{i}\,\varepsilon$; see (14.11)) shows that the integral over k_1 is non-zero only in the interval $0 < \beta_1 < 1$. Similar analysis of the integral over k_2 yields $0 < \alpha_2 < 1$. These conditions ensure that the non-integer powers β_1^ℓ and $(1 - \beta_1)^{\ell'}$ are properly defined and do not introduce redundant complexity into these integrals.

Now we can see already that F contains the integral $A_{\ell\ell'}$, i.e. the amplitude of two-reggeon creation, which differs from the usual amplitude $2 \to 2$ only in that the two external lines are reggeons rather than ordinary particles.

The factors $\beta_1\alpha_2$ and $(1 - \beta_1)(1 - \alpha_2)$ play the rôle of the cosine of the scattering angle which would appear at the vertices if ordinary particles with spin had been created. In our case the spins are ℓ and ℓ' and there appear the corresponding powers β_1^ℓ and $(1 - \beta_1)^{\ell'}$.

14.1.2 Analytic properties of the vertex

The energy invariant this amplitude depends on is $\tilde{s} \simeq s\alpha$. Thus the

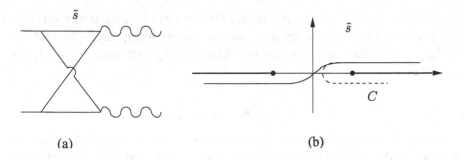

Fig. 14.2.   (a) Two-reggeon production and (b) the integration paths in $\tilde{s}$

integration over α of the Feynman diagram of Fig. 14.2a is equivalent to the integration of the two-reggeon creation amplitude over its energy invariant $\tilde{s}$:

$$\int \mathrm{d}\alpha \to \frac{1}{s} \int \mathrm{d}\tilde{s}\, A_{\ell\ell'}(\tilde{s}, p_2^2, k_\perp^2). \qquad (14.12)$$

As this amplitude emerges from a Feynman diagram, for instance a crossed box of Fig. 14.2a, it has the usual analytic properties (β_1^ℓ and

$(1 - \beta_1)^{\ell'}$ are real): the right- and left-hand cuts lie in the complex $\tilde{s}$ plane and the integration is performed along the contour shown by a solid line in Fig. 14.2b. The integral of $A_{\ell\ell'}$ does not vanish since this diagram has both cuts.

If we consider instead the diagram of Fig. 14.3 without a cross, then it is clear that its contribution is equal to zero since this *planar* diagram has a right cut only.

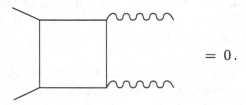

$$= 0.$$

Fig. 14.3.   Planar diagram for the two-particle–two-reggeon transition

The same is true for the pole diagrams with single particle exchange:

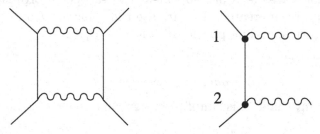

Here there is no left-hand cut either, and the form factors at the vertices 1 and 2 allow us to close the contour on the left, yielding zero. (The integral of $A_{\ell\ell'}$ over $\tilde{s}$ does not contain anything asymptotic since $\alpha = \mathcal{O}(1/s)$, $\tilde{s} = \mathcal{O}(1)$.)

14.1.3 Factorization

It is clear that such a separation of variables in the integral of F does not depend on details of the structure of the internal particle lines that link together the four-point functions of Fig. 14.1.

This result is a consequence of pure kinematics. The momenta in the left-hand loop integral over k_1 have Sudakov variables β_i of the order of unity (along p'_1) and $\alpha_i \sim m^2/s \ll 1$ (along p'_2). Therefore the scalar products kk_i, on which the functions entering this diagram depend, contain only the dependence on β_i and α:

$$2kk_i \sim 2(\alpha_i p'_2 + \beta_i p'_1)(\alpha p'_2 + \beta p'_1) \sim s(\alpha_i \beta + \beta_i \alpha) \sim s\alpha\beta_i,$$

$$\beta_i \sim 1, \qquad \alpha \sim \beta \sim \alpha_i \sim \frac{m^2}{s}.$$

Now the entire result can be written in the form of

$$F(s,t) = \frac{i\pi}{2} \int \frac{d\ell \; d\ell'}{2i\pi \; 2i\pi} \xi_\ell \xi_{\ell'} \int \frac{d^2k}{(2\pi)^2} J_\ell(k_\perp) J_{\ell'}((q-k)_\perp)$$
$$\times s^{\ell+\ell'-1} N_{\ell\ell'}^2(k_\perp, (q-k)_\perp), \quad (14.13)$$

where N is the integral of the imaginary part of $A_{\ell\ell'}$ along the right cut (contour C in Fig. 14.2b):

$$N_{\ell\ell'} = \int_C \frac{d\tilde{s}}{2\pi i} A_{\ell\ell'}(\tilde{s}, k_\perp, q_\perp).$$

Hence, the vertex function $N_{\ell\ell'}$ is real.

14.2 Partial wave amplitude of the Mandelstam branching

Now let us write down the partial wave corresponding to the amplitude F:

$$F = -\frac{1}{4i} \int dj \xi_j \phi_j s^j, \qquad \phi_j = \frac{2}{\pi} \int_{s_0}^\infty s^{-j-1} F_1(s,t) ds, \quad (14.14)$$

where F_1 is the imaginary part of F.

Since $N_{\ell\ell'}$ is real, all complexity arises from the product of the signature factors, i.e. F_1 is proportional to the integral of $\gamma_{\ell\ell'}$:

$$\gamma_{\ell\ell'} = \operatorname{Im} i \xi_\ell \xi_{\ell'}.$$

If one writes the signature factor in the form of

$$\xi_\ell = \frac{1}{\zeta_\ell} \exp\left\{ -i\frac{\pi}{2}\left(\ell + \frac{1-P}{2}\right)\right\},$$
$$\zeta_\ell = \left\{ \begin{array}{ll} \sin(\pi\ell/2), & P = +1, \\ \cos(\pi\ell/2), & P = -1, \end{array} \right\} \quad (14.15)$$

where $P = +1$ ($P = -1$) corresponds to positive (negative) signature, then

$$\gamma_{\ell\ell'} = \frac{1}{\zeta_\ell \zeta_{\ell'}} \cos\left[\frac{\pi}{2}(\ell + \ell' + 1) - \frac{P_\ell + P_{\ell'}}{2}\right]. \quad (14.16)$$

Thus we get for F_1

$$F_1 = \frac{\pi}{2} \int \frac{d\ell}{2\pi i} \int \frac{d\ell'}{2\pi i} \int \frac{d^2k}{(2\pi)^2} J_\ell(k) J_{\ell'}(q-k) N_{\ell\ell'} N_{\ell'\ell} s^{\ell+\ell'-1} \gamma_{\ell\ell'}, \quad (14.17)$$

and for ϕ_j this gives

$$\phi_j = \int \frac{d\ell}{2\pi i} \int \frac{d\ell'}{2\pi i} \int \frac{d^2k}{(2\pi)^2} J_\ell(k) J_{\ell'}(q-k) \gamma_{\ell\ell'} \frac{N_{\ell\ell'} N_{\ell'\ell}}{j+1-\ell-\ell'}, \quad (14.18)$$

and one should remember that j lies to the right of the singularities:

$$\mathrm{Re}\, j \geq \mathrm{Re}\,\ell + \mathrm{Re}\,\ell' - 1, \quad \mathrm{Re}\,\ell > \alpha(k^2), \quad \mathrm{Re}\,\ell' > \alpha'((q-k)^2),$$

that is the pole $\ell = j + 1 - \ell'$ lies to the right of the integration contour over ℓ (similarly the pole $\ell' = j + 1 - l$ lies to the right of the integration contour over ℓ'). After integration over ℓ and ℓ' the expression for ϕ_j takes on a simpler form:

$$\phi_j = \int \frac{\mathrm{d}^2 k}{(2\pi)^2} \frac{\gamma_{\alpha\alpha'} N_{\alpha\alpha'} N_{\alpha'\alpha}}{j + 1 - \alpha(k^2) - \alpha'((q-k)^2)}. \tag{14.19}$$

In getting this expression the functions J_ℓ and $J_{\ell'}$ have been taken in the pole approximation:

$$J_\ell = \frac{1}{\ell - \alpha(k^2)}, \quad J_{\ell'} = \frac{1}{\ell' - \alpha'((q-k)^2)}.$$

The formula (14.19) can be represented in the form of

$$\phi_j = \int \frac{\mathrm{d}^2 k}{(2\pi)^2} \frac{\mathrm{d}\ell}{2\pi\mathrm{i}} \frac{\mathrm{d}\ell'}{2\pi\mathrm{i}} (2\pi\mathrm{i})\delta(\ell + \ell' - j - 1) J_\ell(k) J_{\ell'}(q - k)\gamma_{\ell\ell'} N_{\ell\ell'} N_{\ell'\ell},$$

$$\tag{14.20}$$

which resembles the Feynman integral for a non-relativistic particle with the two-dimensional momentum $k_\perp$ and energy $\omega = \ell - 1$. The factor $\delta(\ell + \ell' - j - 1)$ ensures energy conservation:

$$\omega = \omega_1 + \omega_2; \quad \omega = j - 1; \quad \omega_1 = \ell - 1; \quad \omega' = \ell' - 1.$$

In terms of ω one can write the expression

$$f_{\omega,q} = \int \frac{\mathrm{d}\omega_1}{2\pi\mathrm{i}} \int \frac{\mathrm{d}^2 k}{(2\pi)^2} G_{\omega_1}(k) G_{\omega-\omega_1}(q-k) N^2 \gamma \tag{14.21}$$

which corresponds to the graph

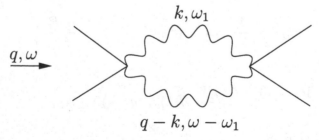

In order to reach full analogy with the Feynman integration, the contour of integration over ω along the imaginary axis can be deformed in general to run along the dashed line:

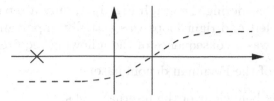

All that has been done so far is valid assuming that there are no singularities of $N_{\ell\ell'}$ with respect to ℓ and ℓ' near $j \sim \alpha_1 + \alpha_2 - 1$. The amplitude of the graph

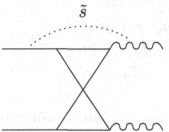

does not have these singularities, due to the rapid drop in $\tilde{s}$.

I do not want to say that I have calculated the whole amplitude. What I have really calculated is only its singular part, the contribution of the branching due to a two-reggeon state. Cases in which reggeon creation contains an exchange of many particles ($\tilde{s}$ is large) are also possible, and they will be considered in the next lecture. We will see that in the region of large $\tilde{s}$ the function N has singularities near j close to unity. So for the time being we assume $N_{\ell\ell'}$ to be smooth functions of ℓ and ℓ'.

Now I want to show how to write the contributions of states containing a larger number of reggeons. The next singularity will be obtained if we insert a branching into f (which, as we already know, is present already in the elastic amplitude). In other words, I will consider the graph of Fig. 14.4.

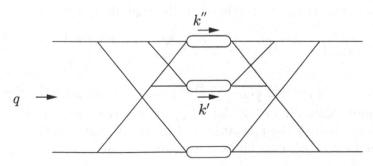

Fig. 14.4. Triple scattering graph generating three-reggeon branching

When I was considering the graph Fig. 14.1, only a separation of integrations on the left and right loops was actually important for me. And that separation was a consequence of the following requirements:

- smallness of the Feynman denominators;
- asymptotic behaviour of the internal blobs.

It is clear that similar arguments will also apply to the new graph Fig. 14.4 (since for the whole upper block f all these requirements remain valid), and we obtain the new branching

$$F = \frac{\pi}{2} \int \frac{\mathrm{d}\ell_1}{2\pi\mathrm{i}} \frac{\mathrm{d}\ell_2}{2\pi\mathrm{i}} \frac{\mathrm{d}\ell_3}{2\pi\mathrm{i}} \int \frac{\mathrm{d}^2 k''}{(2\pi)^2} \frac{\mathrm{d}^2 k'}{(2\pi)^2} J_{\ell_1}(k') J_{\ell_2}(k'') J_{\ell_3}(q - k' - k'')$$
$$\times N_{\ell_1 \ell_2 \ell_3} N_{\ell_3 \ell_2 \ell_1} s^{\ell_1 + \ell_2 + \ell_3 - 2}. \qquad (14.22)$$

In terms of partial waves this formula will correspond to the graph

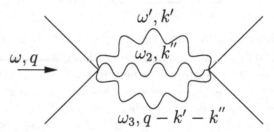

$$\phi_{\omega, q} = \frac{1}{3!} \int \frac{\mathrm{d}\omega_1 \mathrm{d}\omega_2 \mathrm{d}\omega_3}{((2\pi)^3 \mathrm{i})^2} \mathrm{d}^2 k_1 \mathrm{d}^2 k_2 \mathrm{d}^2 k_3 \delta(\omega_1 + \omega_2 + \omega_3 - \omega) \delta\left(q - \sum_i k_i\right)$$
$$\times J_{\omega_1}(k_1) J_{\omega_2}(k_2) J_{\omega_3}(k_3) \gamma_{\omega_1 \omega_2 \omega_3} N_{\omega_1 \omega_2 \omega_3} N_{\omega_3 \omega_2 \omega_1}. \qquad (14.23)$$

The factor $1/3!$ accounts for the identical reggeons. The contribution of all possible branchings can be written in this way.

The only non-triviality is related to the sign alternation of γ:

$$\gamma_{\omega_1 \omega_2 \ldots \omega_n} = (-1)^{n-1} \sin \frac{\pi}{2} \left(j - \sum \frac{P_i - 1}{2} \right) \prod_i \frac{1}{\xi_i}. \qquad (14.24)$$

The factor $(-1)^{n-1} \sin \frac{\pi}{2}(j - \sum \frac{P_i-1}{2})$, which determines the sign of the partial wave discontinuity of the Regge singularity (reggeon threshold), arises as a result of the appearance of an additional integration over k_i when the number of branchings is increased by unity,

$$\frac{\mathrm{d}^4 k_i}{(2\pi)^4 \mathrm{i}} \simeq \mathrm{i} \frac{\mathrm{d}\alpha_i}{2\mathrm{i}\pi} \frac{\mathrm{d}\beta_i}{2\mathrm{i}\pi} \frac{\mathrm{d}^2 k_{i\perp}}{(2\pi)^2} \frac{|s|}{2},$$

and one integration over ℓ with a negative sign,

$$(-1)\frac{\mathrm{d}\ell}{4\mathrm{i}}.$$

The integration over $\alpha_i/(2\mathrm{i}\,\pi)$ and $\beta_i/(2\mathrm{i}\,\pi)$ consists of the integration of the imaginary parts of reggeon creation amplitudes and therefore gives real values. This means that the phase of the nth branching is contained only in the expression

$$(-1)^n \mathrm{i}^{\,n-1}\xi_{\ell_1}\xi_{\ell_2}\cdots\xi_{\ell_n}.$$

Since the partial wave is expressed in terms of an integral of the imaginary part of F (see formula (14.2)), it is proportional to

$$
\begin{aligned}
\gamma_{\ell_1\ldots\ell_n} &= (-1)^n \mathrm{Im}\,\mathrm{i}^{\,n-1}\prod_j \frac{1}{\xi_j}\exp\left[-\frac{\mathrm{i}\,\pi}{2}\left(\sum_j \ell_j - \sum_j \frac{P_j-1}{2}\right)\right] \\
&= (-1)^n \mathrm{Im}\,\prod_j \frac{1}{\xi_j}\exp\left[-\frac{\mathrm{i}\,\pi}{2}\left(\sum_j^n \ell_j - n + 1 - \sum_j \frac{P_j-1}{2}\right)\right] \\
&= (-1)^n \prod_j \frac{1}{\xi_j}\sin\frac{\pi}{2}\left(\sum_j^n \ell_j - n + 1 - \sum_j \frac{P_j-1}{2}\right).
\end{aligned}
$$

But because *energy* is conserved, i.e.

$$\omega = j - 1 = \sum_i^n \omega_i, \qquad \omega_i = \ell_i - 1,$$

we have

$$\sum_i^n \ell_i - n + 1 = j.$$

Therefore

$$\sin\frac{\pi}{2}\left(\sum_j^n \ell_j - n + 1 - \sum_j^n \frac{P_j-1}{2}\right) = \sin\frac{\pi}{2}\left(j - \sum_i^n \frac{P_i-1}{2}\right).$$

This leads to the expression for $\gamma_{\omega_1\ldots\omega_n}$ given in equation (14.24).

The factor

$$\sin\frac{\pi}{2}\left(j - \sum_i^n \frac{P_i-1}{2}\right)$$

reflects the fact that the contribution of moving branch points for physical values of j is zero.

The sign alternation is in fact essential only for the case of the vacuum pole, when $\ell_i = 1$, $P_i = 1$ and $j = 1$ (the amplitude corresponding to the exchange of a vacuum reggeon is purely imaginary).

How could one calculate the singularities of the two-dimensional Feynman integral for a partial wave amplitude?

$$\phi_\omega(q) = \frac{1}{n!} \int \prod_i^n \frac{d\omega_i}{2i\,\pi} \frac{d^2k_i}{(2\pi)^2} \frac{1}{\omega_i - \varepsilon(k_i^2)} N_{\omega_1...\omega_n}^2 \gamma_{\omega_1...\omega_n}$$

$$\times (2i\,\pi)\delta\left(\omega - \sum_i \omega_i\right)(2\pi)^2\delta\left(q - \sum_i k_i\right).$$

We shall calculate these integrals assuming (as was mentioned above) that all singularities in ω_i are in the reggeon propagators $(\omega_i - \varepsilon(k_i^2))^{-1}$, whereas all the remaining functions are smooth, so in what follows we will replace them by constants and in $\gamma_{\omega_1...\omega_n}$ take into account the factor $(-1)^{n-1}$. Then this integral will be

$$\frac{(-1)^{n-1}}{n!} N^2 \int \frac{\prod_i^{n-1} d^2k_i}{(2\pi)^{2(n-1)}} \frac{\sin\frac{\pi}{2}(j - \sum_i \frac{P_i-1}{2})}{\omega - \sum_i \varepsilon(k_i^2)}. \tag{14.25}$$

Since I am working in the physical region where all $\varepsilon_i(k_{i\perp}^2) < 0$, this expression for $\omega > 0$ does not have any singularity. The singularity arises when $\omega \leq 0$ under the condition that $\sum_i k_i = q$, i.e. to determine the position of the reggeon singularity it is necessary to find an extremum of the denominator in (14.25):

$$\nabla_{k_i} \left\{ \sum_j \varepsilon(k_j^2) + \lambda\left(\sum_j k_j - q\right)\right\} = 0,$$

which gives $k_{j,\text{extr.}} = q/n$. (Here I am studying the more interesting case of identical reggeons.)

Writing k_i in the form of

$$k_i = \frac{q}{n} + x_i, \qquad \sum_i x_i = 0,$$

we obtain for $\sum_i \varepsilon(k_i^2)$ the following quadratic form:

$$\sum_{i=1}^n \varepsilon(k_i^2) = n\varepsilon\left(\frac{q^2}{n^2}\right) + \left(\varepsilon' + 2\frac{q^2}{n^2}\varepsilon''\right)\sum_i x_{1i}^2 + \varepsilon'\sum_i x_{2i}^2.$$

Here x_{1i} are components along $\boldsymbol{q}$ and x_{2i} are transverse to $\boldsymbol{q}$.

To find the singularities of such an integral, it is sufficient to calculate its discontinuity which is much easier than to compute the whole integral:

$$\delta\phi_\omega = \pi\frac{(-1)^{n-1}}{n!}N^2 \int \frac{\prod_i^n dx_{1i}}{(2\pi)^{n-1}} \frac{\prod_i^n dx_{2n}}{(2\pi)^{n-1}} \cos\frac{\pi\omega}{2}$$
$$\times \delta(\omega - \omega_n - c_1 I_1 - c_2 I_2)\,\delta\left(\sum x_i\right). \quad (14.26)$$

Here we used the following notation:

$$\left.\begin{array}{ll} \omega_n = n\varepsilon\left(\dfrac{q^2}{n^2}\right), & I_1 = \displaystyle\sum_i^n x_{1i}^2, \quad I_2 = \displaystyle\sum_i^n x_{2i}^2, \\[3mm] c_1 = \varepsilon'\left(\dfrac{q^2}{n^2}\right), & c_2 = c_1 + \dfrac{2q^2}{n^2}\varepsilon''\left(\dfrac{q^2}{n^2}\right). \end{array}\right\} \quad (14.27)$$

Due to the positiveness of $I_{1,2}$ the integral (14.26) differs from zero only when $\omega \leq \omega_n$, i.e. the singularities of ϕ_ω in the ω plane are situated on the real axis at $\omega < n\varepsilon(q^2/n^2)$:

The value of the discontinuity can be easily calculated and has a simple geometrical meaning: it is the surface of the $(2n-2)$-dimensional ellipsoid

$$\frac{(-\omega + \omega_n)^{n-2}(n-1)\pi^{n-1}}{n!\,(c_1 c_2)^{\frac{n-1}{2}}}.$$

Thus we obtain for $\delta\varphi_j$

$$\delta\varphi_j = \pi(-1)^{n-1}(c_1 c_2)^{-\frac{n-1}{2}}\frac{n-1}{(n!)^2(4\pi)^{n-1}}(\omega_n - \omega)^{n-2}N^2 \cos\frac{\pi\omega}{2}. \quad (14.28)$$

This discontinuity corresponds to the logarithmic singularity at any n:

$$\varphi_j \sim (\omega_n - \omega)^{n-2}\ln(\omega_n - \omega).$$

An interesting picture appears in the ω plane if we have vacuum poles with $\alpha(0) = 1$. In this case

$$\varepsilon(q^2) = -\alpha' q^2, \qquad q^2 = -t$$

and the nth branching is situated at

$$\omega_n = -\alpha' \frac{q^2}{n}. \tag{14.29}$$

From (14.29) we see that for $t = 0$ all the singularities are positioned at $\omega = 0$. For $t \neq 0$, $\omega = 0$ is a point of accumulation of the branchings. The pattern of singularities for $t < 0$ is displayed in Fig. 14.5a. The

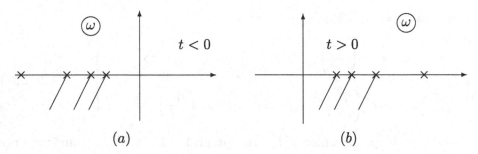

Fig. 14.5.   For $\alpha(0) = 0$ higher branchings accumulate at $\omega = 0$

pole is the leftmost singularity, and the branching points accumulate to $\omega = 0$ with $n \to \infty$. This means that in the physical region of s-channel scattering it is necessary to take into account the higher branchings. This results in specific oscillations of the cross section in momentum transfer – a manifestation of the sign-alternating character of the discontinuity of the multi-reggeon singularity (the factor $(-1)^{n-1}$).

At $t > 0$ the picture in the ω plane is reversed, as shown in Fig. 14.5b. Now the pole is the rightmost singularity.

15

Interacting reggeons

In the previous lectures we have found the rules of how to calculate multi-reggeon contributions to the scattering amplitude:

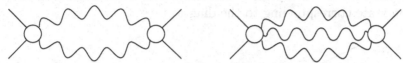

The Green functions $\mathcal{G}(\omega, k)$ of reggeons entering the diagram depend on $\omega = j - 1$ and on the two-dimensional momentum k. Furthermore, ω can be interpreted as an energy; the cuts in the ω plane run along the left-hand real axis.

We have investigated the singularities of $\varphi_\omega(q)$ in the ω plane for each of these diagrams. We have assumed that the vertex $N_{\omega_1\omega_2}(k_1, k_2)$ of the transition of two particles into two reggeons with momenta k_1 and k_2 has no singularities with respect to ω_i. $N_{\omega_1\omega_2}(k_1, k_2)$ itself represents the integral of the absorptive part of the particle–reggeon collision amplitude:

$$N \sim \int A^{(1)}_{\omega_1\omega_2}(s_1, k_1, k_2)\mathrm{d}s_1. \tag{15.1}$$

The simplest contribution to this amplitude,

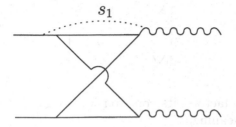

at large s_1 behaves as $1/s_1^2$ and therefore the corresponding integral (15.1) for $N_{\omega_1\omega_2}$ is well convergent.

For small energies s_1 one can restrict oneself to considering the contributions to the absorptive part of the particle–reggeon scattering amplitude that contain a small number of particles in the intermediate s-channel state (for instance two particles as in the above crossed box diagram). Recall that taking into account only *one-particle* s-channel states in the creation amplitudes of two, three, etc. reggeons

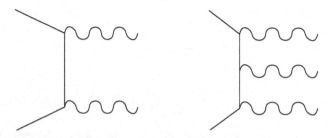

would yield zero. As we have discussed, the contribution of the two-reggeon state corresponding to the diagram

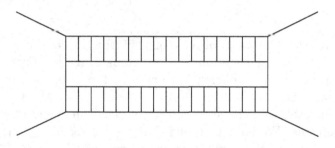

is zero, due to the absence of a left-hand cut in $A^{(1)}_{\omega_1\omega_2}$, and also due to the presence in the reggeon–particle–particle vertices of the form factors which decrease at large particle virtualities $\sim s_1$. Note that if we were to consider the particle–reggeon graph

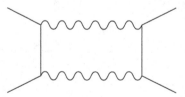

without such form factors, its contribution would not vanish, but it has merely a formal meaning.

If we now consider large s_1, then in the absorptive part of $A^{(1)}_{\omega_1\omega_2}$ we have to take into account the states with large numbers of particles:

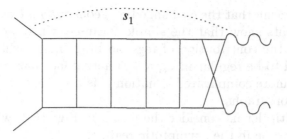

Then, the ladder-type diagrams make a large contribution of the order of s_1^β to the particle–reggeon scattering amplitude. Therefore, restricting ourselves to the region $s' \sim s \gg s_1 \sim m^2$ for extracting the asymptotics cannot be justified (for the notation see Fig. 15.1).

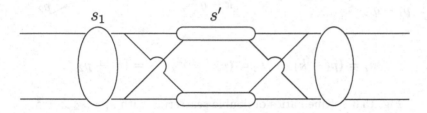

Fig. 15.1.   Reggeon (s') and particle–reggeon transition blocks (s_1)

Considering the integration region $s_1 \sim s' \sim s$, it is natural to introduce the reggeon graphs shown in Fig. 15.2. These graphs can be looked upon

Fig. 15.2.   One-reggeon–two-reggeon transition diagrams

as describing *decays* of the reggeon, its instability. At large s the branch points as well as the Regge pole can give a contribution to the vertex N_{ω_1,ω_2}, and this leads to the reggeon diagrams

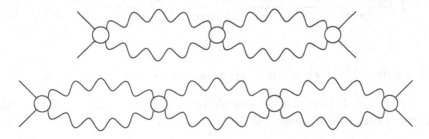

These graphs mean that the exchanged reggeons *interact* with each other. These arguments show that the simple formulæ which we have written previously for the contribution of reggeon branchings, where N_{ω_1,ω_2} has been supposed to be regular in ω_i, are in actual fact not valid.

In the real, more complicated situation it is necessary to study a whole series of reggeon graphs.

To begin with, let us consider the graph in Fig. 15.3, where each one of the blobs can be in the asymptotic regime.

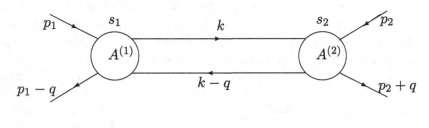

$$s_1 = (p_1 - k)^2, \quad s_2 = (p_2 + k)^2; \quad s = (p_1 + p_2)^2.$$

Fig. 15.3.   Kinematics of double scattering with $s_1 \sim s_2 \gg m^2$

We are going to find the asymptotic contribution which comes from the regions $s_1 \gg m^2$ and $s_2 \gg m^2$. The contribution we are looking for is analogous, in the case of the usual particles, to the photon–vector-meson transition:

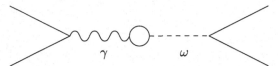

As before, we expand the momentum k in the Sudakov variables:

$$k = \alpha p_2' + \beta p_1' + k_\perp. \tag{15.2}$$

The amplitude corresponding to the graph of Fig. 15.3 is given by the following expression:

$$\int \frac{|s/2|\,\mathrm{d}\alpha\,\mathrm{d}\beta}{(2\pi)^4\mathrm{i}} \frac{A^{(1)}(s_1, k^2, (q-k)^2)\,A^{(2)}(s_2, k^2, (q-k)^2)}{(m^2 - k^2)(m^2 - (q-k)^2)}, \tag{15.3}$$

where

$$k^2 = \alpha\beta s + k_\perp^2, \quad s_1 \simeq -\alpha s, \quad s_2 \simeq \beta s, \quad \alpha\beta s \sim m^2, \quad k_\perp^2 \sim m^2. \tag{15.4}$$

We see from (15.4) that it is impossible to make both s_1 and s_2 of the order of s, since this would mean that $\alpha \sim 1$ and $\beta \sim 1$ and hence $k^2 \gg m^2$.

From condition (15.4) one finds that

$$s_1 s_2 \sim s\, m^2. \qquad (15.5)$$

Let us assume that $A^{(1)}(s_1)$ behaves as $s_1^{\ell_1}$ for $s_1 \gg m^2$ and $A^{(2)}(s_1)$ is proportional to $s_2^{\ell_2}$ for $s_2 \gg m^2$. If $\ell_2 > \ell_1$, then the leading contribution $\sim s^{\ell_2}$ to the integral will come from the region $\alpha \sim m^2/s$, $\beta \sim 1$, which corresponds to taking into account only the second pole:

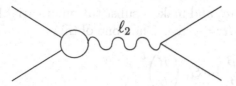

The situation when ℓ_1 is close to ℓ_2 is especially interesting. Under this condition another region of the variables α and β dominates in (15.3):

$$\left.
\begin{aligned}
\frac{m^2}{s} &\ll |\alpha| \ll 1, & \frac{m^2}{s} &\ll \beta \ll 1, \\
s_1 &= -\alpha s \gg m^2, & s_2 &= \beta s \gg m^2.
\end{aligned}
\right\} \qquad (15.6)$$

This region gives to (15.3) more than each of the two contributions of separate Regge poles. It is not difficult to verify that both poles, $1/(m^2 - k^2 - \mathrm{i}\varepsilon)$ and $1/(m^2 - (q-k)^2 - \mathrm{i}\varepsilon)$, lie in the α plane on the same side of the integration contour (below when $\beta > 0$ and above when $\beta < 0$):

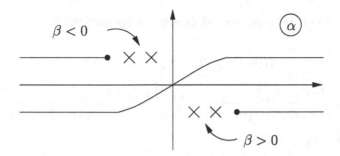

Therefore one can enclose the contour of integration around one of the cuts of $A^{(i)}$, situated in the half-plane where those poles are absent.

As a result the absorptive part of the amplitude $A(s_1)$ enters in the integral (15.3) along the left-hand or right-hand cut and we arrive at the following expression:

$$F = 2\mathrm{i} \int\limits_{\beta>0} \frac{\mathrm{d}^2 k\, \mathrm{d}(\alpha s/2)\mathrm{d}\beta}{(2\pi)^4 \mathrm{i}} \frac{A_1^{(1)}(s_1, k^2, (q-k)^2) A^{(2)}(s_2, k^2, (q-k)^2)}{(m^2 - k^2 - \mathrm{i}\varepsilon)(m^2 - (q-k)^2 - \mathrm{i}\varepsilon)}$$

$$+ 2\mathrm{i} \int\limits_{\beta<0} \frac{\mathrm{d}^2 k\, \mathrm{d}(\alpha s/2)\mathrm{d}\beta}{(2\pi)^4 \mathrm{i}} \frac{A_2^{(1)}(s_1, k^2, (q-k)^2) A^{(2)}(s_2, k^2, (q-k)^2)}{(m^2 - k^2 - \mathrm{i}\varepsilon)(m^2 - (q-k)^2 - \mathrm{i}\varepsilon)}.$$

$$(15.7)$$

Let us consider the first integral. We substitute $A_1^{(1)} \sim (\mathrm{Im}\, \xi_{\ell_1}) s_1^{\ell_1} g_1 g_1'$ and $A^{(2)} \sim \xi_{\ell_2} s_2^{\ell_2} g_2 g_2'$ and make a substitution of variables, that is instead of α we introduce $R = -s\alpha\beta$. Then one obtains

$$-2\mathrm{i} \int_{\beta>0} \frac{\mathrm{d}^2 k\, \mathrm{d}R\, \mathrm{d}\beta}{2(2\pi)^4 \mathrm{i}\, \beta}\, \mathrm{Im}\, \xi_{\ell_1} \left(\frac{R}{\beta}\right)^\ell$$

$$\times \frac{\xi_{\ell_2}(\beta s)^{\ell_2} g_1 g_1' g_2 g_2'}{(m^2 + R - k_\perp^2 - \mathrm{i}\varepsilon)(m^2 + R - (q-k)_\perp^2 - \mathrm{i}\varepsilon)}. \qquad (15.8)$$

And now we integrate over β in the essential region (15.6). This gives

$$\int\limits_{\beta>0} \frac{\mathrm{d}^2 k\, \mathrm{d}R}{2(2\pi)^4 \mathrm{i}} \frac{-2\mathrm{i}\, \mathrm{Im}\, \xi_{\ell_1} \cdot \xi_{\ell_2} R^{\ell_1} s^{\ell_2}\, g_1\, g_1'\, g_2\, g_2'}{(m^2 + R - k_\perp^2 - \mathrm{i}\varepsilon)(m^2 + R - (q-k)_\perp^2 - \mathrm{i}\varepsilon)} \int\limits_{m^2/s}^{1} \frac{\mathrm{d}\beta}{\beta} \beta^{\ell_2 - \ell_1}$$

$$= -2\mathrm{i} \int\limits_{\beta>0} \frac{\mathrm{d}^2 k\, \mathrm{d}R}{2(2\pi)^4 \mathrm{i}}\, \mathrm{Im}\, \xi_{\ell_1}\, \frac{s^{\ell_2} - s^{\ell_1}}{\ell_2 - \ell_1}$$

$$\times \frac{\xi_{\ell_2} R^{\ell_1}\, g_1\, g_1' g_2\, g_2'}{(m^2 + R - k_\perp^2 - \mathrm{i}\varepsilon)(m^2 + R - (q-k)_\perp^2 - \mathrm{i}\varepsilon)}.$$

This formula can be represented in a more convenient form:

$$F = \xi_{\ell_2} \frac{s^{\ell_2} - s^{\ell_1}}{\ell_2 - \ell_1}\, g_1(q^2)\, r(q, \ell)\, g_2(q^2), \qquad (15.9\mathrm{a})$$

$$r(q, l) = -2\mathrm{i} \int \frac{\mathrm{d}^2 k\, \mathrm{d}\left(\frac{R}{2}\right)}{(2\pi)^4 \mathrm{i}} \frac{g_1'\, (\mathrm{Im}\, \xi_{\ell_1})\, R^{\ell_1}\, g_2'}{(m^2 + R + k^2 - \mathrm{i}\varepsilon)(m^2 + R + (q-k)^2 - \mathrm{i}\varepsilon)}. \quad (15.9\mathrm{b})$$

Note that for $\ell_2 \to \ell_1$

$$\int_{m^2/s}^{1} \frac{\mathrm{d}\beta}{\beta} \beta^{\ell_2 - \ell_1} \longrightarrow \ln s,$$

that is for $\ell_2 = \ell_1 = \ell$ the contribution of the region under consideration is not s^ℓ but $s^\ell \ln s$, which is $\ln s$ times larger.

We find now the partial wave amplitude ϕ_j which corresponds to the amplitude F calculated above (cf. (15.9)):

$$\phi_j \sim \frac{2}{\pi} \int s^{-j-1} F^{(1)} \mathrm{d}s, \qquad (15.10)$$

where $F^{(1)}$ is the absorptive part of F, hence

$$\phi_j = g_1 \frac{1}{j - \ell_1} r_{\ell_1 \ell_2}(q^2) \frac{1}{j - \ell_2} g_2. \qquad (15.11)$$

The expression (15.11) can be set in correspondence with the diagram

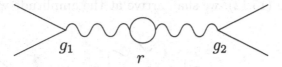

Note that a reggeon with one signature cannot make a transition to a reggeon of the opposite signature.

Indeed, since the amplitude $A^{(2)}(s_2)$ and the absorptive part of the amplitude $A_2^{(1)}(s_1)$ enter the second term in (15.7) with energies $s_2 < 0$ and $s_1 < 0$, respectively, whereas they enter the first term with $s_2 > 0, s_1 > 0$, then, in the case of Regge poles having opposite signatures, they cancel each other.

It is evident that a modification of the intermediate state in the t-channel shown in Fig. 15.3, for instance

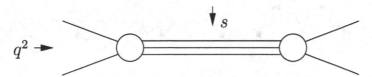

while retaining the conditions $s \gg m^2$, $s_1 \gg m^2$, $s_2 \gg m^2$ ($|q^2| \lesssim m^2$), would change only the explicit form of $r_{\ell_1 \ell_2}(q^2)$.

This discussion can be easily generalized to the case when one inserts a reggeon branching instead of a Regge pole for the asymptotics of the amplitude. To this end, one has to keep the integration over $k_\perp$ in the amplitude, corresponding to the two-reggeon exchange, as the external one, and substitute $s^{\ell_2 + \ell_1 - 1}$ for s^{ℓ_2}:

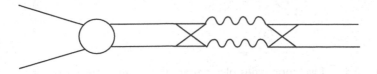

Expression (15.11) will then be replaced by the formula

$$\phi_j = \frac{g_1}{j - \ell_1(q^2)} \int \frac{\mathrm{d}^2 k}{(2\pi)^2} r_{\ell_1 \ell_2 \ell_3} \frac{1}{j - \ell_1(k^2) - \ell_2((q-k)^2) + 1} N_{\ell_1 \ell_2},$$

(15.12)

which corresponds to the graph

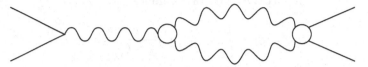

Using in turn the asymptotics for $A^{(1)}$ which corresponds to the partial wave amplitude (15.12), we shall arrive at the amplitude whose diagram is

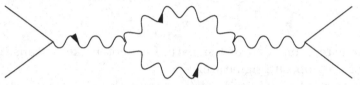

Such reasoning leads to more and more complicated diagrams, which take into account contributions of branchings whose singularities are close to each other:

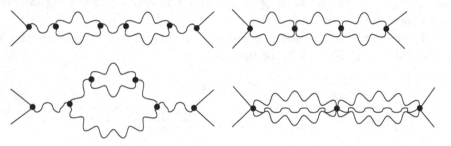

From this discussion it is still not clear whether, for example, the diagrams of Fig. 15.4 correspond to some actual high energy processes. But if one turns to the Feynman diagram then it is possible to find two re-

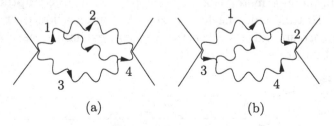

Fig. 15.4.   Two topologies of reggeon exchange between two reggeons

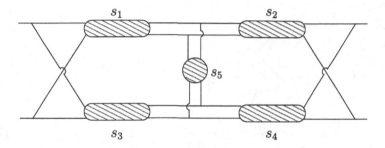

gions in the Feynman integral where all the blobs appear to be in our asymptotical regime.

Then one can relate the graph of Fig. 15.4a with the kinematical region $s_1 s_5 \sim m^2 s_3$ and that of Fig. 15.4b with $s_3 s_5 \sim m^2 s_1$ by deriving the corresponding partial wave amplitudes.

In reggeon graphs there are no closed loops such as appear for example in QED: all the arrows in Fig. 15.4 point in one direction. This is because in the reggeon case the Green function has only one cut in the plane of the energy ω. The latter, in its turn, is explained by the absence of singularities of the partial wave amplitude in the right half-plane of the complex angular momentum.

Let us look now at the two-reggeon diagrams for the vertex Λ of $2 \to 2$ reggeon transition:

The summing of these contributions under the condition that all the vertices in the graphs are constants permits one to evaluate the leading singular part of the vertex Λ:

$$\Lambda = \frac{\lambda}{1 + \lambda \ln(\omega - \omega_2)}. \qquad (15.13)$$

Here λ is the renormalized vertex of $2 \to 2$ reggeon transition, which does not contain two-reggeon intermediate states.

In a similar way we can find the leading singular part of the vertex N of the transition of two particles to two reggeons:

$$N = \frac{N^*}{1 + \lambda \ln(\omega - \omega_2)}, \qquad (15.14)$$

where N^* is the renormalized vertex for the particle–reggeon transition which does not contain two-reggeon singularities.

And finally we turn to the contribution of a two-reggeon singularity to the two-particle scattering amplitude:

$$\phi_\omega = \frac{N^{*2} \ln(\omega - \omega_2)}{1 + \lambda \ln(\omega - \omega_2)} + C = -\frac{1}{\lambda} \cdot \frac{N^{*2}}{1 + \lambda \ln(\omega - \omega_2)} + C'. \quad (15.15)$$

Here C and C' are the parts of the scattering amplitude having no two-reggeon singularity. Now, using (15.13)–(15.15) we can write down the unitarity condition for the non-two-reggeon singularity. Since according to (15.13)

$$\Lambda^{-1} = \lambda^{-1} + \ln(\omega - \omega_2),$$

the discontinuity of $1/\Lambda$ is easily calculated:

$$\delta\Lambda^{-1} = \pi, \qquad (15.16a)$$

$$\delta\Lambda = -\pi\Lambda\Lambda^*. \qquad (15.16b)$$

From the above formulæ, for $\delta\phi$ and δN we obtain

$$\delta\phi = -\pi N N^*, \qquad (15.17)$$

$$\delta N = -\pi N \Lambda^*. \qquad (15.18)$$

16

Reggeon field theory

Now we shall continue our discussion of the situation arising when the reggeon diagrams have several singularities located close to each other. In this case it is necessary to take into account their mutual influence. We shall consider the most interesting case of the vacuum pole and associated vacuum reggeon branch points.

As was shown above, if one denotes the trajectory of the vacuum pole by $\alpha(k^2)$ $(k^2 = -t)$, then the position of the branch points is given by the formula

$$j_n = n\alpha\left(\frac{k^2}{n^2}\right) - n + 1. \qquad (16.1)$$

Let us measure k^2 in units of $(\alpha')^{-1}$. Then at small k^2 we have

$$\alpha(k^2) = 1 - k^2, \qquad j_n = 1 - \frac{k^2}{n} \qquad (16.2)$$

and hence at small k^2 the vacuum pole and all branch points are located near the point $j = 1$ (see Fig. 14.5).

Let us consider the contribution of various reggeon diagrams to the partial wave amplitude $f_j(-k^2)$. We shall be interested in small momentum transfers, since this is the case when all singularities of reggeon diagrams essential for the theoretical description of the high energy behaviour of the total cross section are located near by. In addition we restrict ourselves to the region $j \simeq 1$, where $\omega = j - 1$ is small. It is the point $\omega = 0$ where singularities of $f_j(-k^2)$ accumulate.

The simplest reggeon pole diagram is shown in Fig. 16.1. It gives a contribution to the partial amplitude equal to

$$\frac{\pi}{2}f_j(-k^2) = g_1 g_2 G_0(\omega, k), \qquad G_0 = \frac{1}{\omega + k^2}, \qquad (16.3)$$

Fig. 16.1.   Regge pole diagram

where g_1 and g_2 are the couplings of the reggeon to particles 1 and 2, respectively.

The contribution of this graph to the scattering amplitude $A(s,t)$ can be found using its representation in the form of the Sommerfeld–Watson integral:

$$A(s,t) = s \int_{-i\infty}^{i\infty} \frac{d\omega}{4i} e^{\omega\xi} f_j(-k^2) \left\{ i + \tan\left(\frac{\pi}{2}\omega\right) \right\},$$

$$\xi = \ln s, \quad \omega = j - 1, \quad t = -k^2. \tag{16.4}$$

Using this expression we find that the pole graph contribution to the amplitude at small k^2 is equal to

$$A^{\text{pole}}(s,t) = sg_1g_2 \left(i - \frac{\pi}{2}k^2 \right) e^{-\xi k^2}. \tag{16.5}$$

Let us consider now the reggeon graphs with two vacuum poles in the intermediate state (see Fig. 16.2). The contribution of these graphs to

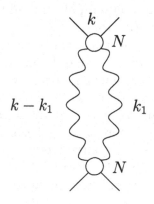

Fig. 16.2.   Two-reggeon exchange

the partial wave amplitude can be represented as

$$\frac{\pi}{2}f_j(-k^2) = -\int \frac{d^2k_1}{(2\pi)^2}\frac{d\omega_1}{2i\,\pi}\, N^2(\omega_1,k_1,\omega-\omega_1,k-k_1)$$
$$\times\, G_0(\omega_1,k_1)\,G_0(\omega-\omega_1,k-k_1), \qquad (16.6)$$

where N is the amplitude of the transition of two particles to two reggeons. The negative sign is due to the product of the two pole amplitudes (16.5), where we neglected small corrections coming from the real part of the signature factor. In (16.6) the essential region of integration is $\omega_1 \sim k_1^2 \sim \omega \sim k^2$.

We calculate first the contribution of the regular part of the amplitude N which does not have singularities at small ω_i, k_i. It corresponds to the diagram in Fig. 16.2, where N has been replaced by a constant:

$$\frac{\pi}{2}f_j(-k^2) =$$

$$= -N^2 \int G_0(\omega_1,k_1)G_0(\omega-\omega_1,k-k_1)\frac{d\omega_1 d^2k_1}{(2\pi)^3 i}. \qquad (16.7)$$

It follows from expression (16.7) for small ω, k^2 that

$$f_j(-k^2) \sim \ln\left(\omega + \frac{k^2}{2}\right) \qquad (16.8)$$

hence at small $\omega \sim k^2 \to 0$ the contribution to $f_j(-k^2)$ from the two-reggeon branching is less than that of the pole graph ($\ln\omega \ll (1/\omega)$, $\omega \to 0$). The same relation is valid for the contribution of these graphs to the asymptotics of the amplitude $A(s,t)$. Indeed, using (16.4) one can verify that the corresponding amplitude has the asymptotics $(s/\xi)e^{-\xi k^2/2}$ ($\xi \gg 1$) which is much smaller than the pole contribution (16.5).

16.1 Enhanced reggeon diagrams

Let us find now what contribution arises from the singular parts of the amplitude N. Taking into account only the pole terms in N, i.e. calculating the diagram in Fig. 16.3a, we find that at $\omega \sim k^2 \to 0$ this contribution is $(g^2r^2/\omega^2)\ln(1/\omega)$, where r is the coupling of one reggeon with two reggeons (see Fig. 16.3b).

The form of the ω-dependence of $f_j(-k^2)$ can be easily understood if one takes into account that each reggeon has a propagator $1/\omega$ and the

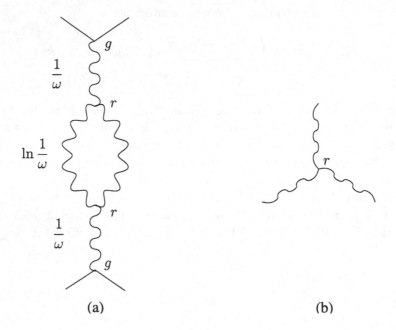

$$\ln\frac{1}{\omega}$$

$$\frac{1}{\omega}$$

(a)   (b)

Fig. 16.3.   Enhanced reggeon diagram (a) and the three-reggeon coupling (b)

reggeon loop contribution is $\ln(1/\omega)$. Thus the *enhanced* graph contribution of Fig. 16.3a exceeds that of the non-enhanced graph of Fig. 16.2 and is bigger than the pole graph (Fig. 16.1). It is easy to see that the enhanced diagram

gives an even larger contribution of the order of $(g^2r^4/\omega^3)\ln^2\omega$. Consequently all such enhanced diagrams turn out to be significant when one calculates the partial wave amplitude near the point $j = 1$.

Let us consider now the reggeon diagrams which have three reggeons in the intermediate state. The simplest one is shown in Fig. 16.4.

Fig. 16.4. Non-enhanced reggeon diagram with exchange of three reggeons

The contribution of this diagram is given by the integral

$$f_j^{(3)}(-k^2) \sim N'^2 \int \frac{\mathrm{d}\omega_1 \mathrm{d}\omega_2 \mathrm{d}^2 k_1 \mathrm{d}^2 k_2}{(2\pi)^6} G_0(\omega_1, k_1) G_0(\omega_2, k_2)$$
$$\times G_0(\omega - \omega_1 - \omega_2, k - k_1 - k_2). \qquad (16.9)$$

Instead of the direct calculation of this integral it is more convenient to compute its absorptive part, which is given by the formula

$$\mathrm{Im}\, f_j^{(3)}(-k^2) \sim N'^2 \int \frac{\mathrm{d}\omega_1 \mathrm{d}\omega_2 \mathrm{d}^2 k_1 \mathrm{d}^2 k_2}{(2\pi)^6} \delta(\omega_1 + k_1^2)\delta(\omega_2 + k_2^2)$$
$$\times \delta[\omega - \omega_1 - \omega_2 + (k - k_1 - k_2)^2]$$
$$\sim N'^2 \cdot \left(\omega + \frac{k^2}{3}\right) \vartheta\left(-\omega - \frac{k^2}{3}\right). \qquad (16.10)$$

Using this expression and dispersion relation over ω between real and imaginary parts of f_j one can find

$$f_j^{(3)}(-k^2) \sim eN'^2 \cdot \left(\omega + \frac{k^2}{3}\right) \ln\left(\omega + \frac{k^2}{3}\right), \qquad (16.11)$$

i.e. at small ω, $k^2 \to 0$ the non-enhanced diagram, corresponding to a three-reggeon branching, gives a small contribution of the order of $\omega \ln(1/\omega)$.

Let us consider now the enhanced diagrams of Fig. 16.5a,b.

In the diagram of Fig. 16.5a the number of integrations is the same as in the non-enhanced diagram of Fig. 16.4, but there are two more Green functions, together contributing a factor of $1/\omega^2$. So the diagram of Fig. 16.5a at $\omega \sim k^2 \to 0$ gives the contribution $\sim (N^2 r^2/\omega) \ln(1/\omega)$. The diagram of Fig. 16.5b contains two additional Green functions as compared with Fig. 16.5a and therefore it behaves at small ω as $(g^2 r^4/\omega^3) \ln(1/\omega)$.

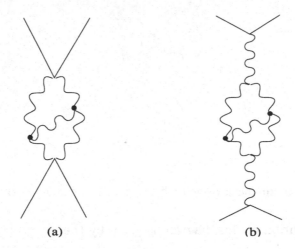

Fig. 16.5.   Enhanced diagrams with three reggeons in the intermediate state

All examples considered above show that in order to obtain a large contribution to the partial wave amplitude at $\omega \to 0$ it is necessary to insert reggeon poles in any possible place. If the reggeon line is inserted into the interior of a diagram, then there appear

1. two new vertices 'r' (see Fig. 16.3b), that is a factor of r^2,

2. three reggeon propagators, i.e. a factor of $1/\omega^3$,

3. an additional integration equivalent to including a factor of ω^2 in the numerator.

Therefore as a result of such insertion the reggeon diagram acquires an additional factor of r^2/ω (for example one can compare the diagrams in Fig. 16.3a and Fig. 16.5b). If the reggeon diagram contains some vertex different from r, for example the vertex λ in Fig. 16.6a, then inserting inside it the vacuum pole (see Fig. 16.6b), we obtain a diagram which contains r^2/ω instead of λ, thus giving a large contribution at $\omega \to 0$.

Thus we conclude that the maximum contribution to the partial wave amplitude at $\omega \to 0$ arises from the diagrams which contain only three-reggeon vertices r (Fig. 16.3b). All such diagrams are of the order of $(g^2/\omega)(r^2/\omega)^n$, if one is interested only in powers of $1/\omega$ and does not, for the sake of simplicity, take account of less singular terms such as $\ln \omega$. For instance the diagram of Fig. 16.3a differs from the pole graph by a factor of r^2/ω, the diagram of Fig. 16.5b contains an extra factor of $(r^2/\omega)^2$ and so on. As a result of summing these diagrams we obtain a series in powers of r^2/ω.

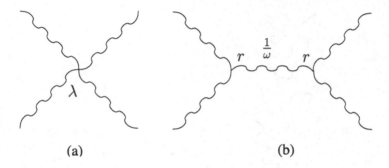

Fig. 16.6.   Interactions between four reggeons

If the three-reggeon coupling is small ($r^2 \ll 1$) and we are interested in the region of small ω (but satisfying the condition $\omega \gg r^2$), then the expansion parameter $r^2/\omega \ll 1$ is small. In this case perturbation theory is applicable, and to calculate the partial wave amplitude it is sufficient to take into account only a few of the first reggeon diagrams. But in the region of smaller ω (when $\omega \sim r^2$) perturbation theory cannot be applied since all diagrams containing three-reggeon vertices r would be of the same order.

If one supposes that not only is the three-reggeon vertex r small but all many-reggeon vertices are small as well (for example, λ of Fig. 16.6a), then all diagrams containing at least one such vertex with more than three reggeons at one point will give relatively small contributions and can be neglected.

For instance the contribution of the diagram in Fig. 16.7a is of the order of $(g^2 r^2 \lambda/\omega^2) \ln^2 \omega$, and that in Fig. 16.7b is of the order of $(g^2 \lambda_1^2/\omega) \ln \omega$. If one does not account for powers of $\ln \omega$ then these diagrams have an order of $(g^2/\omega)(r^2/\omega)\lambda$ and $(g^2/\omega)\lambda_1^2$, respectively. So at $r^2/\omega \sim 1, \lambda \ll 1, \lambda_1 \ll 1$ these diagrams turn out to be small compared with those which contain only three-reggeon vertices and have an order of g^2/ω (see Fig. 16.7).

Hence in this case, to calculate a partial wave amplitude in the region $\omega \sim r^2$, it is necessary to sum up all diagrams containing only three-reggeon vertices r, and not to take account of other diagrams.

As a result of such a summation the exact three-reggeon vertex Γ_2

$$\Gamma_2$$

could become a small quantity of the order of ω.

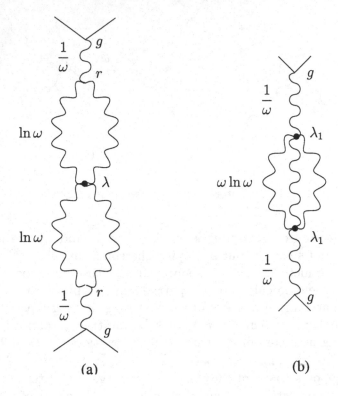

Fig. 16.7.   Enhanced diagrams including point-like four-reggeon interactions

If this is the case then with decreasing ω, all diagrams containing only $\Gamma_2 \propto \omega$ vertices would be small. For instance if in the diagram of Fig. 16.7a one should replace r by the exact vertex $\Gamma_2 \sim \omega$, then this diagram would become a quantity of the order of $(g^2/\omega)(\Gamma_2^2/\omega)\lambda \sim g^2\lambda$, i.e. it would be small compared with the pole diagram g^2/ω. A more rigorous consideration shows that one must take into account terms of the order of $\ln\omega$ in addition to those of the order of $1/\omega$. Then one can formulate the following rule:

> in order to calculate the partial wave amplitude at $\omega \to 0$, it is necessary to sum up all diagrams containing three- and four-reggeon vertices displayed in Fig. 16.8.

Diagrams with five or more reggeons turn out to be small at $\omega \to 0$ and can be omitted.

16.2 Effective field theory of interacting reggeons

Thus the calculation of the partial wave amplitude results in the problem of finding the exact Green function of a non-relativistic particle in the two-

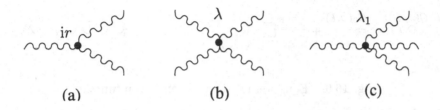

Fig. 16.8.   Basic three- and four-reggeon interaction vertices

dimensional space with the interaction described by the non-renormalized vertices of Fig. 16.8. The Hamiltonian for the reggeon interactions has the form

$$H = \Psi^+ \nabla^2 \Psi + i\,r(\Psi^+\Psi^+\Psi + \Psi^+\Psi\Psi) + \lambda\Psi^+\Psi^+\Psi\Psi$$
$$+ \lambda_1(\Psi^+\Psi^+\Psi^+\Psi + \Psi^+\Psi\Psi\Psi), \tag{16.12}$$

where $\Psi = \sum_k a_k \exp(i\,kx)$ is a non-relativistic field operator in the two-dimensional space.

If the exact Green function is known, then the partial wave amplitude of Fig. 16.1 at $\omega \to 0$, $k^2 \to 0$ is given by the formula

$$\frac{\pi}{2} f_j(-k^2) \;=\; g_1\, g_2\, G(\omega, k^2). \tag{16.13}$$

Note that the constants r, λ and λ_1 are real, so the three-reggeon vertex of Fig. 16.8a is purely imaginary and the Hamiltonian (16.12) is non-Hermitian.

That the triple vertex is imaginary is related to the signature factors in the reggeon diagrams. As we have seen in the previous lectures, it is necessary to associate the factor (-1) to the intermediate state with an even number of vacuum poles and the factor $(+1)$ to that with an odd number of vacuum poles. For instance for the diagram in Fig. 16.3a one associates the factor (-1) which appears from the internal loop. If we want to consider this contribution as the usual Feynman one, not keeping in mind the signature factors, then the factor (-1) should be shared between the two vertices r, hence one should multiply each of them by an imaginary unit. Similar reasons applied to the diagrams containing the vertices λ and λ_1 show that the latter remain real. The non-Hermitian property of the Hamiltonian will be essential in the investigation of the properties of the Green function $G(\omega, k)$.

16.3 Equation for the Green function G

For the exact Green function $G(\omega, k)$ one can write down the Dyson equation shown graphically in Fig. 16.9. The wavy lines represent the exact

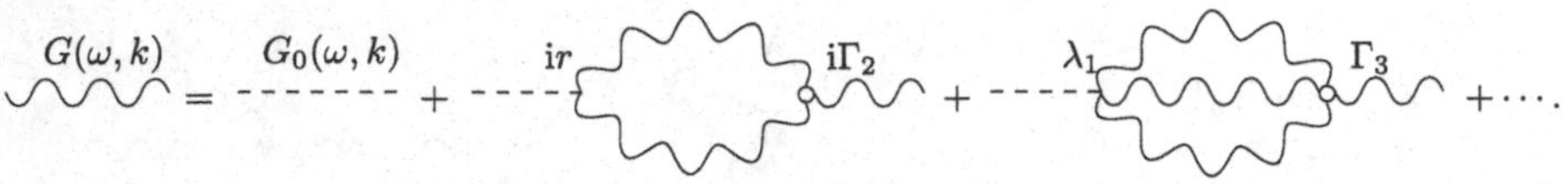

Fig. 16.9.   Equation for the reggeon Green function

Green functions $G(\omega, k)$, the dashed lines correspond to the Green function of a free reggeon, $G_0 = 1/(\omega + k^2)$.

The integral equations for the vertex parts Γ_2 and Γ_3 are as usual represented by infinite series. The first terms of the series for Γ_2 are shown in Fig. 16.10.

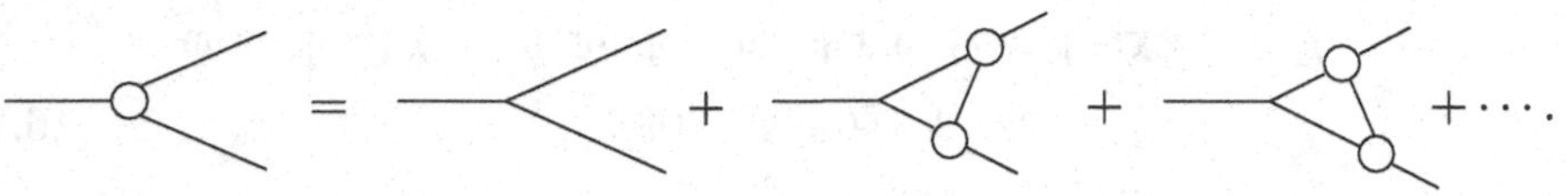

Fig. 16.10.   Equation for the exact three-reggeon vertex Γ_2

Unfortunately the equation for the reggeon Green function, Fig. 16.9, is rather formal and cannot be used to determine $G(\omega, k)$ or $\Sigma(\omega, k)$, which is the self-energy part entering $G(\omega, k)$. Actually, our aim is to account properly for all singularities in $\Sigma(\omega, k)$ at small ω and k^2. However in the integrals of the equation of Fig. 16.9, the essential region of integration corresponds to large internal ω' and k'. This region of integration can lead to the appearance of extra singularities in $\Sigma(\omega, k)$ at small ω and k^2 which are related to properties of the vertex Γ but not to the structure of the equation itself. These singularities are indeed fictitious since, according to the equation, the corresponding singular terms are proportional to Γ, whereas from the unitarity condition for $\Sigma(\omega, k)$ it follows that $\text{Im} \Sigma \sim \Gamma^2$. Hence in the equation for $\Sigma(\omega, k)$ a striking cancellation of singular terms must occur and this fact cannot be seen from the equation itself.

16.4 Equation for the vertex function Γ_2

The equation for the vertex function Γ_2 (Fig. 16.10) does not suffer these drawbacks. If on the r.h.s. of this equation we insert constants instead of exact vertices, then the integrals remain convergent. Hence in these integrals the region of large internal frequencies ω and momenta k^2 is inessential, the problem of calculating Γ_2 becomes self-consistent and the fictitious singularities do not appear. Nevertheless in this equation some care is needed as well. For, apart from the singular part of Γ_2, which arises from the region of small internal frequencies, there appear singular terms coming from the regions where some variables are small (of the order of external frequencies) whereas the others are large.

Let us consider for example the diagram

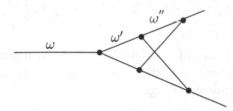

In the region of integration $\omega'' \sim 1 \gg \omega$, $\omega' \sim \omega$, and hence in all internal lines containing ω'' one can neglect ω and ω' as compared with ω'', i.e. shrink the corresponding lines to points (see Fig. 16.11a). In the same way for $\omega' \gg \omega$, $\omega'' \sim \omega$ we arrive at the diagram of Fig. 16.11b. Both these diagrams contain the singular factor $\ln \omega$ due to the reggeon

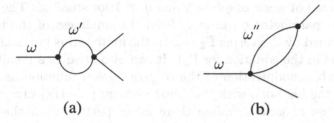

(a)   (b)

Fig. 16.11.   Reduced diagrams generating effective four-reggeon interactions

loop. But similar diagrams appear directly if one takes into account the vertices λ and λ_1 from the Hamiltonian (Fig. 16.8b,c). Therefore the above mentioned regions of integration will show up in the renormalization of the vertices λ and λ_1.

Let us leave aside for a while the problem of taking account of the vertices λ and λ_1 and restrict ourselves to a theory with only three-reggeon vertices (Fig. 16.8a). Consider in more detail the properties of Γ_2 in such a theory at $\omega \to 0$. As has been shown above, for a small three-reggeon coupling constant r and sufficiently small frequencies ω (but satisfying the condition $\omega \gg r^2$) we obtain the expression for the vertex part $\Gamma_2 \equiv r \cdot \gamma_2$ in the form of a convergent series in powers of r^2/ω. For $\omega \sim r^2$ all terms of this series become of the same order. When ω is further decreased, the series diverges as a result of an inadequate choice of the Green functions of the zeroth approximation as $G = G_0$ and $\gamma_2 = 1$. We can rearrange terms and represent Γ_2 in the form of a series of diagrams built up of exact Green functions and vertices as in Fig. 16.10. Comparing various terms of this new series, for instance the first two diagrams on the r.h.s.

of Fig. 16.10, one can convince oneself that there arises a new expansion parameter η instead of r^2/ω:

$$\eta = G^3 \cdot \Gamma_2^2 \cdot \omega \bar{k}^2 = \frac{r^2}{\omega} \cdot (\omega G)^3 \gamma_2^2 \frac{\bar{k}^2}{\omega}. \qquad (16.14)$$

Here $\bar{k}^2$ means the order of magnitude of internal momenta that are significant in the diagram: $\omega \sim \alpha(\bar{k}^2) - 1 \simeq \alpha' k^2$. The parameter η plays the rôle of a renormalized charge.

16.5 Weak and strong coupling regimes

One must distinguish three possibilities depending on the magnitude of η for $\omega \to 0$, as follows.

1. The case of *weak coupling*, when $\eta \ll 1$ for small ω. The smallness of the parameter η can arise from the smallness of the interaction described by the vertex $\Gamma_2 \sim \omega$ (if the finite terms in the $\omega \to 0$ limit cancel in the equation for Γ_2). It can also arise as a result of a considerable modification of the reggeon Green function as compared with the bare one with the linear vacuum pole trajectory $\alpha(\bar{k}^2)$. In the case of *weak coupling* there arises perturbation theory in the new parameter η.

2. The case of *strong coupling*, when $\eta \sim 1$. In this case there is no perturbation theory, and all terms of the series are of the same order of magnitude.

3. The case of *superstrong coupling*, when $\eta \gg 1$, i.e. no rearrangement of the terms can cause the series of reggeon diagrams to be convergent. Recall that $\omega \gg r^2$ the series has been convergent. So the divergence of the series when $\omega \leq r^2$ indicates the presence of a new singularity at small frequencies $\omega \sim r^2$. This singularity is of a different nature from that of the singularities which have been assumed from the very beginning, i.e. a vacuum pole and multi-reggeon states. So in the case of *superstrong coupling* the problem becomes not self-consistent.

If one assumes that there are no singularities other than the vacuum pole and associated multi-reggeon contributions, then the problem of *superstrong coupling* need not be considered. The cases of *weak* and *strong* couplings will be discussed in the following lecture.

16.6 Pomeron Green function and reggeon unitarity condition

Let us return now to the question of definition of the reggeon Green function. The exact Green function $G(\omega, k)$ can be written in the following form:

$$G(\omega, k) = \frac{1}{\omega + k^2 + \Sigma(0,0) - \Sigma(\omega, k)}, \qquad (16.15)$$

where $\Sigma(\omega, k)$ is the self-energy of the reggeon. The constant term $\Sigma(0,0)$ in the denominator of the Green function ensures the condition $\alpha(0) = 1$, i.e. the presence of the pole in the Green function at $\omega = 0, k^2 = 0$. The appearance of this term can be explained from different points of view. One can consider $\Sigma(0,0)$ as a modification of the initial vacuum pole trajectory chosen so that after accounting for the mutual influence of the pole and multi-reggeon states the trajectory of the renormalized vacuum reggeon passes through the point $\omega = 0$ at $k^2 = 0$. Alternatively one can consider $\Sigma(0,0)$ as the regular part extracted from $\Sigma(\omega, k)$, which comes from the integration over the region of large frequencies and momenta $\omega' \sim 1, k'^2 \sim 1$ and which is not determined by the reggeon diagram technique.

Let us consider now how one can calculate the singular part of the self-energy $\Sigma(\omega, k)$. As was noted above we cannot use the Dyson-type equation (Fig. 16.9) for this purpose. Instead we may deduce the integral equation for $d\Sigma/d\omega$. From the point of view of reggeon diagrams this equation is analogous to the equation for the vertex part Γ_2. The value of the function at small external ω will be determined by the integration over the region of small internal frequencies ω'. Therefore the difficulties discussed above characteristic for the function Σ do not appear here.

To calculate $\Sigma(\omega, k)$ one can use the reggeon unitarity condition. The function $\Sigma(\omega, k)$ satisfies it at $t > 0$, i.e. $k^2 < 0$. (The problem for this method is a possibility that the whole analyticity picture may drastically change in the process of crossing into the region $t < 0$ ($k^2 > 0$), due to accumulation of singularities at $t = 0$.)

At $t > 0$ the quantity $\operatorname{Im} \Sigma(\omega, k)$ can be represented as an infinite series of diagrams of the type

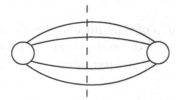

In this diagram all reggeons in the intermediate state are *real*, i.e. located on their mass shell ($\omega + k^2 = 0$). The Green function of a reggeon near the pole $\omega = -k^2$ has the form $Z(k^2)/(\omega + k^2)$, where $Z(k^2)$ is an

unknown renormalization function. Therefore a factor $Z\delta(\omega + k^2)$ enters the unitarity condition from the propagator of each reggeon. As a result the contribution to $\operatorname{Im}\Sigma$ from the diagram with n reggeons in the intermediate state takes the following form:

$$(\operatorname{Im}\Sigma)_n \sim \frac{1}{\omega}\Gamma_n^2 Z^n (\bar{k}^2)^{n-1}, \tag{16.16}$$

where Γ_n is the vertex for the transition of one reggeon to n reggeons. The contribution of n reggeons to the unitarity condition (16.16) can be put in the equivalent form

$$(\operatorname{Im}\Sigma)_n \sim \Gamma_n^2 G^n \omega^{n-1}(\bar{k}^2)^{n-1}. \tag{16.17}$$

Let us estimate the order of magnitude of these quantities as a function of n. For the state $n = 2$, the contribution to the unitarity condition differs from the parameter η considered above by the factor $1/G$:

$$(\operatorname{Im}\Sigma)_2 \sim \Gamma_2^2 G^2 \omega(\bar{k}^2) \sim \frac{\eta}{G}. \tag{16.18}$$

We consider now the state with $n = 3$. If one assumes that there is only one bare vertex with $n = 2$, then the vertex Γ_3 can be represented as

Hence, $\Gamma_3 \sim (\Gamma_2)^2 G$ and

$$(\operatorname{Im}\Sigma)_3 \sim \Gamma_3^2 G^3 \omega^2(\bar{k}^2)^2 \sim \Gamma_2^4 G^5 \omega^2(\bar{k}^2)^2 \sim \frac{\eta^2}{G}. \tag{16.19}$$

In the same way we can convince ourselves that for the n-reggeon intermediate state

$$\Gamma_n \sim (\Gamma_2)^{n-1} G^{n-2}, \qquad (\operatorname{Im}\Sigma)_n \sim \frac{\eta^{n-1}}{G}.$$

Consequently with the help of the unitarity condition one obtains for $\operatorname{Im}\Sigma$ and for the singular part of the self-energy Σ the expression in the form of an expansion in powers of the parameter η introduced earlier:

$$\Sigma \sim \frac{\eta}{G} + \frac{\eta^2}{G} + \cdots + \frac{\eta^n}{G} + \cdots . \tag{16.20}$$

This permits us to define the notions of *weak* and *strong couplings*, given above, according to another principle, as follows.

1. If $\eta \ll 1$ (*weak coupling*), then $\Sigma \ll 1/G \sim \omega + k^2$. Therefore the equivalent definition of the *weak coupling* is the condition

$$\Sigma(\omega, k) \ll \omega + k^2, \qquad \text{weak coupling.}$$

2. If $\eta \sim 1$, then $\Sigma \sim 1/G$, that is $G \sim 1/\Sigma$. But this is possible for $\Sigma \sim \omega + k^2$ or $\Sigma \gg \omega + k^2$. The first of these possibilities is not interesting, as it means that $\Sigma(\omega, k)$ does not have singularities at all, i.e. it is a polynomial. Therefore as a definition of the *strong coupling* one can use the condition

$$\Sigma(\omega, k) \gg \omega + k^2, \qquad \text{strong coupling.}$$

17

The structure of weak and strong coupling solutions

In the previous lecture we have investigated the reggeon Green function

where the solid (wavy) lines represent particles (reggeons). We have constructed three types of interactions:

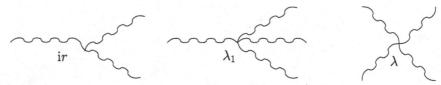

We have seen that there are two essentially different cases:

1. weak coupling, when $\omega + k^2 \gg \Sigma$;

2. strong coupling, when $\omega + k^2 \ll \Sigma$.

17.1 Weak coupling regime

To begin with consider the weak coupling case.

17.1.1 The Green function

The reggeon Green function G satisfies the equation

$$G(\omega - \omega', k - k'). \tag{17.1}$$

208

Here the dashed line denotes the free reggeon Green function

$$G_0 = \frac{1}{\omega + k^2}.$$

The black points mark the exact vertices Γ_2 and Γ_3 that describe the coupling of the reggeon to two and three reggeons, respectively, taking into account the interaction. Obviously we need to know their properties in order to study the equation.

Let us consider $\text{Im}\,\Sigma$ at $k^2 < 0$. Then the rightmost singularity in the ω plane is a pole, followed by the two-, three-, etc.-reggeon cuts. The unitarity condition permits us to express $\text{Im}\,\Sigma$ in terms of the contributions of these cuts:

$$\text{Im}\,\Sigma = \frac{1}{n!}\sum_n \left(-\frac{1}{4\pi^2}\right)^{n-1} \int \prod_i \mathrm{d}^2 k_i \,\mathrm{d}\omega_i \,\Gamma_n^2$$

$$\times \delta\!\left(\omega - \sum_i \omega_i\right) \delta\!\left(k - \sum_i k_i\right) \prod_i \delta(\omega_i + k_i^2). \qquad (17.2)$$

For the two-reggeon contribution all δ-functions can be integrated out and there remains an angular integration only. Therefore $(\text{Im}\,\Sigma)_2 \sim \Gamma_2^2$. The weak coupling regime implies $\Gamma_2^2 \ll \omega, k^2$ (recall that $k^2 \sim \omega$).

For the three-reggeon contribution we obtain an estimate (neglecting a possible dependence of Γ_3 on momenta) $(\text{Im}\,\Sigma)_3 \sim (-\omega - k^2/3)\Gamma_3^2$. Here the weak coupling means that $\Gamma_3 \ll 1$.

17.1.2 $P \to PP$ vertex

Consider now the equation for Γ_2. Although a closed equation does not exist, one can write it down in the form of an expansion in all types of irreducible diagrams. First we will consider the $1 \to 2$ reggeon transitions only. Then we obtain the graphic equation

$$(17.3)$$

Here the second and third graphs on the r.h.s. differ by exchange of the external lines. For instance, if one denotes the external variables by ω, ω_1 and ω_2, then the integral corresponding to the graph

$$(17.4a)$$

has the form

$$\int \frac{d\omega' d^2 k'}{(\omega' + k'^2)(\omega - \omega' + (k - k')^2)(\omega' - \omega_2 + (k' - k_2)^2)}. \quad (17.4b)$$

In the integral for the third graph in (17.3) one has to substitute $\omega_2 \leftrightarrow \omega_1$ and $k_2 \leftrightarrow k_1$.

Assume that $r^2 \ll 1$ and consider the different regions of values of the internal variables ω'.

For $\omega' \gg r^2$ one can set $\Gamma_2 = 1$ in the diagrams, since in this case perturbation theory can be applied (the pole and the branchings are far away from each other). This region corresponds to the tail of convergence of the integrals and its contribution can be omitted when the external variables ω_i are small.

There remain the regions $\omega' \sim r^2$ and $\omega' \to 0$.

The region $\omega'_i \sim r^2$ for $\omega_i \to 0$ gives a regular contribution, i.e. a constant and polynomials. The former renormalizes the vertex. If we want to have $\Gamma_2 \to 0$, this constant should cancel the initial (bare) coupling constant r, thus leaving in the sum $\Gamma_2 \simeq a\omega + bk^2$.

The region $\omega' \to 0$ gives singular terms. Consider first the case when all internal ω' and external ω_i have the same order of smallness. Then the diagram

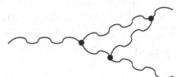

gives a contribution to the imaginary part of the order of

$$\int \frac{\Gamma_2^3}{\omega'^3} d^2 k' d\omega' \sim \frac{\Gamma_2^3}{\omega} \sim \Gamma_2 \frac{\Gamma_2^2}{\omega}. \quad (17.5a)$$

The contribution of the diagram

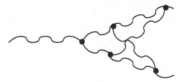

is of the order of

$$\frac{\Gamma_2^3}{\omega} \cdot \frac{\Gamma_2^2}{\omega^3} \cdot \omega^2 \sim \frac{\Gamma_2^5}{\omega^2} = \Gamma_2 \left(\frac{\Gamma_2^2}{\omega}\right)^2. \quad (17.5b)$$

Due to the condition $\Gamma_2^2 \ll \omega$ the correction terms (17.5) are smaller than Γ_2. Therefore, if other contributions are absent, then the r.h.s. and l.h.s. of the equation (17.3) can match only if $\Gamma_2 = a\omega + b(k_1^2 + k_2^2) + ck^2$. Then the singular terms give $\omega^2 \ln \omega$, $\omega^3 \ln \omega$ and so on.

17.1.3 Induced multi-reggeon vertices

In reality the situation is more complicated, due to the possiblity of over-lapping integration regions when some of the internal variables ω'_i are $\mathcal{O}(r^2)$ and the others are smaller.

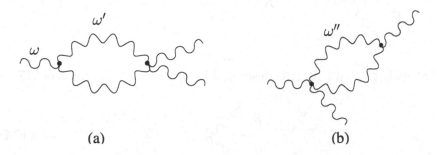

(a)   (b)

Fig. 17.1.   Induced multi-reggeon interactions

Consider for instance the diagram

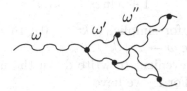

Suppose that $\omega \sim \omega' \ll \omega'' \sim r^2$. Then one can skip ω' and ω in the denominators of the Green functions as compared with ω'' and obtain the effective vertex of Fig. 17.1a. If, on the contrary, $\omega' \sim r^2 \gg \omega'' \sim \omega$, then the diagram of Fig. 17.1b appears. Thus the overlapping regions produce higher types of interactions. These new types of interactions have not been in the initial Hamiltonian. Nevertheless they should be taken into account.

The diagram of Fig. 17.1a gives a contribution to the imaginary part of the order of $\Gamma_2 \omega^{-2} \lambda \omega^2$ (here ω^{-2} comes from the Green functions and ω^2 from phase space). Hence its contribution to the vertex function is of the order of $\Gamma_2 \lambda \ln \omega$, i.e. it exceeds Γ_2. The diagrams with higher numbers of loops give even larger contributions. Therefore it is necessary to sum all ladder-type diagrams

Similarly for Γ_3 and for Λ the diagrams of the following types have to be resummed:

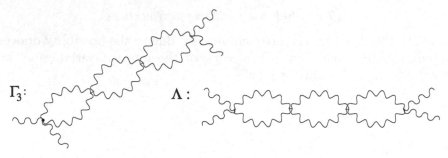

Γ_3: Λ:

Let us begin with Λ. It is described by the sum of diagrams of (17.6a).

$$+ \quad\quad + \quad\quad + \cdots (17.6a)$$

Collecting them we obtain

$$\Lambda = \frac{\lambda}{1 + \lambda \ln \omega} \sim \frac{1}{\ln \omega}, \qquad (17.6b)$$

where we have taken for simplicity $k^2 = 0$. In the general case it is necessary to substitute $\omega \to \omega + k^2/2$.

To find $\Gamma_{2,3}$ it is convenient to write down the unitarity condition for the ladder diagrams. For Γ_2 we have

$$\delta\Gamma_2 = \quad\quad = -\pi\Gamma_2\Lambda^*. \qquad (17.7)$$

For Γ_3 we obtain

$$\delta\Gamma_3 = \quad\quad = -3\pi\Gamma_3\Lambda^*, \qquad (17.8)$$

where the factor of 3 takes account of the fact that there are three types of diagrams, which differ from each other by permutations of external lines.
Similarly for Λ

$$\delta\Lambda = -\pi\left[\Lambda\Lambda^* + 4\Gamma_3\Gamma_3^*\right], \qquad (17.9)$$

where the two terms in square brackets arise, respectively, from the diagrams

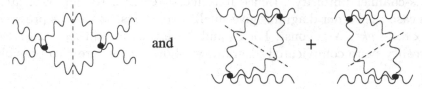

The factor of 4 in (17.9) is related to the contributions with an exchange of external lines.

Since we know that $\Gamma_3 \to 0$, the second term in (17.9) is a small correction. If it is omitted, then we obtain the already known result $\Lambda \sim 1/\ln \omega$ of (17.6b).

Thus $\Gamma_3 \sim 1/\ln^3 \omega$, were the cubic power arises due to the factor of 3 in (17.8). For Γ_2 we obtain $\Gamma_2 \sim \omega/\ln \omega$. Here the factor ω appears due to the cancellation of the bare term in the sum of the diagrams of (17.3).

It is interesting to note that when one does such a summation, polynomial terms (not accompanied by $\ln \omega$) cancel out, though they have been present in Γ_2 at the beginning.

Thus we have a set of solutions, in which everything has been taken into account:

$$\Lambda \sim \frac{1}{\ln \omega}, \quad \Gamma_2 \sim \frac{\omega}{\ln \omega}, \quad \Gamma_3 \sim \frac{1}{\ln^3 \omega}.$$

This seems to be the final answer. However besides Λ, Γ_2 and Γ_3 we would also need to know the quantity N which describes the coupling of reggeons to particles. For N we obtain the sum of diagrams

In this figure the first term on the r.h.s. is a constant N_0, the contribution of the second one is small due to the smallness of Γ_2, whereas the third diagram gives $N_0 \lambda \ln \omega$ which exceeds N_0. Therefore it is again necessary to sum up all such diagrams. After doing this we obtain

$$\delta N = -\pi N \Lambda^*, \qquad N \simeq \frac{\text{const}}{\ln \omega}. \tag{17.10}$$

17.1.4 Vanishing of multi-reggeon couplings

Everything seems to be fine with our weak coupling solution, but in fact it is not. Unfortunately this answer contradicts the *s-channel unitarity* (we will prove this elsewhere).

Why has this happened? Because of our assumption that $\lambda \neq 0$, which results in $\Lambda \sim 1/\ln\omega$ and so on. And now we see that $\lambda \neq 0$ contradicts the s-channel unitarity. Hence it is more realistic to put $\lambda \propto \omega$. In this case the ladder diagrams give small corrections, so that as previously $\Gamma_2 \propto a\omega + bk^2$, $N \simeq$ const. For Γ_3 and Λ we then obtain $\Gamma_3 \sim \omega$, $\Lambda \sim \omega$. In reality more complicated diagrams as shown below are also possible:

If the couplings of two reggeons to three and one reggeon to four are non-zero, then due to such diagrams $\Gamma_3 \sim \omega \ln\omega$, $\Lambda \sim \omega \ln\omega$, that is the corrections would violate the initial solution in the $\omega \to 0$ limit. Thus, for the weak coupling regime to be consistent, *all* multi-reggeon interaction vertices should be vanishing in the $\omega \propto k^2 \to 0$ limit.

To conclude our discussion of the weak coupling regime let us see now why the constant term r in Γ_2 can be, in principle, cancelled for arbitrary r. The equation for Γ_2 has, symbolically, the following structure:

$$(17.11)$$

Of course, for a cancellation of the constant the series should be alternating. This is the case.

Now consider the different regions of integration over internal variables (for simplicity we restrict ourselves to a consideration of the second term only). The region $\omega' \gg r^2$ gives a small contribution due to the convergence of the integrals. The region $\omega' \ll r^2$ is inessential too, since here Γ_2 is small. The remaining region is $\omega' \sim r^2$, where $\Gamma_2 \sim 1$. But here $r^2 \frac{1}{\omega'^3}\omega'^2 = \frac{r^2}{\omega'} \sim 1$. Therefore the contribution of this region is indeed able to cancel the constant for any r. We see however that this cancellation imposes certain restrictions on the behaviour of Γ_2 in the overlap region, i.e. at $\omega \sim r^2$.

17.2 Problems of the strong coupling regime

So far we have considered the weak coupling case. The strong coupling regime with non-Hermitian Hamiltonian runs into difficulties. Here we discuss only objections against the strong coupling.

As was described in Lecture 16, in this case $G = -1/\Sigma$, and the vertex function Γ_2 and Σ depend on the parameter $r^2 \Gamma_2^2 G^3$. Therefore due to its invariance under the replacement $r^2 \to -r^2$ and $G \to -G$, the equation for $G' = -G$ has the same form as in the case of the Hermitian Hamiltonian. Let us consider the dependence of G on ω, putting for simplicity $k^2 = 0$.

In the problem with the Hermitian Hamiltonian the Green function obeys a dispersion relation in ω of the type of the Källen–Lehmann representation. Therefore G^{-1} and its derivative with respect to ω are positive.

In the problem with an anti-Hermitian Hamiltonian, G^{-1} and its derivative with respect to ω must be negative in the region of small $\omega > 0$, i.e. in the region of strong coupling.

What is the meaning of this condition? In the Hermitian case, contributions of all branchings to $\Sigma = -G^{-1}$ have the same sign, whereas in the anti-Hermitian case their signs alternate. This result means that the contribution of the three-reggeon branching, in spite of the *opposite* sign, cannot compensate the two-reggeon contribution. Such behaviour of G at small ω evidently leads to a negative total cross section. Indeed, we have found that G near $\omega = 0$ and the two-reggeon branching have the same sign. But the contribution of this branching to the total cross section is negative.

In our discussion we have tacitly assumed that the problem of strong coupling has an unambiguous solution. However, it is possible that the solution is ambiguous and should be chosen according to the condition of matching to the region $\omega \gtrsim r^2$, where $G^{-1} = 0$ at $k^2 = 0$. Although this possibility cannot be excluded in principle, the consideration of concrete examples shows that either the cross section is still negative or there exist several poles near $\omega = 0$, contrary to the initial assumption of a single pomeron. Thus the solution is not self-consistent. This has the following meaning. One can remove the negative cross section by introducing strong coupling in the three-reggeon system, in order to make its contribution dominant. But the strong interaction yields bound states, i.e. additional poles. This reasoning shows that the difficulty, described above, is presumably of a deeper character.

Appendix A

Space-time description of the hadron interactions at high energies

V. N. Gribov

Here we consider the strong and electromagnetic interactions of hadrons in a unified way. It is assumed that there exist point-like particles (*partons*) in the sense of quantum field theory and that a hadron with large momentum p consists of $\sim \ln(p/\mu)$ partons which have restricted transverse momenta, and longitudinal momenta which range from p to zero. The density of partons increases with the increase of the coupling constant. Since the probability of their recombination also increases, an equilibrium may be reached. In this lecture we will consider consequences of the hypothesis that the equilibrium really *occurs*. We demonstrate that it leads to constant total cross sections at high energies, and to the Bjorken scaling in the deep inelastic ep scattering. The asymptotic value of the total cross sections of hadron–hadron scattering turns out to be universal, while the cross sections of quasi-elastic scattering processes at zero angle tend to zero.

The multiplicity of the outgoing hadrons and their distributions in longitudinal momenta (rapidities) are also discussed.

Introduction

In this lecture we will try to describe electromagnetic and strong interactions of hadrons in the same framework which follows from general quantum field theory considerations without the introduction of quarks or other exotic objects.

We will assume that there exist point-like constituents in the sense of quantum field theory which are, however, strongly interacting. It is convenient to refer to these particles as *partons*. We will not be interested in the quantum numbers of these partons, or the symmetry properties of their interactions. We will assume that, contrary with the perturbation theory,

216

the integrals over the transverse momenta of virtual particles converge as in the $\lambda\varphi^3$ theory. It turns out that within this picture a common cause exists for two seemingly very different phenomena: the Bjorken scaling in deep inelastic scattering, and the recent theoretical observation that all hadronic cross sections should approach the same limit (provided that the Pomeranchuk pole exists).

The material is organized as follows. In the first section we discuss the propagation of the hadrons in the space as a process of creation and absorption of the virtual particles (partons) and formulate the notion of the parton wave function of the hadron. The second section describes momentum and coordinate parton distributions in hadrons. In the third section we consider the process of deep inelastic scattering. It is shown that from the point of view of our approach the deep inelastic scattering satisfies the Bjorken scaling, and, in contrast with the quark model, the multiplicity of the produced hadrons is of the order of $\ln(\nu/\sqrt{q^2})$. The fourth section is devoted to the strong interactions of hadrons and it is shown that in the same framework the total hadron cross sections have to approach asymptotically the same limiting value. In the last section we discuss the processes of elastic and quasi-elastic scattering at high energies. It is demonstrated that the cross sections of the quasi-elastic scattering processes at zero angle tend to zero at asymptotically high energies.

Let us discuss how one can think of the space-time propagation of a physical particle in terms of virtual particles which are involved in the interaction with photons and other hadrons. It is well known that the propagation of a real particle is described by its Green function, which corresponds to a series of Feynman diagrams of the type shown in Fig. A.1.

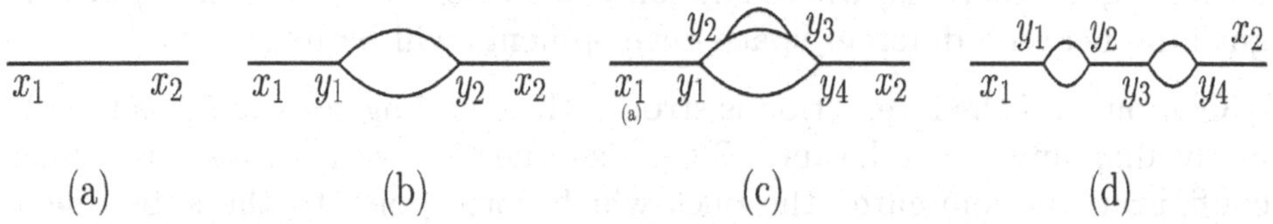

$$
\begin{array}{cccc}
x_1 \qquad x_2 & x_1\ y_1 \qquad y_2\ x_2 & x_1\ y_1 \qquad\qquad y_4\ x_2 & x_1 \qquad\qquad\quad x_2 \\
 & & y_2 \quad y_3 & y_1\ y_2 \qquad y_3\ y_4 \\
(\text{a}) & (\text{b}) & (\text{c}) & (\text{d})
\end{array}
$$

Fig. A.1

(for simplicity, we will consider identical scalar particles). The Feynman diagrams, having many remarkable properties, have, nevertheless, a disadvantage compared to the old-fashioned perturbation theory. Indeed, they do not show how a system evolves with time in a given coordinate reference frame. For example, depending on the relations between the time coordinates x_{10}, y_{10}, x_{20} and y_{20}, the graph in Fig. A.1(b) corresponds to different processes as shown in Fig. A.2.

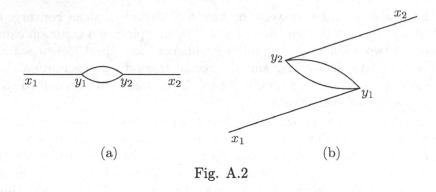

(a)

(b)

Fig. A.2

Similarly, the diagram Fig. A.1(c) corresponds to the processes shown in Fig. A.3.

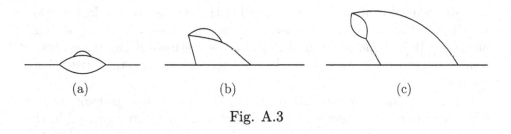

(a)

(b)

(c)

Fig. A.3

In quantum electrodynamics, where explicit calculations can be carried out, this complicated correspondence is of little interest. However, for strong interactions, where explicit calculations are impossible, distinguishing between different space developments will be useful.

Obviously, if the interaction is strong (the coupling constant λ is large), many diagrams are relevant. The first question which arises is which configurations dominate: the ones which correspond to the subsequent decays of the particles – the diagrams Fig. A.2(a) and Fig. A.3(a) – or those which correspond to the interaction of the initial particle with virtual "pairs" created in the vacuum. It is clear that if the coupling constant is large and the momentum of the incoming particle is small (see below), configurations with "pairs" dominate (at least if the theory does not contain infinities). Indeed, if $x_{20} - x_{10}$ is small, then in the case of configurations without "pairs" the integration regions corresponding to each correction will tend to zero with an increase of the number of corrections.

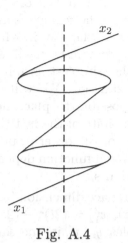

Fig. A.4

At the same time, for the configurations containing "pairs" the region of integration over time will remain infinite. Hence, if the retarded Green function $G^r(y_2 - y_1)$ does not have a strong singularity at $x_{20} - x_{10} \to 0$, the contribution of the configurations without "pairs" will be relatively small if the coupling constant is large. Even the graphs of the type Fig. A.1(d) are determined mainly by configurations like those in Fig. A.4.

This means that if we observe a low energy particle at any particular moment of time (the cut in the diagram in Fig. A.4), we will see a few partons which are decay products of the particle, and a large number of virtual "pairs" which will interact with these partons in the future.

What happens if a particle has a large momentum in our coordinate reference frame? To analyse the space-time evolution of a fast particle we have to consider a space-time interval $(x_1 - x_2)^2$ such that $(x_1 - x_2)^2 \sim 1/\mu^2$, and $t_2 - t_1 \sim E/\mu^2 \gg 1/\mu$. Here μ is the mass of the particle, E its energy. In this case, $\vec{x}_1 - \vec{x}_2 = \vec{v}(t_2 - t_1)$, $(x_2 - x_1)^2 = (\mu^2/E^2)(t_2 - t_1)^2 \sim 1/\mu^2$. For such intervals the relation between the configurations with and without "pairs" changes. Configurations corresponding to a decay of one parton into many others start to dominate, while the rôle of configurations with "pairs" decreases.

The physical origin of this phenomenon is evident. A fast parton can decay, for example, into two fast partons which, due to the energy–time uncertainty relation, will exist for a long time (of the order of E/μ^2), since

$$\Delta E = \sqrt{\mu^2 + \vec{p}^2} - \sqrt{\mu^2 + \vec{p}_1^2} - \sqrt{\mu^2 + (\vec{p} - \vec{p}_1)^2}$$

$$\sim \frac{\mu^2}{2|\vec{p}|} - \frac{\mu^2}{2|\vec{p}_1|} - \frac{\mu^2}{2|\vec{p} - \vec{p}_1|}.$$

Each of these two partons can again decay into two partons and this will continue up to the point when slow particles, living for a time of the order of $1/\mu$, are created. After that the fluctuations must evolve in the reverse direction, i.e. the recombination of the particles begins.

On the other hand, due to the same uncertainty relation, creation of virtual "pairs" with large momenta in vacuum is possible only for short

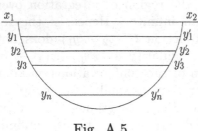

Fig. A.5

time intervals of the order of $1/p$. Hence, it affects only the region of small momentum partons. The way in which this phenomenon manifests itself can be seen using the simplest graph in Fig. A.5 as an example. We will observe that it is possible to place here many emissions in spite of the fact that the interval x_{12}^2 is of the order of unity $(1/\mu^2)$, and the Green function depends only on the invariants.

For the sake of simplicity, let us verify this for one space dimension ($y_i = (t_i, z_i)$). Suppose that $x_1 = (-t, -z)$ and $x_2 = (t, z)$, $x_{12}^2 = (2t)^2 - (2z)^2$. Then $t = z + x_{12}^2/8z$. Let us choose the variables y_i, y_i' in the same way $y_i = (-t_i, -z_i)$, $y_i' = (t_i', z_i')$, and consider the following region of integration in the integral, corresponding to the diagram in Fig. A.5:

$$1 < z_n < z_{n-1} < \ldots < z_1 < t,$$
$$1 < z_n' < z_{n-1}' < \ldots < z_1' < t,$$
$$z_i \sim z_i', \quad t_i = z_i + \frac{y_i^2}{2z_i}, \quad t_i' = z_i' + \frac{y_i'^2}{2z_i'}.$$

The integrations over $\mathrm{d}^2 y_1 \ldots \mathrm{d}^2 y_n \mathrm{d}^2 y_1' \ldots \mathrm{d}^2 y_n'$ can be replaced by integrations over y_i^2, $y_i'^2$ and $z_i \equiv y_{iz}$, $z_i' \equiv y_{iz}'$:

$$\mathrm{d}^2 y_i = \frac{1}{2} \mathrm{d} y_i^2 \frac{\mathrm{d} z_i}{z_i}, \qquad \mathrm{d}^2 y_i' = \frac{1}{2} \mathrm{d} y_i'^2 \frac{\mathrm{d} z_i'}{z_i'}.$$

It is easy to see that in this region of integration the arguments of all Green functions, $(y_i - y_i')^2$, $(y_i - y_{i+1}')^2$, $(y_i' - y_{i+1}')^2$, are of the order of unity, and the integrals do not contain any small factors. All these conditions for y_i can be satisfied simultaneously for a large number of emissions: $n \sim \ln t$. Indeed, if we write z_n in the form $z_n = C^n$, all conditions will be fulfilled for $n \sim \ln t / \ln C$, $C \geq 1$. Obviously, one can consider a more complicated diagram than Fig. A.5 by including interactions of the virtual particles. On the other hand, configurations containing vacuum "pairs" play a minor rôle. Moving backward in time is possible only for short time intervals (see Fig. A.6).

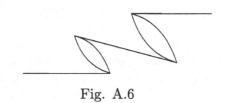

Fig. A.6

Hence, we come to the following picture. A real particle with a large momentum p can be described as an ensemble of an indefinite number of partons of the order of $\ln(p/\mu)$ with momenta in the range from p to zero, and several vacuum pairs with small momenta which in the future can interact with the target.

The observation of a slow particle during an interval of the order of $1/\mu^2$ does not tell us anything about the structure of the particle since we cannot distinguish it from the background of the vacuum fluctuations, and we can speak only about the interaction of particles or about the spectrum of states. On the contrary, in the case of a fast particle we can speak about its structure, i.e. about the fast partons which do not mix with the vacuum fluctuations.

As a result, in a certain sense a fast particle becomes an isolated system which is only weakly coupled to the vacuum fluctuations. Hence, it can be described using a quantum-mechanical wave function or an ensemble of wave functions, which determine probabilities of finding certain numbers of partons and their momentum distribution. Such a description is not invariant, since the number of partons depends on the momentum of the particle, but it can be considered as covariant. Moreover, it may be even invariant, if the momentum distribution of the partons is homogeneous in the region of momenta much smaller than the maximal one, and much larger than μ.

Indeed, under the transformation from one reference frame to another in which the particle has, for example, a larger momentum, a new region emerges in the distribution of partons; in the old region, however, the parton distribution remains unchanged. One usually describes hadrons in terms of the quantum mechanics of partons in the reference frame which moves with an infinite momentum, because in this case all partons corresponding to vacuum fluctuations have zero momenta, and such a description is exact.

Such a reference frame is convenient for the description of the deep inelastic scattering. However, it is not as good for describing strong interactions, where the slow partons are important. In any case, it appears useful to preserve the freedom in choosing the reference frame and to use the covariant description. This allows a more effective analysis of the accuracy of the derivations.

A.1 Wave function of the hadron. Orthogonality and normalization

The preceding considerations allow us to introduce the hadron wave function in the following way. Let us assume, as usual, that for $t \to -\infty$ the hadron can be represented as a bare particle (the parton). After a sufficiently long time the parton will decay into other partons and form a stationary state which we call a hadron. Diagrams corresponding to this process are shown in Fig. A.7.

Let us exclude from the Feynman diagrams those configurations (in the sense of integrations over intermediate times) which correspond to vacuum pair creation.

For the theory $\lambda\varphi^3$ such a separation of vacuum fluctuations corresponds to decomposing φ into positive and negative frequency parts $\varphi = \varphi^+ + \varphi^-$ and replacing

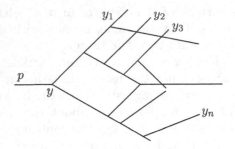

Fig. A.7

$\varphi^3 = (\varphi^+ + \varphi^-)^3$ by $3(\varphi^{-2}\varphi^+ + \varphi^-\varphi^{+2})$. The previous discussion shows that the ignored term $\varphi^{+3} + \varphi^{-3}$ would mix only partons with small momenta.

It is natural to consider the set of all possible diagrams with a given number of partons n at the given moment of time as a component of the hadron wave function $\Psi_n(t, \vec{y}_1, \ldots, \vec{y}_n, p)$. Similarly, we can determine the wave functions of several hadrons with large momenta provided the energy of their relative motion is small compared with their momenta. The latter condition is necessary to ensure that slow partons are not important in the interaction. The Lagrangian of the interaction remains Hermitian even after the terms corresponding to the vacuum fluctuations are omitted. As a result, the wave functions will be orthogonal, and will be normalized in the usual way:

$$\sum_n \int \Psi_n^{b*}(\vec{y}_1 \ldots, \vec{y}_n, p_b) i \overset{\leftrightarrow}{\partial} \Psi_n^a(\vec{y}_1 \ldots, \vec{y}_n, p_a) \frac{d^3 y_1 \ldots d^3 y_n}{n!}$$

$$= (2\pi)^3 \delta(\vec{p}_a - \vec{p}_b)\, \delta_{ab}, \quad \text{(A.1)}$$

or similarly in the momentum space, after separating $(2\pi)^3 \delta(\vec{p} - \sum \vec{k}_i)$

$$\sum_n \frac{1}{n!} \int \Psi_n^{b*}(\vec{k}_1 \ldots, \vec{k}_n, \vec{p}) \Psi_n^a(\vec{k}_1 \ldots, \vec{k}_n, \vec{p}) \frac{d^3 k_1 \ldots d^3 k_n}{2k_{10} \ldots 2k_{n0}} \frac{\delta(p - \sum k_i)}{(2\pi)^{3n-1}}$$

$$= \delta_{ab}. \quad \text{(A.2)}$$

For the momentum range $k_i \gg \mu$, the wave functions coincide with those calculated in the infinite momentum frame. In this reference frame they do not depend on the momentum of the system (except for a trivial factor). This can be easily proven by expanding the parton momenta

$$\vec{k}_i = \beta_i \vec{p} + \vec{k}_{i\perp}, \quad \text{(A.3)}$$

and writing the parton energy in the form

$$\varepsilon_i = \sqrt{\vec{k}_i^2 + m^2} = \beta_i p + \frac{m^2 + k_{i\perp}^2}{2p\beta_i}. \quad \text{(A.4)}$$

Note now that the integrals which determine Ψ_n, corresponding to Fig. A.7, can be represented in the form of the old-fashioned perturbation theory where only the differences between the energies of the intermediate states and the initial state $E_k - E$ enter, and the momentum is conserved. Hence, the terms linear in p cancel in these differences, and consequently

$$E_k - E = \frac{1}{2p} \left(\sum_i \frac{m^2 + k_{i\perp}^2}{\beta_i} - m^2 \right). \qquad (A.5)$$

Each consequent intermediate state in Fig. A.7 in the $\lambda\varphi^3$ model differs from the previous one by the appearance or disappearance of one particle. The factor $\frac{1}{2k_{i0}} = \frac{1}{2p} \frac{1}{\beta_i}$, which comes from the propagator of this particle, cancels $2p$ in (A.5). Hence, there remain only integrals over $\mathrm{d}^2 k_{i\perp} \frac{\mathrm{d}\beta_i}{\beta_i}$, and the resulting expression does not depend on p:

$$\sum_n \frac{1}{n!} \int \Psi_n^{b*}(k_{i\perp}, \beta_i) \Psi_n^a(k_{i\perp}, \beta_i) \prod \frac{\mathrm{d}^2 k_{i\perp}}{2(2\pi)^3} \frac{\mathrm{d}\beta_i}{\beta_i} (2\pi)^3 \delta \left(1 - \sum \beta_i \right)$$
$$= \delta_{ab}. \quad (A.6)$$

For slow partons, where the expansion (A.4) is not correct, the dependence on momentum p does not disappear, and contrary to the case of the system moving with $p = \infty$, this dependence cuts off the sum over the number of partons.

A.2 Distribution of the partons in space and momentum

The distribution of partons in longitudinal momenta can be characterized by the rapidity:

$$\eta_i = \frac{1}{2} \ln \frac{\varepsilon_i + k_{iz}}{\varepsilon_i - k_{iz}}, \qquad (A.7)$$

where k_{iz} is the component of the parton momentum along the hadron momentum.

$$\eta_i \simeq \ln \frac{2\beta_i p}{\sqrt{m^2 + k_{i\perp}^2}}. \qquad (A.8)$$

As is well known, this quantity is convenient since it simply transforms under the Lorentz transformations along the z direction: $\eta_i' = \eta_i + \eta_0$, where η_0 is the rapidity of the coordinate system.

The determination of the parton distribution over η is based on the observation that in each decay process $k_1 \to k_2 + k_3$ shown in Fig. A.7 the momenta $\vec{k}_2$ and $\vec{k}_3$ are, in the average, of the same order. This means that in the process of subsequent parton emission and absorption the rapidities of the partons change by a factor of the order of unity. At the same time the overall range of parton rapidities is large, of the order of $\ln(2p/m)$. This implies that in the rapidity space we have short range forces.

Let us consider the density of the distribution in rapidity

$$\varphi(\eta, k_\perp, p) = \sum_n \frac{1}{n!} \int |\Psi_{n+1}(k_\perp, \eta, k_{\perp 1}, \eta_1, \ldots, k_{\perp n}, \eta_n,)|^2 (2\pi)^3$$

$$\times \delta\left(\vec{p} - \vec{k} - \sum \vec{k_i}\right) \prod \frac{\mathrm{d}k_i \mathrm{d}\eta_i}{2(2\pi)^3} \qquad (A.9)$$

in the interval $1 \ll \eta \ll \eta_p$ (see Fig. A.8).

The independence of φ from p for these values of η means that Ψ depends only on the differences $\eta_i - \eta_p$. If $\varphi = \varphi(\eta - \eta_p, k_\perp)$ decreases with the increase of $\eta - \eta_p$, this corresponds to a weak coupling, i.e. to a small probability of the decay of the initial parton. If the coupling constant grows, the number of partons increases and at a *certain* value of the coupling constant an equilibrium is reached, since the probability of recombination also increases. The value of this critical coupling constant has to be such that the recombination probability due to the interaction should be larger than the recombination probability related to the uncertainty principle.

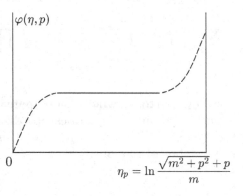

Fig. A.8

The basic hypothesis is that such an equilibrium *does occur* and that due to the short range character of interaction it is *local*. This is equivalent to the hypothesis of the constant total cross sections of interaction at $p \to \infty$. Hence we assume that the equilibrium is determined by the neighbourhood of the point η of the order of unity and it does not depend on η_p. Obviously, this can be satisfied only if $\varphi(\eta, \eta_p, k_\perp) = \varphi(k_\perp)$ does not depend on η and η_p at $1 \ll \eta \ll \eta_p$. According to the idea of Feynman, this situation resembles the case of a sufficiently long one-dimensional matter in which, due to the homogeneity of the space, far from the boundaries the density is either constant or oscillating (for a crystal). In our case the analogue of the homogeneity of space is the relativistic invariance (the shift in the space of rapidities). *For the time being* we will not consider the case of the crystal. According to (A.9), the integral of $\varphi(\eta, \eta_p, k_\perp)$ over η and $k_\perp$ has the meaning of the average parton density which is, obviously, of the order of $\eta_p \sim \ln(2p/m)$.

Generally speaking, we cannot say anything about the parton distribution in the transverse momenta except for one statement: it is absolutely crucial for the whole concept that it must be restricted to the region of the order of parton masses, as in the $\lambda\varphi^3$ theory.

Consider now the spatial distribution of the partons. First, let us discuss parton distribution in the plane perpendicular to the momentum $\vec{p}$. For that purpose it is convenient to transform from $\Psi_n(k_{1\perp}, \eta_1, k_{2\perp}, \eta_2, \ldots, k_{n\perp}, \eta_n)$ to the impact parameter representation $\Psi_n(\vec{\rho}_1, \eta_1, \vec{\rho}_2, \eta_2, \ldots, \vec{\rho}_n, \eta)$:

$$\Psi_n(\vec{\rho}_n, \eta_n) = \int e^{i\sum k_{i\perp}\rho_i} \Psi(k_{i\perp}, \eta_i) \delta\left(\sum k_{\perp i}\right) (2\pi)^2 \prod \frac{\mathrm{d}^2 k_i}{(2\pi)^2}. \quad (A.10)$$

Let us rank the partons in the order of decreasing rapidities. Consider a parton with the rapidity $\eta \ll \eta_p$ and let us follow its history from

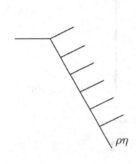

the initial parton. Initially, we will assume that it was produced solely via parton emissions (Fig. A.9). In this case it is clear that if the transversal momenta of all partons are of the order of μ, then each parton emission leads to a change of the impact parameter $\vec{\rho}$ by $\sim 1/\mu$. If n emissions are necessary to reduce the rapidity from η_p to η, and they are independent and random, $\overline{(\Delta\rho)^2} \sim n$. If every emission changes the rapidity of the parton by about one unit, then

Fig. A.9

$$\overline{(\Delta\rho)^2} = \gamma(\eta_p - \eta). \quad (A.11)$$

Hence, the process of the subsequent parton emissions results in a kind of diffusion in the impact parameter plane (see Fig. A.10). The parton distribution in ρ for the rapidity η has the Gaussian form

$$\varphi(\rho, \eta) = \frac{C(\eta)}{\pi\gamma(\eta_p - \eta)} \exp\left[-\frac{\rho^2}{\gamma(\eta_p - \eta)}\right], \quad (A.12)$$

if the impact parameter of the initial parton is considered as the origin. Consequently, the partons with $\eta \simeq 0$ have the broadest distribution, and, hence, the fast hadron is of size

$$R = \sqrt{\gamma\eta_p} \simeq \sqrt{\gamma \ln \frac{2p}{m}}. \quad (A.13)$$

The account of the recombination and the scattering of the partons affects only densities of partons and fluctuations, but does not change the radius of the distribution which can be viewed as the front of the diffusion wave.

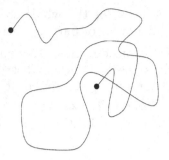

Let us discuss the parton distribution over the longitudinal coordinate. A relativistic particle with a momentum p is commonly considered as a disk of thickness $1/p$. In fact, this is true only in the first approximation of the perturbation theory.

Fig. A.10

In reality, a hadron is a disk with radius $\sqrt{\gamma \ln(2p/m)}$ and thickness of the order of $1/\mu$. Indeed, each parton with a longitudinal momentum k_{iz} is distributed in the longitudinal direction in an interval $\Delta z_i \sim 1/k_{iz}$. Since the parton spectrum exists in the range of momenta from p down to $k_i \sim \mu$, the longitudinal projection of the hadron wave function has the structure depicted in Fig. A.11.

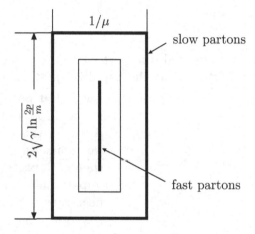

Fig. A.11

Finally, let us consider what is the lifetime of a particular parton.

As we have discussed in the Introduction to this lecture, in a theory which is not singular at short distances, the intervals y_{12}^2 between two events represented by a Feynman diagram are of the order of unity. For a fast particle moving along the z axis, $z_{21} = vt_{21}$ and $y_{12}^2 = t_{21}^2(m^2/p^2)$. Consequently, the lifetime of a fast parton with a momentum k_i is of the order of k_i/μ^2. The arguments presented were based on the $\lambda \varphi^3$ theory which is the only theory providing a cut-off in transverse momenta. Still, the argument should hold for other theories and for particles with spins, if one assumes that in these theories the cut-off of transverse momenta occurs in some way. On the other hand, the $\lambda \varphi^3$ theory cannot be considered as a self-consistent example. Indeed, due to the absence of a vacuum state, the series of perturbation theory do not make sense (series with positive coefficients are increasing as factorials). Hence, the picture we have presented here does not correspond *literally* to any particular field theory. At the same time, it corresponds fully to the main ideas of the quantum field theory and to its basic space-time relations.

A.3 Deep inelastic scattering

It is convenient to consider the deep inelastic scattering of electrons in the frame where the time component of the virtual photon momentum is $q_0 = 0$. In this reference frame the momentum of the photon is equal to $-q_z$ ($q^2 = -q_z^2$), while the momentum of the hadron is $p_z = \omega q_z/2$ ($\omega = -2pq/q^2$). Suppose that q_z is large and $\omega \sim 1$. According to our previous considerations, a fast hadron can be viewed as an ensemble of partons. In this system a photon looks like a static field with wavelength $\sim 1/q_z$.

The main question is, which partons can the photon interact with? We can consider the static field of a photon as a packet with a longitudinal size of the order of $1/q_z$. The interaction time between a hadron with the size $1/\mu$ and such a packet is of the order of $1/\mu$. However, due to the big difference between the parton and photon wavelengths, the interaction with a slow parton is small. Hence, the photon interacts with partons which have momenta of the order of q_z. Partons with such momenta are distributed in the longitudinal direction in the region $1/q_z$. Because of this, the time of the hadron–photon interaction is in fact of the order of $1/q_z$, i.e. much shorter than the lifetime of a parton. This means that the photon interacts with a parton as with a free particle, and so not only the momentum but also the energy is conserved. As a result, the energy–momentum conservation laws select the parton with momentum $q_z/2$, which can absorb a photon:

$$k_{iz} - q_z = k'_{iz}, \quad |k_{iz} - q_z| = k_{iz}.$$

This gives

$$k_{iz} = \frac{q_z}{2}, \quad k'_{iz} = -\frac{q_z}{2}.$$

The cross section of such a process is, obviously, equal to the cross section σ_0 of the absorption of a photon by a free particle, multiplied by the probability of finding a parton with a longitudinal momentum $q_z/2$ inside the hadron, i.e. by the value $\varphi(\eta_{q/2}, \eta_p)$ in (A.9), integrated over $k_\perp$. (The necessary accuracy of fulfilment of the conservation laws allows any $k_\perp \ll q_z$).

As was already discussed, $\varphi(\eta, \eta_p) = \varphi(\eta - \eta_p) \equiv \varphi(\omega)$. Hence, using the known cross section for the interaction of the photon with a charged spinless particle, we obtain for the cross section of the deep inelastic scattering

$$\frac{d^2\sigma}{dq^2 d\omega} = \frac{4\pi\alpha^2}{q^4}\left(1 - \frac{pq}{pp_e}\right)\varphi(\omega), \tag{A.14}$$

where p_e is the electron momentum. If the partons have spins, the situation becomes more complicated, since the cross sections of the interactions between photons and partons with different spins are different. The parton distributions in rapidities for different spins may also be different, leading to the form

$$\frac{\mathrm{d}^2\sigma}{\mathrm{d}q^2\mathrm{d}\omega} = \frac{4\pi\alpha^2}{q^4}\left\{\left(1-\frac{pq}{pp_e}\right)\varphi_0(\omega)\right.$$

$$\left.+\left[1-\frac{pq}{pp_e}+\frac{1}{2}\left(\frac{pq}{pp_e}\right)^2\right]\varphi_{\frac{1}{2}}(\omega)\right\}. \quad (A.15)$$

Let us discuss now a very important question, namely, what physical processes take place in deep inelastic scatterings? To clarify this, we go back to Fig. A.7 determining the hadron wave function. We will neglect the parton recombinations in the process of their creation from the initial parton, i.e. we consider fluctuations of the type shown in Fig. A.9. Suppose that the photon was absorbed by a parton with a large momentum $q_z/2$. As a result, this parton obtained a momentum $-q_z$ and moves in the opposite direction with momentum $-q_z/2$. The process is depicted in Fig. A.12. What will now happen to this parton and to the remaining partons?

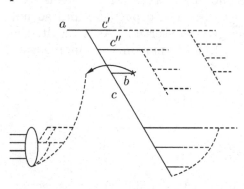

Fig. A.12

Within the framework we are using it is highly unlikely that the parton with momentum $-q_z/2$ will have time to interact with the other partons. The probability of interacting directly with residual partons will be small, because the relative momentum of the parton with $-q_z/2$ and the rest of the partons is large. It could interact with other partons after many subsequent decays which, in the end, could create a slow parton. However, the time needed for these decays is large, and during this time the parton and its decay products will move far away from the remaining partons, thus the interaction will not take place.

Hence, we come to the conclusion that one free parton is moving in the direction $-q_z$. What will we observe experimentally, if we investigate particles moving in this direction? To answer this question, it is sufficient to note that, on average, a hadron with momentum k_z consists of n partons, $n = c\ln(k_z/\mu)$ at $k_z \gg \mu$.

In a sense there should exist an uncertainty relation between the number of partons in a hadron (n) and the number of hadrons in a parton (n_p):

$$n_p n \gtrsim c\ln\frac{k_z}{\mu}, \quad (A.16)$$

where k_z is the momentum of the state.

We came to the conclusion that the parton decays into a large number of hadrons, i.e. in fact the parton is very short-lived, highly virtual. Hence, we have to discuss whether this conclusion is consistent with the assumption that the photon–parton interaction satisfies energy conservation. To answer this question, let us calculate the mass of a virtual parton with momentum k_z, decaying into n hadrons with momenta k_i and masses m_i.

$$M^2 = \left(\sum \sqrt{m_i^2 + k_i^2}\right)^2 - k_z^2 = \left(k_z + \sum_i \frac{m_i^2 + k_{i\perp}^2}{2k_{iz}}\right)^2 - k_z^2$$

$$\simeq k_z \sum_i \frac{m_i^2 + k_{i\perp}^2}{k_{iz}}.$$

If the hadrons are distributed almost homogeneously in rapidities, their longitudinal momenta decrease exponentially with their number, and in the sum only a few terms, corresponding to slow hadrons, are relevant. As a result, $M^2 \sim k_z\mu$, i.e. the time of the existence of the parton is of the order of $1/\mu$, much larger than the time of interaction with a photon $1/q_z$.

Let us discuss now what happens to the remaining partons. Little can be determined using only the uncertainty relation (A.16). This is because the number of partons before the photon absorption was n, after the photon absorption it became $n - 1$ and, consequently, according to the uncertainty relation, the number of hadrons corresponding to this state can range from 1 to n. Hence, everything depends on the real perturbation of the hadron wave function due to the photon absorption.

Consider now the fluctuation shown in Fig. A.12. The photon absorption will not have any influence on partons created after the parton b which absorbed the photon was produced, and which have momenta smaller than b. These fluctuations will continue, and the partons can, in particular, recombine back into the parton c. The situation is different for partons which occurred earlier and have large momenta (c', c''). In this case the fluctuation cannot evolve further the same way, since the parton b has moved in the opposite direction. As a result, it is highly probable that partons c' and c'' will move apart and lose coherence. On the other hand, slow partons which were emitted by c' and c'' earlier and which are not connected with the parton b will be correlated, as before, with each of them. Thus c' and c'' will move in space together with their slow partons, i.e. in the form of hadrons. Hence, it appears that partons flying in the initial direction lead to the production of of the order of $c\ln(\omega q_z/2) - c\ln(q_z/2) = c\ln\omega$ hadrons with rapidities ranging from $\ln(\omega q_z/\mu)$ to $\ln(q_z/\mu)$. This answer can be interpreted in the following way. After the photon is absorbed, a hole is created in the distribution of partons moving in the initial direction.

Contrary to the case of rapidities of partons, we will count the rapidity of the hole not from zero rapidity but from the rapidity $\ln(\omega q_z/\mu)$. In this case the rapidity of the hole is $\ln\omega$. If we now represent the parton hole with rapidity $\ln\omega$ as a superposition of the hadron states, this superposition will contain $\ln\omega$ hadron states.

Let us represent the whole process by a diagram describing rapidity distributions of partons and hadrons. Before the photon absorption the partons in the hadrons are distributed at rapidities between zero and $\ln(\omega q_z/\mu)$, while after the photon absorption a parton distribution is produced which is shown in Fig. A.13.

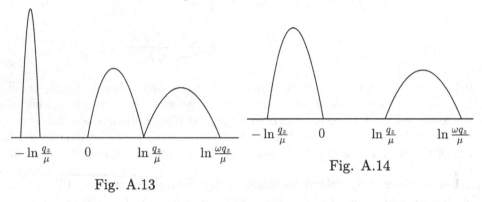

Fig. A.13

Fig. A.14

This parton distribution leads to the hadron distribution shown in Fig. A.14. The total multiplicity corresponding to this distribution is

$$\bar{n} = c\ln\frac{q_z}{\mu} + c\ln\omega = c\ln\frac{\nu}{\mu\sqrt{-q^2}}.$$

This hadron distribution in rapidities in the deep inelastic scattering differs qualitatively from those previously discussed in the literature. It corresponds to $c\ln(\sqrt{-q^2}/\mu)$ hadrons moving in the photon momentum direction, while $\ln\omega$ hadrons are moving in the nucleon momentum direction, with a gap in rapidity between these distributions. The hadron distribution which was obtained in the framework of perturbation theory for superconverging theories like $\lambda\varphi^3$ (Drell and Yan [3]) differs qualitatively from the distribution in Fig. A.14.

In conclusion of this section, it is necessary to point out that the problem of spin properties of the partons exists in this picture even if the partons do not have quark quantum numbers. If, as experiment shows, the cross section σ_t for the interaction of the transversal photons is larger than the cross section for the interaction of the longitudinal photons, σ_l, the charged partons have predominantly spin $1/2$. This means that at least one fermion, for example a nucleon, has to move in the direction of the photon momentum. In other words, in deep inelastic scattering the distribution of the created hadrons in quantum numbers as the function of their rapidities differs essentially from what we are used to in strong interactions. Perhaps this is one of the key predictions of the non-quark parton picture for $\sigma_t \gg \sigma_1$.

A.4 Strong interactions of hadrons

Let us discuss now the strong interactions of hadrons. First, we consider a collision of two hadrons in the laboratory frame. Suppose that a hadron 1 with momentum $\vec{p}_1$ hits hadron 2 which is at rest. Obviously, the parton wave function makes no sense for the hadron at rest, since for the latter the vacuum fluctuations are absolutely essential. However, the hadron at rest can also be understood as an ensemble of slow partons distributed in a volume of the order of $1/\mu$, independent of the origin of the partons. Indeed, it does not matter whether these partons are decay products of the initial parton or the result of the vacuum fluctuations. How can a fast hadron, consisting of partons with rapidities from $\ln(2p_1/\mu)$ to zero, interact with the target which consists of slow partons? Obviously, the cross section of the interaction of two point-like particles with a large relative energy is not larger than $\pi\lambda^2 \sim 1/s_{12} \sim \exp(-\eta_{12})$ (where λ is the wave length in the cms, η_{12} is the relative rapidity). That is why only slow partons of the incident hadron can interact with the target with a cross section which is not too small.
This process is shown in Fig. A.15.

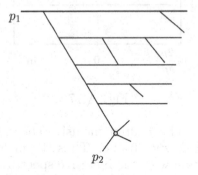

If the slow parton which initiated the interaction was absorbed in this interaction, the fluctuation which led to its creation from a fast parton was interrupted. Hence, all partons which were emitted by the fast parton in the process of fluctuation cannot recombine any more. They disperse in space and ultimately decay into hadrons leading to the creation of hadrons with rapidities from zero to $\ln(2p_1/\mu)$.

Fig. A.15

The interaction between the partons is short range in rapidities. Hence, the hadron distribution in rapidities will reproduce the parton distribution in rapidities. In particular, the inclusive spectrum of hadrons will have the form shown in Fig. A.8, with an unknown distribution near the boundaries. The total hadron multiplicity will be of the order of $\eta_p = \ln(2p_1/\mu)$. If the probability of finding a slow parton in the hadron does not depend on the hadron momentum (this would be quite natural, since with the increase of the momentum the lifetime of the fluctuation is also growing), the total cross section of the interaction will not depend on the energy at high energies.

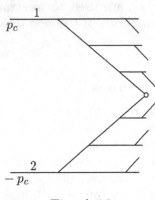

Fig. A.16

Before continuing the analysis of inelastic processes, let us discuss how to reconcile the energy independence of the total interaction cross section at high energies with the observation discussed above according to which the transverse hadron sizes increase with the increase of the energy as $\sqrt{\gamma \ln(2p/\mu)}$.

The answer is that slow partons are distributed almost homogeneously over the disk of radius $\sqrt{\gamma \ln(2p/\mu)}$ (equation (A.11)), while their overall multiplicity during the time of $1/\mu$ is of the order of unity.

Let us see now how the same process will look, for example, in the cms. In this reference frame the interaction will have the form shown in Fig. A.16. Each of the hadrons consists of partons with rapidities ranging from $-\ln(2p_c/\mu)$ to zero and from zero to $\ln(2p_c/\mu)$, respectively. The slow partons interact with cross sections which are not small. As a result, the fluctuations will be interrupted in both hadrons, and the partons will

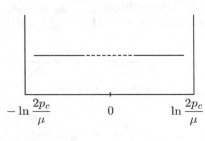

Fig. A.17

fly away in opposite directions, leading to the creation of hadrons with rapidities from $-\ln(2p_c/\mu)$ to $\ln(2p_c/\mu)$. From the point of view of this reference frame the inclusive spectrum must have the form shown in Fig. A.17, with unknown distributions not only at the boundaries but also in the centre, since the distribution of the slow partons in the hadrons and in vacuum fluctuations is unknown.

The hadron inclusive spectrum, however, should not depend on the reference frame. Thus the inclusive spectrum in Fig. A.17 should coincide with the inclusive spectrum in Fig. A.8, and they should differ only by a trivial shift along the rapidity axis, i.e. due to relativistic invariance we know something about the spectra of slow partons and vacuum fluctuations. Let us demonstrate that this comparison of processes in two reference frames leads to a very important statement, namely that at ultra-high energies the total cross sections for the interactions of arbitrary hadrons should be equal. Indeed, we have assumed that the distribution of hadrons reproduces the parton distribution.

From the point of view of the laboratory frame the distribution of partons and, consequently, the distribution of hadrons in the central region of the spectrum is completely determined by the properties (quantum numbers, mass, etc.) of particle 1, and does not depend on the properties of particle 2. On the other hand, from the point of view of the anti-laboratory frame (where the particle 1 is at rest) everything is determined by the properties of particle 2. This is possible only if the distribution of partons in the hadrons with rapidities η much smaller than the hadron rapidity η_p does not depend on the quantum numbers and the mass of the hadron, that is the parton distribution with $\eta \ll \eta_p$ should be universal. From the point of view of the cms the same region is determined by slow partons of both hadrons and by vacuum fluctuations (which are universal), and, consequently, the distribution of slow partons is also universal.

It is natural to assume that the probability of finding a hadron in a sterile state without slow partons tends to zero with the increase of its momentum, or, in other words, to assume that slow partons are always present in a hadron (compare with the decrease of the cross section of the elastic electron scattering at large q^2). In this case, considering the process in the c.m. system, we see that the total cross section of the hadron interaction is determined by the cross section of the interaction of slow partons and by their transverse distribution which is universal. Consequently, the total hadron interaction cross section is also universal, i.e. equal for any hadrons.

This statement looks rather strange if we regard it, for instance, from the following point of view. Let us consider the scattering of a complicated system with a large radius, for example, deuteron–nucleon scattering. As we know, the cross section of the deuteron–nucleon interaction equals the sum of the nucleon–nucleon cross sections, thus it is twice as large as the nucleon–nucleon cross section. How and at what energies can the deuteron–nucleon cross section become equal to the nucleon–nucleon cross section? How is it possible that the density of slow partons in the deuteron turns out to be equal to the density of slow partons in the nucleon? To answer this question, let us discuss the parton structure of two hadrons which are separated in the plane transverse to their longitudinal momenta by a distance much larger than their Compton wavelength $1/\mu$. Suppose that at the initial moment they were point-like particles. Next, independently of each other, they begin to emit partons with decreasing longitudinal momenta. At the same time the diffusion takes place in the transverse plane so that the partons will be distributed in a growing region. The basic observation which we shall prove and which answers our question is that if the momenta of the initial partons are sufficiently large, then during one fluctuation the partons coming from different initial partons will inevitably meet in space (Fig. A.18) in a region of the order of $1/\mu$. They will have similar large rapidities and, hence, will be able to interact with a probability of the order of unity. If such "meetings" take place sufficiently frequently, the probability of the parton interaction will be unity. Consequently, the further evolution and the density of the slow partons which are created after the meeting may not depend on the fact that initially the transverse distance between two partons was large.

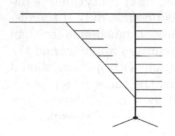

Fig. A.18

In terms of the diffusion in the impact parameter plane this statement corresponds to the following picture. Suppose that the initial partons were placed at points ρ_1 and ρ_2 in Fig. A.19 and that their longitudinal momenta are of the same order of magnitude, i.e. the difference of their rapidities is of the order of unity, while each of the rapidities is large. We will follow the parton starting from point ρ_1, which decelerates via emission of other

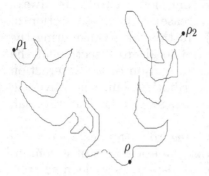

Fig. A.19

partons. As we have seen, its propagation in the perpendicular plane corresponds to diffusion.

The difference of rapidities $\eta_p - \eta$ at the initial and considered moments plays the rôle of time in this diffusion process.

The diffusion character of the process means that the probability density of finding a parton with rapidity η at the point ρ if it started from the point ρ_1 with rapidity η_p is

$$\omega(\vec{\rho}, \vec{\rho}_1, \eta_p - \eta) = \frac{1}{\pi\gamma(\eta_p - \eta)} \exp\left[-\frac{(\vec{\rho} - \vec{\rho}_1))^2}{\gamma(\eta_p - \eta)}\right]. \qquad (A.17)$$

The situation is exactly the same for a decelerating parton which started from the point ρ_2. Thus, the probability of finding both partons at the same point ρ with equal rapidities is proportional to

$$\begin{aligned}
\omega(\rho_{12}, \eta_p - \eta) &= \int \omega(\vec{\rho}, \vec{\rho}_1, \eta_p - \eta)\omega(\vec{\rho}, \vec{\rho}_2, \eta_p - \eta)\mathrm{d}^2\rho \\
&= \frac{1}{2\pi\gamma(\eta_p - \eta)} \exp\left[-\frac{(\vec{\rho}_1 - \vec{\rho}_2))^2}{2\gamma(\eta_p - \eta)}\right]. \qquad (A.18)
\end{aligned}$$

If we now integrate this expression over η, i.e. estimate the probability for the partons to meet at some rapidities, we obtain

$$\int_0^{\eta_p} \omega(\rho_{12}, \eta_p - \eta)\mathrm{d}\eta \simeq \frac{1}{\pi} \ln \frac{2\gamma\eta_p}{\rho_{12}^2}\bigg|_{\eta_p \to \infty} \longrightarrow \infty. \qquad (A.19)$$

This means that if $2\gamma\eta_p \gg \rho_{12}^2$, the partons will inevitably meet. According to (A.19) we get a probability much larger than unity. The reason is that under these conditions the meetings of partons at different values of η are not independent events and therefore it does not make sense to add the probabilities. It is easy to prove this statement directly, for example with the help of the diffusion equation. We will not do this, however. According to a nice analogy suggested by A. Larkin, this theorem is equivalent to the statement that if you are in an infinite forest in which there is a house at a finite distance from you, then, randomly wandering in the forest, you sooner or later arrive at this house. Essentially, the reason is that in the two-dimensional space the region inside of which the diffusion takes place and the length of the path travelled during the diffusion increase with time in the same way. From the point of view of the reference frame in which the deuteron is at rest and is hit by a nucleon in the form of a disk, the radius of which is much larger than that of the deuteron, the statement of the equality of cross sections means that the parton states inside the disk are highly coherent.

It is clear from the above that the cross sections of two hadrons can become equal only when the radius of parton distribution $\sqrt{\gamma \eta_p}$ which is increasing with the energy becomes much larger than the size of both hadrons. Substituting $4 \cdot 0.25/m^2$ for the value of γ (m is the proton mass)[*] we see that the deuteron–nucleon cross section will practically never coincide with the nucleon–nucleon cross section, while the tendency for convergence of cross sections for pion–nucleon, kaon–nucleon and nucleon–nucleon scatterings may be manifested already starting at incident energies $\sim 10^3$ GeV.

A.5 Elastic and quasi-elastic processes

So far we focused on the implications of the picture considered for inelastic processes with multiplicities, growing logarithmically with the energy. However, with a certain probability it can happen that slow partons scatter at very small angles and the fluctuations will not be interrupted in either of the hadrons (for example, if we discuss the process in the cms).

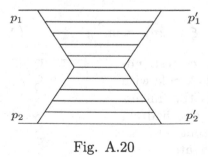

Fig. A.20

In this case small angle elastic or quasi-elastic scattering will take place (Fig. A.20).

First, let us calculate the elastic scattering amplitude. It is well known that the imaginary part of the elastic scattering amplitude can be written in the form

---

[*] It will be demonstrated below that $\gamma = 4\alpha'$, where α' is the slope of the Pomeron trajectory. The current data give $\alpha' \sim 0.25$ GeV^{-2}.

$$A_1(s_{12}) = s_{12} \int d^2\rho_{12}\, e^{i\vec{q}\vec{\rho}_{12}} \sigma(\rho_{12}, s_{12}),\qquad (A.20)$$

where s_{12} is the energy squared in the c.m. system, ρ_{12} is the relative impact parameter, $\sigma(\rho_{12}, s_{12})d^2\rho_{12}$ the total interaction cross section of particles at the distance ρ_{12} and $\vec{q}$ is the momentum transferred. In order to calculate $\sigma(\rho_{12}, s_{12})$ it is sufficient to notice that, according to (A.12), the probability of finding a slow parton with rapidity η_1 at the impact parameter ρ'_1 which originated from the first hadron with an impact parameter $\vec{\rho}_1$ is

$$\varphi_1\left(\vec{\rho}_1, \vec{\rho'_1}, \eta_1, \eta_{pc}\right) \frac{C(\eta_1)}{\pi\gamma\eta_{pc}} \exp\left[-\frac{(\vec{\rho}_1 - \vec{\rho'_1})^2}{\gamma\eta_{pc}}\right]. \qquad (A.21)$$

The probability or finding a parton originating from the second hadron at impact parameter ρ'_2 is

$$\varphi_2\left(\vec{\rho}_2, \vec{\rho'_2}, \eta_2, \eta_{pc}\right) \frac{C(\eta_2)}{\pi\gamma\eta_{pc}} \exp\left[-\frac{(\vec{\rho}_2 - \vec{\rho'_2})^2}{\gamma\eta_{pc}}\right]. \qquad (A.22)$$

The total cross section of the hadron interaction which is due to the interaction of slow partons is equal to

$$\sigma(\rho_{12}, s_{12}) = \int d\eta_1 d\eta_2 d^2\rho'_{12}\sigma(\eta_1, \eta_2, \rho'_{12})C(\eta_1)C(\eta_2)$$

$$\times \int \frac{d^2\rho}{(\pi\gamma\eta_{pc})^2} \exp\left[-\frac{(\vec{\rho} - \vec{\rho}_1)^2}{\gamma\eta_{pc}} - \frac{(\vec{\rho} - \vec{\rho}_2)^2}{\gamma\eta_{pc}}\right].$$

We have taken into account that $\rho'_1 = \rho + \rho'_{12}/2$, $\rho'_2 = \rho - \rho'_{12}/2$, and that the dependence on ρ'_{12} can be neglected in the exponential factor.

After carrying out the integration over ρ, we obtain

$$\sigma(\rho_{12}, s_{12}) = \frac{\sigma_0}{2\pi\gamma\eta_{pc}} \exp\left[-\frac{(\vec{\rho}_1 - \vec{\rho}_2)^2}{2\gamma\eta_{pc}}\right]. \qquad (A.23)$$

Inserting (A.22) into (A.20), we get

$$A_1 = s_{12}\sigma_0 e^{-\frac{7}{4}q^2\xi}, \qquad \xi = 2\eta_{pc} = \ln\frac{s_{12}}{\mu^2}. \qquad (A.24)$$

We obtained the scattering amplitude corresponding to the exchange by the Pomeranchuk pole with slope $\alpha' = \gamma/4$, $\sigma_0 = g^2$, where g is the universal coupling constant of the Pomeron and hadron. The amplitude (A.24) is usually represented by the diagram in Fig. A.21, where a propagator of the form $\exp(-\alpha' q^2\xi)$ corresponds to the Pomeron. In the impact parameter space this propagator has the form (A.22).

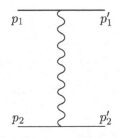

Fig. A.21

Let us discuss the physical meaning of σ_0 in more detail. For this purpose, let us calculate the zero angle scattering amplitude at $\vec{q} = 0$, without using the impact parameter representation. The probability of finding a parton with rapidity η and a transverse momentum $k_\perp$ is described by (A.9). This expression at $\eta \ll \eta_p$ corresponds to the diagram in Fig. A.22. The wavy line represents integration over parton rapidities from η_p to zero. This figure reflects the hypothesis that the calculation of $\varphi(\eta, k_\perp, \eta_p)$ for sufficiently large η_p and $\eta \ll \eta_p$ leads to an expression for φ which is factorized in the same way as the Pomeron contribution to the scattering amplitude. This is because the parton distribution in this region is independent of the properties of the hadron as well as the values of η, η_p. Compared with the diagram in Fig. A.7, Fig. A.22 indicates

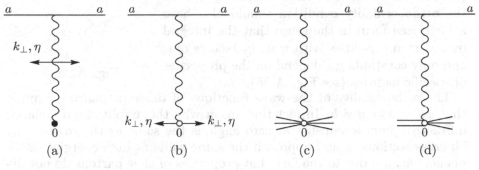

Fig. A.22

that the calculation of $\varphi(\eta, k_\perp, \eta_p)$ is similar to the calculation of the inclusive cross section due to the Pomeron exchange. The only difference is that the coupling of the hadron with the Pomeron should be replaced by unity, since a hadron always exists in a Pomeron state. If $\eta \sim 1$, $\varphi(\eta, k_\perp)$ corresponds to the diagram in Fig. A.22(a), which shows that $\varphi(\eta, k_\perp)$ depends on η. Similarly, it is possible to determine the probability of finding several slow partons (Fig. A.22(b)), and even the density matrix of slow partons. In this case the amplitude of elastic hadron–hadron scattering in the cms is determined by the diagram of Fig. A.23 and the

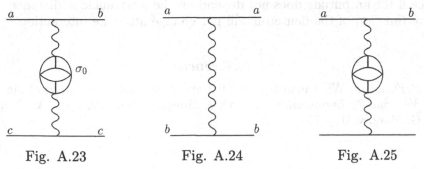

Fig. A.23   Fig. A.24   Fig. A.25

value of σ_0 is determined solely by the interaction of slow partons. Now let us consider the quasi-elastic scattering, corresponding to the Pomeron exchange (Figs. A.24, A.25) at zero transverse momentum. While the probability of finding the parton in hadron a is determined in (A.8) by the integral of the wave function squared, the analogous quantity for the amplitude of the inelastic diffractive process (Fig. A.25) will lead to the integral of the product of the parton functions of different hadrons. They are orthogonal to each other and it is almost obvious that the amplitude for the inelastic diffractive process at zero angle should vanish for this reason. Indeed, the orthogonality condition of (A.6) has the same structure as the imaginary part of the amplitude. Thus, if at high energies the amplitude factorizes (as it should do for the Pomeron exchange), then the orthonormality condition should also have a factorized form in the sense that the integral over parton rapidities with $\eta \ll \eta_p$ factors out, and only constants g_{ab} depend on the properties of specific hadrons (see Fig. A.26).

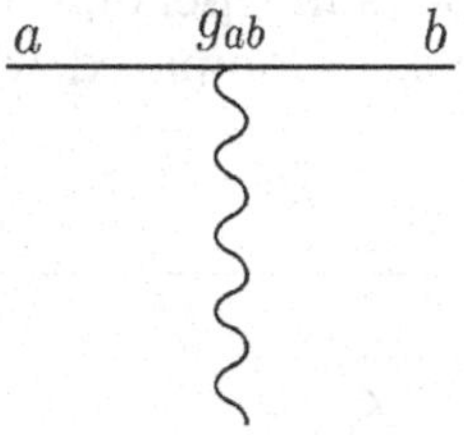

$a \qquad g_{ab} \qquad b$

Fig. A.26

The orthogonality of the wave functions of different hadrons implies that $g_{ab} = 0$ at $a \neq b$. In fact the reason why the amplitude of inelastic diffractive process vanishes at zero angle is the same as the reason why all cross sections should approach the same value at high energies. Both phenomena are due to the fact that properties of slow partons do not depend on the properties of hadrons to which they belong. We can illustrate this again using the example of quasi-elastic dissociation of the composite system – e.g. deuteron. Let us consider the interaction of a fast nucleon with a deuteron. As we discussed in the previous section, at very large energies partons from different nucleons will always interact with each other independently of the distance between nucleons. This will lead to the production of the spectrum of slow partons which does not depend on the relative distance between nucleons in the deuteron. This means that the amplitude of the nucleon–nucleon interaction will not depend on the internucleon distance as well. Thus, if nucleons inside the deuteron remain intact after the interaction, then the deuteron will not dissociate either, since if the amplitude does not depend on the inter-nucleon distance, the wave function of the deuteron will not change after the interaction.

References

[1] R. Feynman, What neutrinos can tell about partons, *Proceedings of the Neutrino-72 Europhysics Conference*, Hungary, June 1972, ed. A. Frenkel, G. Marx, v. II, p. 75.

[2] J. D. Bjorken, *Phys. Rev.* **179** (1969) 1547.

[3] S. D. Drell and T. M. Yan, *Annals of Phys.* **66** (1971) 555.

[4] V. N. Gribov, *Proceedings of Batavia Conference*, 1972.

This lecture was previously published in the following.

Proceedings of the Eighth LNPI Winter School on Nuclear and Elementary Particle Physics, Leningrad, 1973, p. 5 (in Russian).

Proceedings of the First ITEP Winter School, Moscow, 1973, Vol. I, p. 65 (in Russian).

V.N. Gribov, *Gauge Theories and Quark Confinement*, Phasis, Moscow, 2002, pp. 3–27.

Appendix B

Character of inclusive spectra and fluctuations produced in inelastic processes by multi-pomeron exchange

V. A. Abramovski, V. N. Gribov, and O. V. Kancheli

We attempt to determine which absorptive parts of the reggeon graphs for σ_{tot} are not small in the limit as $s \to \infty$, and present on this basis a classification of the asymptotically essential inelastic processes. The only absorptive parts that are not small are those for which each reggeon is cut as a whole or is not cut at all. For the physically interesting case of the n-reggeon exchange, these absorptive parts are expressed explicitly via the n-reggeon contribution to σ_{tot}. The relations obtained show that the main part of the j-plane branch points does not contribute to the inclusive cross sections. The main corrections to the scaling form of the spectrum arise from one-loop contributions to the vertex functions and to the pomeron propagators. Their explicit form is determined by the three-reggeon vertex only. The asymptotic form of the two-particle correlation function ρ_2 in the central region is also determined by the three-reggeon vertex only and decreases logarithmically with the relative rapidity but preserves its positive sign. In the last section we study the fluctuations in the distributions of the produced particles in individual events. For this purpose it is convenient to introduce the concept of the inclusive cross section $\hat{f}$ for a given type of inhomogeneity in the spectrum. The quantities $\hat{f}$ are expressed via the absorptive parts of the diagrams for the pomeron propagator. It is shown that the final particle density distribution is quite inhomogeneous; the distribution with respect to the number of fluctuations with ranges $\geq \lambda$, in rapidity units, has a Poisson form with a mean fluctuation number $\sim \lambda^{-2} \ln s$.

Introduction

The description of the asymptotic behaviour of strong interactions in terms of the Pomeranchuk singularity (the pomeron P) includes a representation of both processes of the diffraction type and inelastic processes, which give the main contribution to the total cross sections. As has been

240

Fig. B.1

well known since the work of Amati, Fubini, and Stanghellini [1], pomeron exchange can be described in the language of Feynman diagrams by the set of ladder diagrams shown in Fig. B.1.

A characteristic property of the inelastic processes at asymptotic energies that are described by these diagrams is that the particles produced have a uniform distribution in the rapidity η – the logarithm of the longitudinal momentum of these particles (in the sense of inclusive cross sections), with the exception of the region of longitudinal momenta near the momenta of the colliding particles; this leads to a logarithmic multiplicity of the particles produced, $\bar{n} = a\xi$.

A second important property of these diagrams is the fact that a uniform density appears in each individual event only after a distribution which is actually non-uniform is averaged over an interval of η greater than the characteristic scale $\lambda_0 \sim a^{-1} \sim 1$ determined by a rung of the ladder. The probability of fluctuations of order λ much greater than λ_0 then falls off exponentially.

It has been realized that these properties do not require that the interaction be described literally in terms of ladder diagrams, but may be a result of a more general phenomenon – the absence of large momentum transfers at all stages of the interaction.

It is well known that, in addition to pomeron exchange, multi-pomeron exchange processes, corresponding to branch points in the complex angular momentum plane, give appreciable contributions to the interaction at high energy.

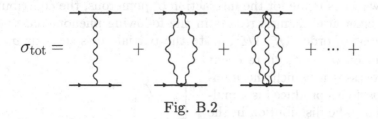

$$\sigma_{\text{tot}} =$$

Fig. B.2

It would be of great interest to determine how allowance for multipomeron exchanges modifies the properties of inelastic processes. In the present work we attempt to analyse this problem. We shall make use of the reggeon diagrammatic technique [2] for the description of multi-pomeron exchanges. The contribution to the total interaction cross section corresponding to the exchange of pomerons can be represented as a series of diagrams of the form shown in Fig. B.2, to which we must add the diagrams that include the mutual interaction of the pomerons.

We shall show that, even if one allows for the exchange of many pomerons and their interactions, it is possible to preserve the first of the above-mentioned properties of inelastic processes – the uniformity of the spectrum (in the sense of inclusive cross sections) and $\bar{n} \sim \xi$.

However, the scale of the averaging for which a uniform density is achieved and the probabilities of fluctuations (in each individual event) turn out to be completely different. This difference from the ladder situation when $\xi \to \infty$ is due entirely to the interaction of reggeons and consists in the fact that a uniform density appears only on averaging over an interval of rapidity $\gtrsim \gamma\sqrt{\xi}$, where γ is determined by the three-pomeron interaction vertex. The probability of fluctuations of order (in the rapidity) $\lambda > \gamma\sqrt{\xi}$ falls off like $\gamma^2\xi/\lambda^2$.

The distribution of large fluctuations is characterized by the following simple property: if we are interested in fluctuations of order $\sim \lambda$, they are separated on the average by a distance $\sim \lambda^2/\gamma^2 \gg \lambda$ in the rapidity. It is of interest to note that, for these large fluctuations, the density of particles is equal to either zero or twice the average density, the second case being encountered twice as frequently as the first.

On the whole, this situation is reminiscent of the behaviour of matter at a second order phase transition, where large fluctuations of the system lead to a state in which separate regions of a substance are in different phases.

The main contents of this work are as follows. Starting from the representation of the total interaction cross section as a set of contributions of reggeon diagrams and assuming that reggeon exchange corresponds to a homogeneous density in the rapidity (in the sense of an inclusive cross section) with an average number of particles $\bar{n} \simeq a\xi$, we show that, if no allowance is made for the interaction of pomerons, the contribution of multi-pomeron exchanges results in the following phenomena: there are corrections of order $1/\xi$, $1/\xi^2$, ... to the partial cross sections σ_n in the main region $n \sim a\xi$; there occur new processes in which the number of particles produced is a multiple of $\bar{n}$. The distribution in the number of particles then has the form [3] shown in Fig. B.3, the cross section being $\sigma_{k\bar{n}} \sim 1/\xi^{k-1}$. We note that, from the point of view of the analogy with gas

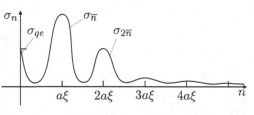

Fig. B.3

models [4, 5], the appearance of oscillations is connected with the finite dimensions of the system (edge effects). It turns out that the corrections to the partial cross sections $\sigma_{\bar{n}}$ and the values of the cross sections $\sigma_{k\bar{n}}$ ($k \neq 1$) are interrelated in such a way that they cancel when the inclusive cross section is evaluated. Let us illustrate this for the case of two-pomeron exchange. This exchange leads to the appearance of new processes with cross sections σ_{qe} and $\sigma_{2\bar{n}}$ and a negative correction $\sigma_{\bar{n}}'$ due to screening (Fig. B.4), where $4\sigma_{qe} = -\sigma_{\bar{n}}' = 2\sigma_{2\bar{n}}$. Then the correction to the inclusive cross section in the central region is

$$\delta \left(\frac{\partial^3 \sigma}{\partial p^3} \right) \sim \partial^3 \sigma_{\bar{n}}' + 2\partial^3 \sigma_{2\bar{n}} = 0. \tag{B.1}$$

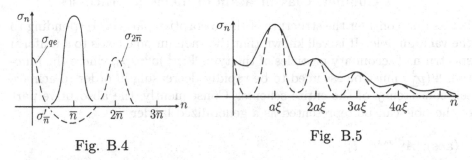

Fig. B.4   Fig. B.5

Thus, if no allowance is made for the interaction of reggeons, the invariant inclusive cross section is

$$f_1(p_\perp, \eta_1, \xi) = (2\pi)^3 2p_0 \frac{\partial^3 \sigma}{\partial p^3}, \tag{B.2}$$

where $\eta = \ln p_0$, the rapidity of the observed particle, is independent of ξ and η to within powers of $1/s$ in the central region.

Allowance for the interaction of reggeons leads to a smoother distribution of σ_n (Fig. B.5). The amplitude of oscillations in the distribution of σ_n is determined by the reggeon interaction constants [3].

The inclusive cross section for the process $p_a + p_b \to p + \{X\}$ in the central region contains the logarithmic corrections

$$f_1 = \sigma_{\text{tot}}(\infty)\Psi(p_\perp^2) \left[1 + \frac{c_1}{\eta_1} + \frac{c_2}{\eta_2} \right], \tag{B.3}$$

where $\eta_1 = \ln(p_a + p)^2$, $\eta_2 = \ln(p_b + p)^2$, and the constants c_i are related to the value of the three-pomeron vertex. If all total cross sections are asymptotically equal [6], f_1 will be universal within the accuracy $1/\eta_i$ and we will then have $c_1 = c_2$.

The correlation function of two particles produced in the central region with rapidities η_1 and η_2 is of the form

$$\rho_2 \simeq \Psi(p_{1\perp}^2) \frac{3\gamma^2}{|\eta_1 - \eta_2|} \Psi(p_{2\perp}^2). \tag{B.4}$$

The last section is devoted to the study of the fluctuations in individual events. In particular, it is shown that the distribution in the number of fluctuations of order $\gtrsim \lambda \gg a^{-1}$ is of the Poisson type, with an average number of fluctuations

$$\overline{m}(\lambda) \sim \gamma^2 \xi / \lambda^2. \tag{B.5}$$

B.1 The absorptive parts of reggeon diagrams in the s-channel. Classification of inelastic processes

Let us first consider the structure of the absorptive parts corresponding to the vacuum pole. It is well known that the vacuum pole leads to a uniform spectrum of secondary particles of the type $\Psi(p_\perp^2) d^3 p / p_0$, where the function $\Psi(p_\perp^2)$ must be assumed to be rapidly decreasing in order to achieve self-consistency of the entire scheme. Consequently, the absorptive part of the pole can be represented as a generalized ladder

$$(\text{abs})_s [A^{(\text{Pow})}(s,t)]$$

$$= (\text{abs})_s \left[\begin{array}{c} \vdots \\ \vdots \end{array} \right] = \sum_n \int d\tau_n |A_{2-n}|^2 = \sum_n \int d\tau_n \left[\begin{array}{c} \vdots \\ \vdots \end{array} \right] \tag{B.6}$$

without specifying the nature of the exchange in the amplitudes $A_{2 \to n}$. The simplest examples of such amplitudes are multi-peripheral ones [1], but this is not essential: the actual amplitudes $A_{2 \to n}$ can also be determined by multi-particle exchange. It is important only that states with large 4-momentum squared be suppressed on the virtual lines. In this case, the intermediate states which appear in the operation $(\text{abs})_s A_{2 \to 2}^{(\text{Pol})}$ will be ordered approximately according to their longitudinal momenta. It is just this fact which is reflected by our method of representing $A_{2 \to n}$ as a "comb".

Adopting such a picture of the absorptive parts for the pole, let us consider the question as to which absorptive parts of the two-reggeon branch cut will be appreciable as $s \to \infty$. The simplest diagram with non-zero third spectral functions has the form

$$A^2 = \quad\quad \simeq \quad\quad \tag{B.7}$$

We see that the dashed line corresponding to the cut of the diagram can be drawn in various ways in calculating $(\text{abs})_s A^2$. Cuts of different types (e. g., l_1, l_2 and l_3 in Fig. B.6) correspond to topologically non-equivalent inelastic processes, where the absorptive parts for most of them (e. g., for l_3 in Fig. B.6) will not be expressed in terms of quantities which are characteristic of the reggeon diagrams.

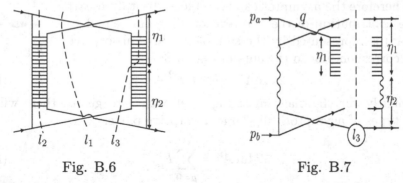

Fig. B.6   Fig. B.7

Let us determine what absorptive parts are appreciable for $\xi = \ln s \to \infty$. It can be seen that the only important cuts of the diagram are those for which the line l which divides the diagram into two parts does not leave the internal part of the reggeon (such as the line l_3 in Fig. B.6). This can be explained qualitatively as follows. The diagrams for the multi-particle amplitudes in which a cut of the type l_3 is made have the form of Fig. B.7. It is clear that $q^2 \sim m^2 e^{\eta_1} \gg m^2$ in the "left-hand" amplitude, since the internal part of the reggeon is significant only for $\eta_1 \gg 1$. Since we are assuming that large q^2 must be suppressed on all the virtual lines, these absorptive parts will be asymptotically small. As to the region of small η_1, it will contribute only to a renormalization of the vertex which enters the absorptive part.[*]

It can be seen that the foregoing property of the absorptive parts refers to arbitrary reggeon diagrams. Cuts of the "sliding" type always lead to a "suspended comb" with a large mass and consequently to large q^2 on the lines in the region where the comb is attached.

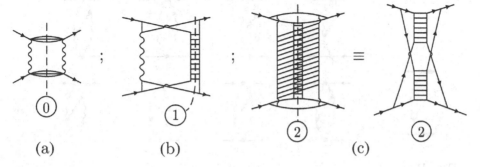

(a)   (b)   (c)

Fig. B.8

---

[*] Note that it is just the contributions of cuts with small η_i which are essential in the cancellation of the leading asymptotic terms in the planar diagrams [7].

Thus, the two-reggeon diagram has absorptive parts of only three types (Fig. B.8), determined by the number of cut reggeons.

Fig. B.9

Therefore the asymptotically complete absorptive part of the contribution of the two-reggeon diagram (equal to $2 \operatorname{Im} A^2$) is represented by the sum of all three absorptive parts corresponding to the cuts in Fig. B.8:

$$2 \operatorname{Im} A^2 = F_0^2 + F_1^2 + F_2^2. \tag{B.8}$$

Similarly, for the diagram of Fig. B.9 with ν reggeons, there will be $\nu + 1$ types of asymptotically large absorptive parts:

$$2 \operatorname{Im} A^\nu = \sum_{\mu=0}^{\nu} F_\mu^\nu, \tag{B.9}$$

where

$$F_\mu^\nu \quad = \quad \tag{B.10}$$

corresponding to μ ($\mu = 0, 1, 2, \ldots$) reggeons being simultaneously cut (the cut reggeons are marked by crosses).

This property of the absorptive parts is readily generalized to arbitrary diagrams with an interaction of the reggeons. We find that there are only a few types of asymptotically large absorptive parts of each reggeon diagram for $A_{2 \to 2}$. For example, in the case of the diagram of Fig. B.10, these parts will be the diagrams of Fig. B.11.

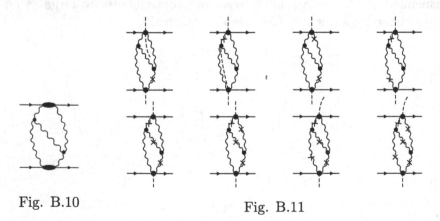

Fig. B.10   Fig. B.11

Let us now consider the question as to what processes correspond to the various absorptive parts of the reggeon diagrams. In the simplest case of the diagram Fig. B.8(a), the absorptive part corresponds to a process in which a small number of particles are produced with momenta close to the momenta of the incident particles – a quasi-elastic process. The absorptive part of the diagram Fig. B.8(b) corresponds to processes that give the main contribution to the cross section like that of the processes described by the absorptive part of the pole term with average multiplicity equal to $a\xi$. It represents a correction to the amplitudes of these basic processes due to screening. The absorptive part of the diagram Fig. B.8(c) corresponds to processes with an average density of particles in a range of rapidity twice as large as for the processes corresponding to the pole term. These absorptive parts lead to processes with the cross sections σ_{qe}, $\sigma'_{\bar{n}}$ and $\sigma_{2\bar{n}}$ (discussed in the Introduction to this Appendix).

Similarly, when more complex diagrams are cut, there occur absorptive parts corresponding to corrections to either the basic processes or processes in which there are no particles in certain ranges of rapidity, while in other ranges the density of particles is a multiple of the density of particles in processes of the basic type.

This result can be described graphically as follows. If we represent the amplitude involving the production of a given number of particles $n \gg 1$ by the set of diagrams

$$\left\{\begin{matrix}\end{matrix} + \begin{matrix}\end{matrix} + \begin{matrix}\end{matrix} + \cdots \right\}_A + \left\{\begin{matrix}\end{matrix} + \begin{matrix}\end{matrix} + \cdots \right\}_B$$

$$+ \left\{\begin{matrix}\end{matrix} + \begin{matrix}\end{matrix} + \cdots \right\}_C + \cdots + \left\{\begin{matrix}\end{matrix} + \begin{matrix}\end{matrix} + \cdots \right\} + \cdots . \quad (B.11)$$

it is obvious that, as functions of the kinematic variables of the particles, the amplitudes corresponding to the classes of diagrams A, B, ... are non-zero in non-overlapping regions of the values of these variables. Therefore processes corresponding to different classes do not interfere when the cross section is evaluated, and the result can be represented symbolically in the form

$$\mathcal{F}_A \mathcal{F}_A^\dagger + \mathcal{F}_B \mathcal{F}_B^\dagger + \cdots . \quad (B.12)$$

B.2 Relations among the absorptive parts of reggeon diagrams

It turns out that in many cases the "large" absorptive parts of a reggeon diagram can be expressed in terms of the value of the contribution of the reggeon diagram itself, so that they differ only in certain combinatorial coefficients (in particular, all the absorptive parts of the diagrams with no interaction between the reggeons are of this type). In other cases, the expression for the absorptive part will also involve new "cut" vertices for the interaction between reggeons.

Consider the diagrams of Fig. B.9. Their contribution to $A_{2\to2}$ can be written in the form of an integral over the two-dimensional transverse momenta of the reggeons [2]:

$$\mathrm{i}\, A^\nu(s, Q^2) = s \int N_\nu[(\mathrm{i}\, D_1)(\mathrm{i}\, D_2)\ldots(\mathrm{i}\, D_\nu)]N_\nu \, \mathrm{d}\Omega_\nu, \qquad (\text{B.13})$$

where

$$\mathrm{d}\Omega_\nu = \frac{1}{\nu!}\delta^{(2)}\left(Q - \sum_i \kappa_i\right)\prod_{i=1}^{\nu}\frac{\mathrm{d}^2\kappa_i}{2(2\pi)^2}$$

is the reggeon phase space, $N_\nu(\kappa_1, \ldots, \kappa_\nu)$ are real vertices for the emission of reggeons, and $D(\xi, \kappa^2)$ are complex reggeon Green functions; comparing a simple pole of positive signature with a reggeon, we have

$$D(\xi, \kappa^2) = \exp(-\alpha'\kappa^2\xi + \alpha(0) - 1)\frac{\exp(-\mathrm{i}\,\pi\alpha(\kappa^2)/2)}{\sin\pi\alpha(\kappa^2)/2}. \qquad (\text{B.14})$$

The quantities N_ν and $\mathrm{d}\Omega_\nu$ entering (B.13) are real and, since we shall be interested in the absorptive parts of A^ν in s, it is convenient to write (B.13) in the symbolic form

$$[-\mathrm{i}\,(\mathrm{i}\, D_1)(\mathrm{i}\, D_2)\ldots(\mathrm{i}\, D_\nu)]. \qquad (\text{B.13a})$$

We have omitted the quantities N_ν here, since they are not changed when the absorptive parts of A^ν are evaluated (see the Appendix to this Appendix).

The calculation of the absorptive parts of the amplitudes A^ν is in essence combinatorial in character. Let us illustrate this for the case of the two-reggeon branch cut:

$$A^2 = [-\mathrm{i}\,(\mathrm{i}\, D_1)(\mathrm{i}\, D_2)]. \qquad (\text{B.15})$$

The quantity F_0^2 is equal to the absorptive part in the case in which the cut passes through the reggeons; we obviously have two possibilities:

$$(\text{B.16})$$

Accordingly, we obtain from (B.15)

$$F_0^2 = [(\mathrm{i}\, D_1)(\mathrm{i}\, D_2)^\dagger] + [(\mathrm{i}\, D_2)(\mathrm{i}\, D_1)^\dagger] = 2[\mathrm{Re}\, D_1 \,\mathrm{Re}\, D_2 + \mathrm{Im}\, D_1 \,\mathrm{Im}\, D_2]. \qquad (\text{B.16a})$$

The absorptive parts of F_1^2 correspond to a "cut" of one of the reggeons; we have four possibilities:

$$1 \left\{ \begin{matrix} \ \end{matrix} \right\} 2 \ + \ 2 \left\{ \begin{matrix} \ \end{matrix} \right\} 1 \ + \ 1 \left\{ \begin{matrix} \ \end{matrix} \right\} 2 \ + \ 2 \left\{ \begin{matrix} \ \end{matrix} \right\} 1 \qquad (\text{B.17})$$

From (B.15) we obtain

$$F_1^2 = [(-\mathrm{i}\,\delta D_1)(\mathrm{i}\,D_2)^\dagger] + [(\mathrm{i}\,D_2)(-\mathrm{i}\delta D_1)]$$
$$+ [(\mathrm{i}\,D_1)(-\mathrm{i}\,\delta D_2)] + [(-\mathrm{i}\,\delta D_2)(\mathrm{i}\,D_1)^\dagger] = -8[\operatorname{Im} D_1 \operatorname{Im} D_2],$$
$$(\text{B.17a})$$

where $\delta D = 2\mathrm{i} \operatorname{Im} D$ is the discontinuity of the amplitude D across its right-hand cut. If the two reggeons are cut at the same time, we have only a single possibility (Fig. B.12). From (B.15) we find

$$F_2^2 = [(-\mathrm{i}\,\delta D_1)(-\mathrm{i}\,\delta D_2)] = 4[\operatorname{Im} D_1 \operatorname{Im} D_2]. \qquad (\text{B.18})$$

Combining (B.16a), (B.17a) and (B.18), we have

$$F_0^2 + F_1^2 + F_2^2 = 2[\operatorname{Re} D_1 \operatorname{Re} D_2 - \operatorname{Im} D_1 \operatorname{Im} D_2]. \qquad (\text{B.19})$$

This obviously coincides with the expression for $2 \operatorname{Im} A^2$ which is found directly from (B.15). The relation (B.19) shows that the sum $F_0^2 + F_1^2 + F_2^2$ actually "saturates" the quantity $2 \operatorname{Im} A^2$. This is not a trivially obvious result – the Feynman diagrams indicate (see the preceding section) that A^2 could also have absorptive parts of other types. Our argument that such absorptive parts are asymptotically small was based on the fact that there would be large q_i^2 on the virtual lines in this case. But it is in fact the cut-off of large q_i^2 which led to a structure of the integrand in (B.13) which is factorizable in D_i^2 and for which (B.8) is satisfied exactly.

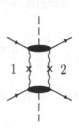

Fig. B.12

In an analogous way, one obtains the absorptive parts of F_μ^ν, corresponding to a cut of μ reggeons in the contribution (B.13). Let $\mu \neq 0$ at first; then, from a cut of μ reggeons in (B.13), there appears a factor

$$\prod_{\beta=1}^{\mu}(-\mathrm{i}\,\delta D_\beta) = \prod_{\beta=1}^{\mu}(2 \operatorname{Im} D_\beta),$$

while each reggeon which is not cut can appear to the right or left of the cut; this gives a factor

$$\prod_{\gamma=\mu+1}^{\nu} [(\mathrm{i}\,D_\gamma) + (\mathrm{i}\,D_\gamma)^\dagger] = \prod_{\gamma=\mu+1}^{\nu} [-2 \operatorname{Im} D_\gamma].$$

Hence

$$F_\mu^\nu = \sum_{(\mu)} \prod_{\beta=1}^{\nu} (-1)^{\nu-\mu} (2 \operatorname{Im} D_\beta), \tag{B.20}$$

where the summation is taken over all possible sets of cut reggeons.
  Since

$$\operatorname{Im} D_\beta = \pm \exp[\xi(\alpha_\beta(0) - 1 - \alpha' \chi_\beta^2)],$$

all the terms in (B.20) are identical. Since we can choose μ reggeons from the set of ν reggeons in $C_\nu^\mu = \nu!/\mu!(\nu-\mu)!$ ways, we finally obtain ($\mu \neq 0$)

$$F_\mu^\nu = (-1)^{\nu-\mu} C_\nu^\mu \prod_{\beta=1}^{\nu} (2 \operatorname{Im} D_\beta). \tag{B.21}$$

The expression for F_0^ν can be written in the form

$$F_0^\nu = \prod_{\beta=1}^{\nu} [(\mathrm{i}\, D_\beta) + (\mathrm{i}\, D_\beta)^\dagger] - \prod_{\beta=1}^{\nu} (\mathrm{i}\, D_\beta) - \prod_{\beta=1}^{\nu} (\mathrm{i}\, D_\beta)^\dagger,$$

where the last two terms take into account the fact that all the reggeons cannot appear on one side of the cut.
  Finally, we have

$$F_0^\nu = (-1)^\nu \prod_{\beta=1}^{\nu} (2 \operatorname{Im} D_\beta) + 2 \operatorname{Im} \left[-\mathrm{i} \prod_{\beta=1}^{\nu} (\mathrm{i}\, D_\beta) \right]. \tag{B.22}$$

As above, we obtain from (B.21) and (B.22)

$$\sum_{\mu=1}^{\nu} \mu F_\mu^\nu(s, t) = 2 \operatorname{Im} \left[(-\mathrm{i}) \prod_{\beta=1}^{\nu} (\mathrm{i}\, D_\beta) \right], \tag{B.23}$$

which coincides with $2 \operatorname{Im} A^\nu$. We note that from (B.21) follow the relations

$$\left. \begin{aligned} \sum_{\mu=1}^{\nu} \mu F_\mu^\nu(s, t) &= \left[(-1)^\nu \prod_{\beta=1}^{\nu} (2 \operatorname{Im} D_\beta) \right] \sum_{\mu=1}^{\nu} (-1)^\mu \mu C_\nu^\mu = 0, \\ \sum_{\mu=2}^{\nu} \mu(\mu - 1) F_\mu^\nu(s, t) &= 0, \\ \cdots\cdots\cdots\cdots\cdots\cdots\cdots & \\ \sum_{\mu=m}^{\nu} \mu(\mu - 1) \ldots (\mu - m + 1) F_\mu^\nu(s, t) &= 0, \end{aligned} \right\} \tag{B.24}$$

which are of great importance in calculating the corrections to the inclusive cross sections.

The generalization to arbitrary reggeon diagrams is obvious. In fact, the contribution of each diagram can be written in the form of an integral of a product of the D_i and the vertex functions, similar to (B.13). The integration then goes over the energy variables for the individual reggeons. Clearly, we can always isolate a "complex" part of the integrand of the type (B.13a). Further, the absorptive parts can be determined by analogy with our considerations for the case in which there is no interaction of the reggeons. However, there is one essential difference. The cut generally passes through a number of vertices for the interaction between reggeons, and it is necessary to know what happens to these vertices when they are "cut".

The vertices N_ν are not changed when they are cut, for arbitrary diagrams in perturbation theory. This means, firstly, that the exact vertices N_ν that take into account the interaction between the reggeons are also not changed when cut and, secondly, that $\Gamma_{1\to\nu}$, the vertices for the transition of a single reggeon into ν reggeons (see Fig. B.13), are not changed when cut. All the remaining vertices, namely $\Gamma_{\nu_1\to\nu_2}$ with $\nu_1, \nu_2 \geq 2$ (Fig. B.14), are in general changed when cut.

Fig. B.13   Fig. B.14

This is shown by an analysis of the Feynman diagrams for $\Gamma_{\nu_1\to\nu_2}$, as well as by arguments of the following type. Let us consider the simplest diagrams for $\Gamma_{2\to2}$ in reggeon perturbation theory (the constant r is small):

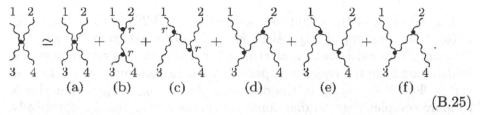

$$(B.25)$$

We see that, if all the reggeons are cut (or at least 1 and 2, or 3 and 4), all of the diagrams (a)–(f) contribute to the cut $\Gamma_{2\to2}$. But if, for example, the cut passes between the reggeons, then of the diagrams that are drawn only (a) and (b) contribute; the contributions of the remaining diagrams will be asymptotically small. The relation between the values of the cut vertices $\Gamma_{2\to 2}$ will therefore depend on the value of r.

Thus, in order to calculate the absorptive parts of arbitrary reggeon diagrams in the general case, it is also necessary to know the values of the cut vertices $\Gamma_{\nu_1\to\nu_2}$ for $\nu_1, \nu_2 \geq 2$.

B.3 Inclusive cross sections

It is convenient to describe the inclusive processes $p_a + p_b \to p + \{X\}$, $p_a + p_b \to p_1 + p_2 + \{X\}$ etc. in terms of the invariant inclusive cross sections

$$f_1(p) = (2\pi)^3 2p_{10}\left(\frac{\partial^3\sigma}{\partial p^3}\right), \quad f_2(p_1,p_2) = (2\pi)^6 2p_{10}p_{20}\left(\frac{\partial^6\sigma}{\partial p_1^3\partial p_2^3}\right), \quad \text{etc.}$$

For the arguments of f_1 we shall choose the quantities $p_\perp^2$, $\eta_1 = \ln(p_a+p)^2$, $\eta_2 = \ln(p_b + p)^2$, $\eta_1 + \eta_2 = \xi$, and we shall confine our analysis to the central region, where $\eta_1, \eta_2 \to \infty$. Then the main contribution to f_1 which does not fall off with energy comes from the diagram of Fig. B.15 [8, 9] and has the form

$$f(p_\perp^2, \infty, \infty) = g_a(0)\Psi(p_\perp^2)g_b(0). \tag{B.26}$$

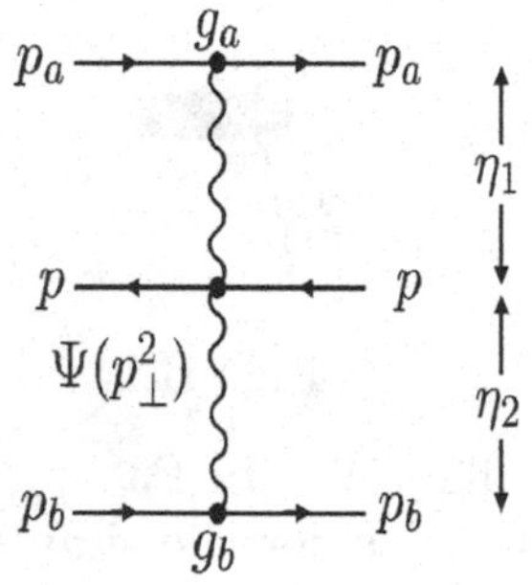

Fig. B.15 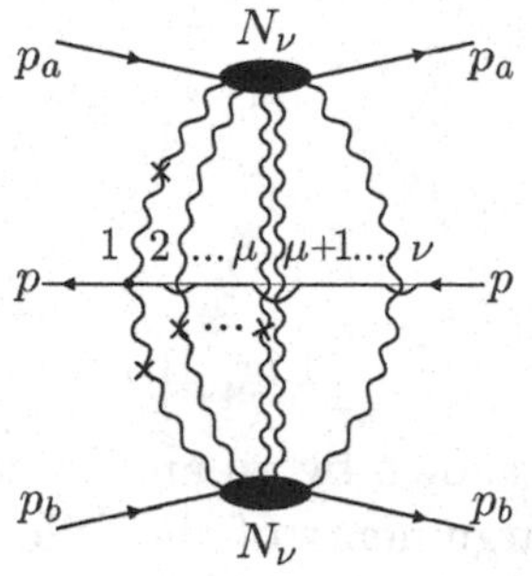   Fig. B.16

The meaning of the quantities appearing in (B.26) is clear from Fig. B.15. The diagram of Fig. B.15 for f_1 is obtained from the reggeon diagram for forward elastic scattering after taking its absorptive part and "extracting from the reggeon" a particle with momentum p, with the new vertex $\Psi(p_\perp^2)$ appearing in the diagram. An analogous procedure leads to more complex reggeon diagrams: we choose one of the asymptotically large absorptive parts of the diagram for $A_{2\to 2}$ and join the vertex $\Psi(p_\perp^2)$ to one of the "cut" lines. Since the other lines in this reggeon diagram can be either cut or not cut, we must also sum over all these possibilities.

Thus, we arrive at reggeon diagrams for f_1 (Fig. B.16) in which, of the ν extracted reggeon lines, μ are cut. Actually, we are considering the general case, since the vertices N_ν may themselves contain arbitrary reggeon diagrams. In order to obtain the contribution of the diagram of Fig. B.16 to f_1, we must perform the operation

$$(\text{abs})_s[-\mathrm{i}\,(\mathrm{i}\,D_1)(\mathrm{i}\,D_2)\ldots(\mathrm{i}\,D_\nu)] \tag{B.27}$$

on the Green functions of the isolated reggeons and, for one of the cut lines, make the substitution

$$\operatorname{Im} D \to (\operatorname{Im} D)\Psi(p_\perp^2)(\operatorname{Im} D). \tag{B.28}$$

Since we can make this substitution for each cut line, an additional factor μ appears in this contribution. Then, taking into account (B.21), we obtain an expression for the contribution of the diagram of Fig. B.16 to f_1:

$$(f_1)^{\nu,\mu} = (-1)^{\nu-\mu}\mu C_\nu^\mu$$

$$\times \int d\Omega_\nu N_\nu \left[(2\operatorname{Im} D_1\Psi(p_\perp^2)\operatorname{Im} D_1)(2\operatorname{Im} D_2)(2\operatorname{Im} D_3)\ldots(2\operatorname{Im} D_\nu)\right] N_\nu. \tag{B.29}$$

It was assumed here that the vertices N_ν do not change when they are cut. We obtain from this the relation

$$\sum_{\mu=1}^{\nu}(f_1)^{\nu,\mu} = 0, \quad \nu \ge 2, \tag{B.30}$$

analogous to (B.24), from which it follows that the diagrams of Fig. B.16 do not contribute to f_1 at all. There remains only the contribution of the diagram of Fig. B.15 with the exact Green functions and vertices[†] (we shall consider this contribution in the following section). If we imagine that the reggeons do not interact with each other, then f_1 is given by (B.26) and the corrections to f_1 which fall off like powers of $1/\eta_i$ are generally absent – the first non-vanishing corrections to (B.26) are due to the non-vacuum trajectories in the diagram of Fig. B.15; but at the same

---

[†] It is interesting to note that this implies at once that it is not possible to have eikonal models [10] with a bare trajectory $\alpha(0) > 1$. By virtue of the above-mentioned cancellation in f_1, only the pole term would remain, i.e., $f_1(\xi,\eta,p_\perp^2) \sim \Psi(p_\perp^2)^{\alpha(0)-1}$ in the central region. Let us now choose in the sum rule associated with the conservation of energy,

$$\sigma_{\text{tot}}^{-1}\int d^2p_\perp \int_{\eta_1}^{\eta_2} d\eta\, e^{\eta-\xi} f_1(p_\perp^2,\eta,\xi) \ge 1,$$

an upper limit $\eta_2 = c\xi$ such that $1 > c > 2-\alpha(0)$. Then, on the one hand, we remain in the central region and, on the other hand, we find a violation of the sum rule if $\alpha(0) - 1 > 0$.

time there are logarithmic corrections in the total cross sections, owing to the diagrams of Fig. B.9. In fact, the interaction between the reggeons is not equal to zero. This leads to additional contributions to f_1 (apart from those of Figs. B.15 and B.16), which correspond to the situation in which the "observed" particle is "extracted" from the reggeon interaction vertex. This mechanism of particle production leads, for example, to the fact that the diagrams for f_1 shown in Fig. B.18 are obtained from the diagram of Fig. B.17 for $A_{2\to2}$. The contributions of these diagrams to f_1 are small ($\sim 1/\eta_1^2\eta_2, 1/\eta_1\eta_2^2$).

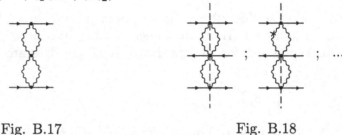

$$\text{Fig. B.17} \qquad\qquad\qquad \text{Fig. B.18}$$

Let us now consider the question as to which diagrams will be important for the double inclusive cross sections $f_2(p_1, p_2)$. The arguments of f_2 are

$$\eta_1 = \ln(p_a + p_1)^2, \quad \eta_2 = \ln(p_1 + p_2)^2, \quad \eta_3 = \ln(p_2 + p_b)^2,$$
$$\eta_1 + \eta_2 + \eta_3 = \xi, \quad p_{1\perp} \quad \text{and} \quad p_{2\perp}.$$

When all the η_i are large, the main contribution to f_2 comes from the diagrams of Fig. B.19 and is of the form

$$g_a \Psi(p_{1\perp}^2)\Psi(p_{2\perp}^2)g_b = \frac{1}{\sigma_{\text{tot}}^{(ab)}} f_1(p_1)f_2(p_2). \tag{B.31}$$

We shall now ascertain which diagrams will give corrections to (B.31). Consider first the diagrams of Fig. B.20 for it, which are analogous to those of Fig. B.16, in which we must sum over the number of cut reggeons, just as in Fig. B.16.

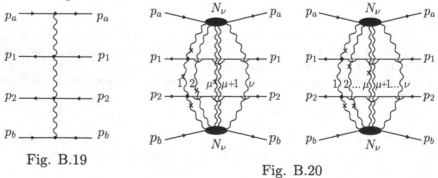

$$\text{Fig. B.19} \qquad\qquad\qquad\qquad \text{Fig. B.20}$$

Using (B.24), it is easy to see that the contributions of all these diagrams (with the exact N_ν) cancel, as in (B.30), apart from the pole term of Fig. B.19 and the two-pomeron contribution of Fig. B.21.

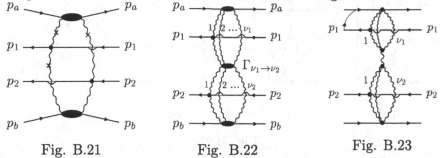

Fig. B.21   Fig. B.22   Fig. B.23

For f_2 there are also the diagrams of Fig. B.22, in which the reggeons interact "between" particles 1 and 2.

Since the vertices $\Gamma_{\nu_1 \to \nu_2}$ with $\nu_1, \nu_2 \geq 2$ are in general changed when they are cut, the diagrams of Fig. B.22 contribute to f_2, except for the cancelling diagrams of Fig. B.23.

Analogously with the case of f_1, there also remain the contributions of the diagrams in which the "observed" particles 1 and 2 are emitted at the point of interaction of the reggeons (Fig. B.24).

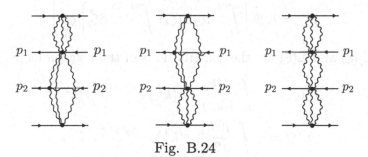

Fig. B.24

B.4 Main corrections to the inclusive cross sections in the central region

As we have seen in the preceding section, the contribution to f_1 which does not fall off as $\eta_1, \eta_2 \to \infty$ and the main corrections to it are determined by the diagram of Fig. B.15 with the exact vertices g and the exact vacuum Green functions D. In setting ourselves the task of finding the main corrections to $f_1(\infty)$ of order $1/\eta_i$, we may confine ourselves to the terms of g that are linear and the terms of D that are quadratic in the three-reggeon vertex r (we recall that $r = 0$ at $k_{i\perp} = 0$ [11]). It is therefore sufficient for us to determine the total contribution to f_1 from the diagrams of Fig. B.25.

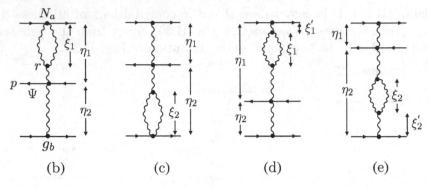

(b)   (c)   (d)   (e)

Fig. B.25

The contributions of the diagrams Figs. B.25(b–e) to f_1 are of the form

$$
\left.\begin{aligned}
f^{(b)} &= \left[\int_1^{\eta_1} \mathrm{d}\xi_1 \Sigma_a'(\xi_1)\right] \Psi g_b, \\[2mm]
f^{(c)} &= g_a \Psi \left[\int_1^{\eta_2} \mathrm{d}\xi_2 \Sigma_b'(\xi_2)\right], \\[2mm]
f^{(d)} &= g_a \left[\int_1^{\eta_1} \mathrm{d}\xi_1 \Sigma(\xi_1) \int_1^{\eta_1-\xi_1} \mathrm{d}\xi_1'\right] \Psi g_b, \\[2mm]
f^{(e)} &= g_a \Psi \left[\int_1^{\eta_2} \mathrm{d}\xi_2 \Sigma(\xi_2) \int_1^{\eta_2-\xi_2} \mathrm{d}\xi_2'\right] g_b,
\end{aligned}\right\} \tag{B.32}
$$

where $\Sigma_i'(\xi)$ and $\Sigma(\xi)$ are the contributions of the reggeon loops,

$$
\begin{aligned}
\Sigma_i'(\xi) &= -\int \frac{\mathrm{d}^2 k_\perp}{4(2\pi)^2} N_i(k_\perp^2)\, e^{-2\alpha' k_\perp^2 \xi}\, r(k_\perp^2), \\[2mm]
\Sigma(\xi) &= -\int \frac{\mathrm{d}^2 k_\perp}{4(2\pi)^2} r(k_\perp^2)\, e^{-2\alpha' k_\perp^2 \xi}\, r(k_\perp^2).
\end{aligned} \tag{B.33}
$$

For small $k_\perp^2$,

$$
r(k_\perp^2) \simeq 2\beta\alpha' k_\perp^2, \quad N_i(k_\perp^2) \simeq N_i.
$$

Hence for large ξ we obtain from (B.26)

$$
\Sigma_i'(\xi) \simeq -\frac{1}{\xi^2} \frac{N_i \beta}{32\pi\alpha'}, \quad \Sigma(\xi) \simeq -\frac{1}{\xi^3} \frac{\beta^2}{16\pi\alpha'}. \tag{B.33a}
$$

The expressions (B.32) for $f^{(b)}, \ldots, f^{(e)}$ must then be renormalized, which corresponds to extracting the polynomial parts in η_i from them. This renormalization is unambiguous, and by combining the extracted polynomials with the contribution of the pole diagram we obtain the renormalized pole term

$$
\tilde{f}^{(a)} = g_a \Psi g_b.
$$

It is simplest to carry out the renormalization of the integrals in (B.32) with the aid of the identities

$$
\left.
\begin{aligned}
&\int_1^{\eta_1} d\xi_1 \Sigma'(\xi_1) = \int_1^{\infty} d\xi_1 \Sigma'(\xi_1) - \int_{\eta_1}^{\infty} d\xi_1 \Sigma'(\xi_1), \\
&\int_1^{\eta_1} d\xi_1 \Sigma(\xi_1) \int_1^{\eta_1-\xi_1} d\xi_1' = \int_1^{\infty} (\eta_1 - \xi_1)\Sigma(\xi_1)d\xi_1 + \int_{\eta_1}^{\infty} d\xi_1 (\xi_1 - \eta_1)\Sigma(\xi_1).
\end{aligned}
\right\} \quad \text{(B.34a)}
$$

Considering the behaviour of $\Sigma(\xi)$ and $N(\xi)$ as $\xi \to \infty$, we note that the last integrals in (B.34a) no longer contain polynomial parts. The renormalized expressions for $\tilde{f}^{(i)}$ take the form

$$
\left.
\begin{aligned}
\tilde{f}^{(a)} &= g_a \Psi g_b, \\
\tilde{f}^{(b)} &= -\left[\int_{\eta_1}^{\infty} d\xi_1 \Sigma_a'(\xi_1)\right] \Psi g_b \simeq N_a \left[\frac{\beta}{32\pi\alpha'\eta_1}\right] \Psi g_b, \\
\tilde{f}^{(c)} &= -g_a \Psi \left[\int_{\eta_2}^{\infty} d\xi_2 \Sigma_b'(\xi_2)\right] \simeq g_a \Psi \left[\frac{\beta}{32\pi\alpha'\eta_2}\right] N_b, \\
\tilde{f}^{(d)} &= g_a \left[\int_{\eta_1}^{\infty} d\xi_1 (\xi_1 - \eta_1)\Sigma(\xi_1)\right] \Psi g_b \simeq g_a \left[\frac{-\beta^2}{32\pi\alpha'\eta_1}\right] \Psi g_b, \\
\tilde{f}^{(e)} &= g_a \Psi \left[\int_{\eta_2}^{\infty} d\xi_2 (\xi_2 - \eta_2)\Sigma(\xi_2)\right] g_b \simeq g_a \Psi \left[\frac{-\beta^2}{32\pi\alpha'\eta_2}\right] g_b.
\end{aligned}
\right\} \quad \text{(B.35a)}
$$

Combining these contributions, we obtain

$$
f_1 = g_a \Psi(p_\perp^2) g_b \left(1 + \frac{c_a}{\eta_1} + \frac{c_b}{\eta_2}\right), \tag{B.36}
$$

$$
c_i = \frac{\beta}{32\pi\alpha' g_i}[N_i - g_i\beta], \quad i = a, b. \tag{B.37}
$$

To the same accuracy as above, i.e., neglecting all terms of higher order in $1/\eta_i$, (B.37) can be further simplified. This follows from the fact that a number of additional conditions are imposed on the vertices for the interaction of the vacuum reggeons with the particles when the total cross sections are constant. First of all, all vertices for diffraction production must reduce to zero when the pomeron momentum tends to zero. The N_i therefore take the eikonal form $N_i = g_i^2$; this implies that

$$
c_i = \frac{\beta}{32\pi\alpha'}(g_i - \beta).
$$

Moreover, it has been shown [6] that, under the same assumptions, all the $g_i(0)$ are equal: $g_i^2 = \sigma_{\text{tot}}(\infty)$, which is independent of the type of colliding hadrons. In this case, all the c_i are equal,

$$c_i = c = \frac{\beta}{32\pi\alpha'}[\sqrt{\sigma_{\text{tot}}(\infty)} - \beta], \tag{B.38}$$

and f_1 takes a symmetric form in η_i:

$$f_1(p_\perp, \eta_1, \eta_2) = \sigma\Psi(p_\perp^2)\left[1 + c\left(\frac{1}{\eta_1} + \frac{1}{\eta_2}\right)\right]. \tag{B.39}$$

It is obvious that f_1 will have a maximum at $\eta_1 = \eta_2 = \xi/2$ if $c < 0$ and a minimum if $c > 0$; in the variables $\xi = \eta_1 + \eta_2, y = (\eta_1 - \eta_2)/2$, we have

$$f_1(p_\perp, y, \xi) = \sigma_{\text{tot}}\Psi(p_\perp^2)\left(1 + \frac{4c\xi}{\xi^2 - 4y^2}\right). \tag{B.40}$$

We note only that the sign of c cannot be determined at present from theoretical considerations.

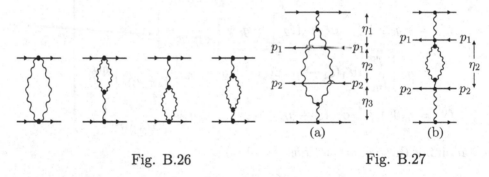

Fig. B.26   Fig. B.27

Let us compare the main asymptotic corrections to f_1 with the corrections of order $1/\xi$ to σ_{tot} which are determined by the diagrams of Fig. B.26. Renormalizing their contributions as in (B.35), we obtain

$$\sigma_{\text{tot}}(\xi) \simeq \sigma_{\text{tot}}(\infty)\left[1 - \frac{1}{32\pi\alpha' g_a g_b}\frac{1}{\xi}(N_a - g_a\beta)(N_b - g_b\beta)\right]$$

$$\rightarrow \sigma_{\text{tot}}(\infty)\left[1 - \frac{(g - \beta)^2}{32\pi\alpha'\xi}\right] \equiv \sigma_{\text{tot}}(\infty)\left(1 - \frac{32\pi\alpha' c^2}{\xi\beta^2}\right). \tag{B.41}$$

Let us now consider the main corrections to the double inclusive cross section f_2, restricting ourselves here to the region $\eta_1, \eta_3 \gg \eta_2 \gg 1$. Thus, we are seeking the main corrections in $1/\eta_2$. It is readily seen that contributions to $f_2 \sim 1/\eta_2$ can come only from the diagrams of Fig. B.27. The contribution of the diagram (a) of Fig. B.27a at large η_2 is easily evaluated and has the form

$$
\begin{aligned}
f_2^a &= g_a \Psi(p_{1\perp}^2) \left[\frac{\beta^2}{4\pi\alpha'} \int_1^{\eta_1} d\xi_1 \int_1^{\eta_2} \frac{d\xi_2}{(\eta_2 + \xi_1 + \xi_2)^3} \right] \Psi(p_{2\perp}^2) g_b \\
&= g_a \Psi(p_{1\perp}^2) \left[\frac{\beta^2}{8\pi\alpha'\eta_2} \right] \Psi(p_{2\perp}^2) g_b.
\end{aligned}
\tag{B.42}
$$

The contribution of the diagram (b) of Fig. B.27 is determined by analogous integrals, and we have the result

$$
f_2^b = -\frac{1}{4} f_2^a.
$$

Then from (B.31) and (B.42) we obtain an expression for the two-particle correlator

$$
\rho_2 \equiv \frac{f_2(p_1, p_2)}{\sigma_{tot}} - \frac{f_1(p_1)}{\sigma_{tot}} \frac{f_1(p_2)}{\sigma_{tot}} = \frac{1}{\eta_2} \frac{3\beta^2}{32\pi\alpha'} \Psi(p_{1\perp}^2)\Psi(p_{2\perp}^2).
\tag{B.43}
$$

It is of interest to note that the sign of ρ_2 for $\eta_2 \gg 1$ is uniquely determined (as positive).

B.5 Fluctuations in the distribution of the density of produced particles

When $\xi \to \infty$, the average number of particles produced in the interaction is large, $\bar{n} \simeq a\xi$. The final state which appears in an individual event can therefore be described by the density of the number of particles $\nu(\eta)$ in rapidity space. The total number of particles produced,

$$
n(\xi) = \int_1^\xi d\eta\, \nu(\eta),
\tag{B.44}
$$

will then fluctuate from event to event. As we have already discussed in the Introduction to this Appendix, the "incorporation" of branch cuts leads to a rather non-trivial structure (of the form of Fig. B.5) in the distribution for the quantity $n(\xi)$. In what follows, we shall attempt to ascertain how frequently the functions $\nu(\eta)$ of a given form appear[‡] and what physical mechanism is responsible for such fluctuations.

[‡] With mathematical rigour, the probabilities that given $\nu(\eta)$ appear are given by the variations of a certain functional $W[\nu(\eta)]$, which in turn can be expressed in terms of the set of all higher inclusive cross sections. However, we shall not dwell on this point here, but shall confine ourselves to the more "transparent" considerations in the text.

The quantity $\overline{\nu(\eta)}$, averaged over many events, can obviously be expressed in terms of the inclusive cross section $f_1(p_\perp, \eta, \xi)$:

$$\overline{\nu(\eta)} = \frac{1}{\sigma_{\text{tot}}} \int f_1(\eta, p_\perp, \xi) \mathrm{d}^2 p_\perp. \tag{B.45}$$

Each cut pomeron of "length ξ_i" contains $a\xi_i$ particles on the average. The average distance between the particles is therefore $\sim a^{-1}$. Clearly, the density $\nu(\eta)$ is meaningful only after averaging over a length which is large in comparison with a^{-1}. This averaging will be implied.

There are two mechanisms which produce a deviation of the function $\nu(\eta)$ from its asymptotic average value $\overline{\nu}(\eta)$. The first is the small distance correlation in the pomeron, which leads to fluctuations in $\nu(\eta)$ with a period of the order of several units of a^{-1}. This correlation is due to the non-vacuum reggeons and, as we have already mentioned, we shall assume that an average is carried out over these fluctuations.[§] The second mechanism is due to the pomeron cuts; it leads to long range fluctuations in $\nu(\eta)$, with periods up to ξ. We shall consider these fluctuations.

What kind of function $\nu(\eta)$ can appear? We saw in Section B.1 that all the inelastic processes associated with the absorptive parts of reggeon diagrams can be divided into topologically non-equivalent classes. We can see at once that to each of them there corresponds a function $\nu(\eta)$, which has a step form and takes integral values at each point (if η is measured in units of a^{-1}) equal to the number of cut reggeons "for the given η".[¶]

For example:

$$\tag{B.46}$$

[§] The fluctuations in $\nu(\eta)$ due to the non-vacuum reggeons have a Poisson-like character (see, e. g., [11]) and fall off exponentially with increasing η.

[¶] The system which we are considering has the properties of a one-dimensional gas placed in a volume ξ. A number of properties of this "Feynman gas" were discussed in the literature (see [4, 5]). The non-exponential fall-off with distance η_2 of the correlation function ρ_2, shows that the case of constant total cross sections corresponds to the situation in which the "Feynman gas" is at the critical point. From the point of view of the gas analogy, the step function $\nu(\eta)$ corresponds to the fact that, at the critical point, the fluctuations divide the system into regions which are in different "phases". In this sense, there exist "phases" for the "Feynman gas" with different densities $\nu = 0, 1, 2, \ldots$ However, if $\alpha(0) < 1$ for the pomeron, the long range fluctuations in $\nu(\eta)$ are cut off like $(\eta_1 - \eta_2)^{-1} \exp[-(1 - \alpha(0))|\eta_1 - \eta_2|]$. The quantity $1 - \alpha(0)$ can then be interpreted as $|T_{\text{crit}} - T|^{1/2}$.

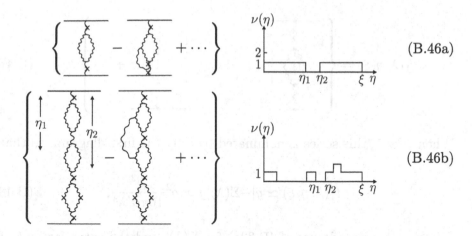

The functions $\nu(\eta)$ which are represented correspond to the absorptive parts of reggeon diagrams with bare pomerons. In fact, it is only when a bare pomeron (with a finite range of correlation in the ladder) is cut that we obtain an η-independent contribution to $\nu(\eta)$ in the interval $\xi_1 < \eta < \xi_2$, where $|\xi_2 - \xi_1|$ is the energy invariant of the cut pomeron.

However, for the bare pomeron we must have $\alpha(0) < 1$ [12]. The probability of finding the homogeneous configuration (B.46) therefore appears as a power of s. This also applies to any configuration of $\nu(\eta)$ with fixed positions of all the steps. It is thus a more important problem to determine the average number of different inhomogeneities in $\nu(\eta)$, corresponding to the dominant events.

We shall first illustrate the method which we employ for the case of the simplest inhomogeneity in $\nu(\eta)$, namely the hole shown in (B.46a). It is convenient to introduce $\hat{f}_1(\lambda, \eta, \xi)$, the inclusive cross section for producing a hole of extent λ and rapidity η, having in mind the following considerations.

The contribution of the diagrams (B.46a) with fixed η_1 and η_2 and with bare trajectories gives the cross section for producing a hole with boundaries η_1 and η_2 on the background of a homogeneous distribution. Let us also consider the other diagrams with bare pomerons which give functions $\nu(\eta)$ having a hole in the interval (η_1, η_2) and taking arbitrary values outside this range (an example of one of these diagrams is shown in (B.46b)). The total contribution of all such diagrams with bare pomerons and fixed η_1 and η_2 it is natural to identify with $\hat{f}_1(\lambda, \eta, \xi)$. But the sum of these diagrams can, on the other hand, be re-expressed in terms of a sum of the reggeon diagrams with the exact renormalized vertices and pomeron Green functions, for which $\alpha(0) = 1$. We will then have

$$\hat{f}_1(\lambda, \eta, \xi) = \left\{ \quad + \quad + \cdots \right\}. \tag{B.47}$$

When $\lambda \gg 1$, this series is dominated by only the first diagram, so that

$$\hat{f}_1(\lambda, \eta, \xi) \simeq g[-\Sigma(\lambda)]g \simeq g^2 \frac{\beta^2}{16\pi\alpha'\lambda^3}, \tag{B.48}$$

where we have made use of (B.33a) for $\Sigma(\lambda)$, and the extra factor (-1) appeared because $\Sigma(\lambda)$ is cut "between" the reggeons. The diagrams (B.47) for $\hat{f}_1$ are naturally compared with the diagram of Fig. B.15 for f_1, and we see that $(\text{abs})\Sigma(\lambda)$ plays the rôle of the vertex Ψ in the diagram of Fig. B.15. Because of this similarity in the structures of the diagrams for f_1 and $\hat{f}_1$, the conclusions of the preceding sections concerning f_1 carry over naturally to $\hat{f}_1$.

Moreover, the function $\hat{f}_1$, which is an inclusive cross section, must satisfy the normalization relation

$$\int \hat{f}_1(\lambda, \eta, \xi)\mathrm{d}\eta = \sigma_{\text{tot}} \langle m(\lambda) \rangle, \tag{B.49}$$

where $\langle m(\lambda) \rangle \mathrm{d}\lambda$ is the average multiplicity of holes in $\nu(\eta)$ with the size $(\lambda, \lambda + \mathrm{d}\lambda)$. Substituting (B.48) into (B.49), we find

$$\langle m(\lambda) \rangle = \xi[(\text{abs})\Sigma(\lambda)] \simeq \xi \frac{\beta^2}{16\pi\alpha'\lambda^3}. \tag{B.50}$$

This relationship is of interest, firstly because only known quantities appear on the right-hand side of (B.50), and secondly because it provides a new *s*-channel meaning for the self-energy part of $\Sigma(\lambda)$. The foregoing considerations obviously carry over directly to other forms of inhomogeneities in the function $\nu(\eta)$.

The diagrams (B.47), but with both cut reggeons in $\Sigma(\lambda)$ (Fig. B.28), give the inclusive cross section for producing a column (in the interval $\eta_1, \eta_1 + \lambda, \nu(\eta) = 2$).

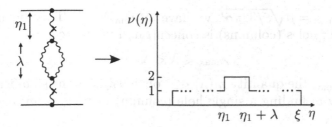

Fig. B.28

Using (B.16a) and (B.18), we obtain at once

$$\langle m(\lambda)\rangle_c = 2\langle m(\lambda)\rangle_h. \tag{B.51}$$

Similarly, the absorptive parts of more complex diagrams for $\Sigma(\lambda)$ give the inclusive cross sections for producing the corresponding inhomogeneities of $\nu(\eta)$.

The inhomogeneities of the hole and column type with $\nu = 2$ obviously dominate in $\nu(\eta)$ for large λ since $\hat{f}_1$ for more complex fluctuations contains a higher power of λ in the denominator.[‖] In what follows, we shall therefore consider only these simplest inhomogeneities of $\nu(\eta)$. The quantity

$$\overline{m}(\lambda) = \int_1^\infty d\lambda' \langle m(\lambda')\rangle \tag{B.52}$$

obviously gives the average multiplicity of holes (columns) with ranges greater than λ. From (B.50) we have

$$\overline{m}(\lambda) = \frac{\xi\beta^2}{32\pi\alpha'\lambda^2}. \tag{B.53}$$

We find from this that the average distances between holes with range $\sim \lambda$ will be

$$\sim \frac{32\pi\alpha'}{\beta^2}\lambda^2 \gg \lambda \quad \text{for} \quad \lambda^2 \gg 1. \tag{B.54}$$

Thus, we see that the holes form a "rarefied gas" for which the average value of the correlator is $\hat{\rho}_2 \sim 1/\lambda^2 \ll 1$. It follows from this that the probability w_m of finding m holes (columns) with ranges $\lambda \gg 1$ in the system has a Poisson distribution with the average number $\overline{m}_h(\lambda)$ $(m_c(\lambda))$:

$$w_m = \frac{e^{-\overline{m}(\lambda)}}{m!}|\overline{m}(\lambda)|^m. \tag{B.55}$$

[‖] We shall not consider here the question of "fine structure" in the columns. The ranges of these inhomogeneities are $\leq \beta\sqrt{\lambda/32\pi\alpha'}$. On "observing" inhomogeneities of $\nu(\eta)$ with ranges $\geq \lambda$, we must therefore average $\nu(\eta)$ over the intervals $0 \lesssim \beta\sqrt{\lambda/32\pi\alpha'}$.

For $\lambda \sim \lambda_{\max} = \beta\sqrt{\xi/32\pi\alpha'}$, we have $\overline{m}(\lambda_{\max}) \sim 1$. This means that the spectrum of holes (columns) is concentrated in the region

$$\lambda_{\max} \gtrsim \lambda \gtrsim \lambda_0 \sim 1. \tag{B.56}$$

For $\lambda \gg \lambda_{\max}$ the quantity $\overline{m(\lambda)} \ll 1$ obviously has the significance of the probability of finding a single hole (column) of range λ in an individual event.

Our discussion shows that, on the average, $\nu(\eta)$ must have (with weight ~ 1) the following structure: there is of the order of one hole (and two columns) with ranges $\sim \lambda_{\max} \sim \beta\sqrt{\xi}/4\sqrt{\pi\alpha'}$, of the order of four holes (and double the number of columns) with ranges $\sim \lambda_{\max}/2$, etc., up to values of λ of the order of the correlation length.

Thus, it is clear that the structure of the real function $\nu(\eta)$ is highly non-uniform and will be only slightly reminiscent of $\nu(\eta) \simeq$ const, which characterizes a simple multi-peripheral chain.**

In conclusion, let us consider what phenomena may be expected in individual multi-particle events at non-asymptotic ξ. When ξ is such that $\lambda_{\max} \sim (\beta/4\sqrt{\pi\alpha'})\sqrt{\xi} \lesssim \lambda_0$, where λ_0 is the correlation length associated with the non-vacuum reggeons ($\lambda_0 \sim 2$), there will be produced mainly events with homogeneous $\nu(\eta)$. But when $(\beta/4\sqrt{\pi\alpha'})\sqrt{\xi}$ becomes greater than λ_0, there will "appear" a hole or column with sizes larger than the correlation length in the individual events (with weight ~ 1). When ξ is increased further, there may appear a second inhomogeneity, etc.

The foregoing behaviour of $\nu(\eta)$ is obviously reminiscent of the observation, which has been discussed for many years (mainly in the literature on cosmic-ray physics), that there are appreciable inhomogeneities in individual multi-particle events; these inhomogeneities are explained (in the same literature) mainly by invoking the hypothesis of fireballs. However, the foregoing considerations show that it may be possible to explain the large inhomogeneities of $\nu(\eta)$ without invoking any new physical ideas.

Appendix

We shall present simple arguments which show that N_ν is not changed for various cuts.

The vertex N_ν arises in calculating the asymptotic form of the Feynman diagrams of Fig. B.29 and can be written in the form [2]

$$N_\nu = \int_{-\infty}^{\infty} \prod_{i=1}^{\nu} \frac{\mathrm{d}(\alpha_i s)\mathrm{d}(\tilde{\alpha}_i s)}{z_i \tilde{z}_i} \delta\left[\sum_{i=1}^{\nu}(\alpha_i + \tilde{\alpha}_i)\right] \left[\prod_{i=1}^{\nu} \frac{\mathrm{d}\beta_i \mathrm{d}^2 p_{i\perp}}{2} g_i \beta_i^{J_i}\right] R, \tag{B.57}$$

** Some of the conclusions of this section coincide with the results contained in a contribution of K. A. Ter-Martirosian to the Batavia conference (September 1972).

where

$$p_i = \alpha_i p_b + \beta_i p_a + p_{i\perp}, \quad k_i = \tilde{\alpha}_i p_b' + \tilde{\beta}_i p_a' + k_{i\perp}, \quad (p_a')^2 \simeq (p_b')^2 \simeq 0,$$
$$z_i = p_i^2 - m^2 + i\varepsilon, \quad \tilde{z}_i = k_i^2 - m^2 + i\varepsilon,$$

J_i are the complex angular momenta of the reggeons, and R is the amplitude for the process shown in Fig. B.30.

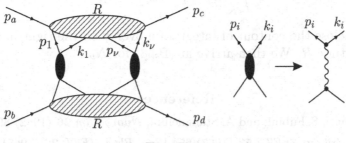

Fig. B.29

The integration in (B.57) is carried out over $\alpha_i, \tilde{\alpha}_i$ with the Feynman rule for avoiding singularities, and over β_i between finite limits ($\beta_i > 0$). It is also important (see [12]) that the factors $\beta_i^{J_i}$ do not lead to any new singularities in the integrals over $\alpha_i, \tilde{\alpha}_i$.

It will be important for us to observe that the amplitude R is integrated over the α variables of all the lines (p_i and k_i) in a completely symmetric manner, independently of the reggeons to which these lines are joined.

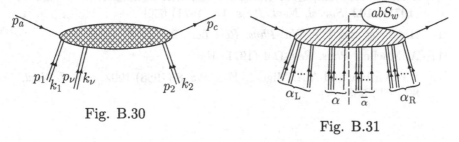

Fig. B.30

Fig. B.31

Let us consider the absorptive part of the diagram of Fig. B.29, when of the ν reggeons μ are cut, μ_1 being to the left of the cut and the remaining number $\mu_2 = \nu - \mu - \mu_1$ to the right.

Then the cut vertex N_ν takes the form

$$N_\nu^{(\mu,\mu_1,\mu_2)} = \int_0^\infty \mathrm{d}\gamma\, \delta(\gamma - (\alpha + \alpha_\mathrm{L})) \int_{-\infty}^\infty \prod_{i=1}^\nu \frac{\mathrm{d}(\alpha_i s)\mathrm{d}(\tilde{\alpha}_i s)}{z_i \tilde{z}_i}$$

$$\times \delta\left(\sum(\alpha_i + \tilde{\alpha}_i)\right) \left[\prod_i \frac{\mathrm{d}\beta_i \mathrm{d}^2 p_{i\perp}}{2} g_i \beta_i^{J_i}\right] [(\mathrm{abs})_W R], \quad (\mathrm{B.58})$$

where $(\text{abs})_W R$ is the absorptive part of R in the variable $W = s\gamma$, corresponding to the re-grouping of the lines shown in Fig. B.31 (α and $\tilde{\alpha}$ are the sets of lines from the cut reggeons).

But the vertex N_ν can itself be written in the same form (B.58). Indeed, we can always multiply the right-hand side of (B.57) by

$$\int_{-\infty}^{\infty} d\gamma \, \delta(\gamma - (\alpha + \alpha_L)),$$

and then close the contour of integration over γ on the right-hand cut of the amplitude R. We then arrive at (B.58) for N_ν.

References

[1] D. Amati, S. Fubini, and A. Stanghellini, *Nuovo Cim* **26** (1962) 896.

[2] V.N. Gribov, *ZhETF* **53** (1967) 654 [*Sov. Phys. JETP* **26** (1968) 414].

[3] V.A. Abramovski and O.V. Kancheli, *ZhETF Pis. Red.* **15** (1972) 559 [*JETP Lett.* **15** (1972) 397].

[4] K. Wilson, Cornell preprint CLNS-131.

[5] J.D. Bjorken, preprint SLAC-PUB-974, 1971.

[6] V.N. Gribov, *Yad. Fiz.* **17** (1973) 603.

[7] S. Mandelstam, *Nuovo Cim.* **30** (1963) 1127, 1148.

[8] A. Mueller, *Phys. Rev.* **D 2** (1970) 2963.

[9] V.A. Abramovski, O.V. Kancheli, and I.D. Mandzhavidze, *Yad. Fiz.* **13** (1971) 1102 [*Sov. J. Nucl. Phys.* **13** (1971) 630].

[10] H. Cheng and T.T. Wu, *Phys. Rev. Lett.* **24** (1970) 1456.

[11] A. Mueller, *Phys. Rev.* **D 4** (1971) 150.

[12] V.N. Gribov and A.A. Migdal, *Yad. Fiz.* **8** (1968) 1002 [*Sov. J. Nucl. Phys.* **8** (1969) 583].

This lecture was previously published in the following.

Yad. Fiz. **18** (1973) 595 [*Sov. J. Nucl. Phys.* **18** (1974) 308].

L. Caneschi (ed.), *Regge Theory of Low p_T Hadronic Interactions*, 1974, pp. 199–208.

V.N. Gribov, *Gauge Theories and Quark Confinement*, Phasis, Moscow, 2002, pp. 67–96.

Appendix C
Theory of the heavy pomeron

V. N. Gribov

The small shrinkage of the diffraction peak is used for studying the pomeron structure in the intermediate non-asymptotic energy region. Even in this region the cross sections are shown to be factorized and basically determined by a pole and by enhanced cuts. The total cross sections rise as $(\ln \ln E)^2$ and the interaction radius as $\ln \ln E$. The general character of inelastic processes is the same as in the asymptotic theory in the case of strong coupling.

C.1 Introduction

A somewhat conflicting situation has developed in describing the strong interactions at high energies on the basis of the complex angular momentum theory and the multi-peripheral model of inelastic processes.

On the one hand, there is a seemingly consistent scheme for describing the total cross sections, elastic and inelastic processes, inclusive spectra etc., which led to a good qualitative description of the experimental data and predicted almost all qualitative phenomena that have been discovered in the last few years, such as the shrinkage of the diffraction peak, scaling, plateau, the nature of diffraction dissociation in the large mass region, a rise in the total cross sections, and others. On the other hand, all attempts to make a quantitative description are handicapped by the fact that the theory contains too many parameters whose choice is mostly arbitrary.

The scheme for describing high energy processes [1, 2] is generally as follows. The energy dependence of the total cross section and the elastic scattering is described as the result of the exchange of one or several pomerons. Each of such exchanges may be described by the reggeon diagrams Fig. C.1. Cross sections of individual inelastic processes are

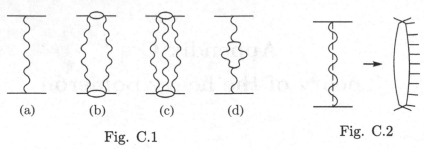

Fig. C.1   Fig. C.2

described by cutting the diagrams (calculating the imaginary part) under the assumption that processes with uniform particle distribution in rapidity correspond to a pole (Fig. C.2). Then all other diagrams describe all kinds of fluctuations in inelastic processes. For instance, the diagram in Fig. C.3(a) describes the fact that both a hole in the density distribution and doubled density are possible over a particular range of rapidity.

Fig. C.1(b) describes the possibility of elastic and quasi-elastic scattering or doubled density throughout the whole range of rapidity. It is obvious that to make a single pomeron exchange a decisive factor, i.e. for the particles produced to have mainly uniform distribution in rapidity, fluctuations should not be anomalously large.

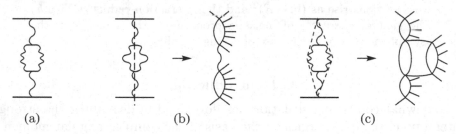

Fig. C.3

Referring to the simple diagrams in Figs. C.1(b) and C.1(c), we see that they comply with this condition because, due to shrinkage of the diffraction peak (the pomeron motion) their contributions decrease with an increase in energy as a power of ξ ($\xi = \ln s/m^2$), that is, the probability of a large fluctuation is small. As to the diagrams of the type shown in Fig. C.3, the situation is not so simple and the result depends upon the assumptions made about the magnitude of the interaction vertices, the problems involved being considered below. But in fact, with the elementary diagrams, too, all is not well. The probability of large fluctuations decreases only due to shrinkage of the diffraction peak determined by the parameter α', the slope of the Pomeranchuk pole,

$$\alpha(K^2) = 1 - \alpha' K^2$$

($K^2 = -t$ is the square of the transferred transverse momentum), which denotes one fourth of the diffusion coefficient of particles on the impact parameter plane.

If $4\alpha'$ were of the order of R^2, where R is the hadron radius, the probabilities of large fluctuations would drop rapidly with an increase in energy and could be ignored in the total cross section calculations. Indeed, experimentally, $4\alpha' \sim 1/m^2$, where m is the nucleon mass and $R^2 \sim \mu^2/4$ is the mass of the π meson; $4\mu^2/m^2 \sim 1/12$, and hence contributions even from the elementary diagrams practically do not increase with an increase in energy. All of these diagrams have to be taken into account, which leads to a large number of unknown parameters. The cause of the existence of such a small parameter is not quite clear. The smallness of the diffusion coefficient may be related to both large mass partons inside a hadron and a fairly strong coupling between partons.

However, regardless of the nature of this phenomenon, it completely changes the character of fluctuations in the inelastic process over the attainable energy range compared with what might be expected, if there were no such small parameter.

Considering that the asymptotic behaviour may occur only if $4\alpha'\xi \gg R^2$, we shall always be within the preasymptotic region $1 < \xi \lesssim R^2/4\alpha'$. On the other hand, experiment suggests that even in this region there are simple regularities corresponding to a neglect of the fluctuations as a first approximation and restriction to one-pomeron exchange. This points to an apparent cancellation of contributions from individual diagrams.

The description suggested in this paper takes into account the smallness of $4\alpha'/R^2$ right from the start and makes it possible to see that such a cancellation really occurs and that even in this region $1 < \xi \lesssim R^2/4\alpha'$ non-enhanced diagrams are small and the total cross sections are described by the Pomeranchuk pole with a slightly corrected Green function. At the same time, from the smallness of the parameters determining the behaviour of the cross sections over the asymptotic region $\xi > R^2/4\alpha'$, the idea comes that if there is a wide region of energies where the peak shrinkage is inessential, there should be at least an approximate theory with a fixed Pomeranchuk pole giving a sensible description both of the total cross sections and of all kinds of inelastic processes. Therefore, to a zero approximation, it should make sense to assume $\alpha' = 0$ and only then consider corrections to this approximation.

C.2 Non-enhanced cuts at $\alpha' = 0$

We consider an elementary two-reggeon cut in Fig. C.4 but take into account the possible interaction between two reggeons.

Let us see how to write it. It is best to do this using not momentum space, but the impact parameter space $\bar{\rho}$. The pomeron Green function

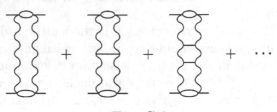

Fig. C.4

$G(K, \xi) = \exp[-\alpha' K^2 \xi]$ in the $\bar{\rho}$ space is of the form

$$G(\bar{\rho} - \bar{\rho}', \xi) = \frac{1}{4\pi\alpha'\xi}\, e^{-(\bar{\rho}-\bar{\rho}')^2/4\alpha'\xi}, \tag{C.1}$$

which at $\alpha' = 0$ gives

$$G(\bar{\rho} - \bar{\rho}', \xi) = \delta(\bar{\rho} - \bar{\rho}'), \tag{C.2}$$

that is, at $\alpha' = 0$ the pomeron has an infinitely large mass and does not move along $\bar{\rho}$. We consider first the pomeron interactions at equal ξ, i.e. the interactions at low relative parton energies. This interaction may be thought of as a particle exchange with the Green function

$$D(\bar{\rho}_1, \bar{\rho}_2, \xi_1, \xi_2) = \delta(\xi_1 - \xi_2)V(\rho_{12}).$$

Let us adopt this kind of interaction. Then the Green function for two pomerons entering into the diagrams in Fig. C.4 may be presented as

$$G(\bar{\rho}_1, \bar{\rho}_1'; \bar{\rho}_2, \bar{\rho}_2'; \xi)$$

$$= \delta(\bar{\rho}_1 - \bar{\rho}_1')\delta(\bar{\rho}_2 - \bar{\rho}_2') - \int_0^\xi V(\rho_{12})G(\bar{\rho}_1, \bar{\rho}_1', \bar{\rho}_2, \bar{\rho}_2', \xi')\mathrm{d}\xi', \tag{C.3}$$

and hence

$$G(\bar{\rho}_1, \bar{\rho}_1'; \bar{\rho}_2, \bar{\rho}_2'; \xi) = \delta(\bar{\rho}_1 - \bar{\rho}_1')\delta(\bar{\rho}_2 - \bar{\rho}_2')G(\rho_{12}, \xi), \tag{C.4}$$

$$G(\rho_{12}, \xi) = e^{-V(\rho_{12})\xi}, \tag{C.5}$$

or in the complex momentum space

$$G(\rho_{12}, \omega) = \frac{1}{\omega + V(\rho_{12})}. \tag{C.6}$$

It is obvious from this fact that if we want the contribution from the cuts to fall and not to rise with increasing ξ, we must assume that $V(\rho_{12}) > 0$. In this case the contribution of the two-reggeon cut is

$$f_2(\bar{K}, \xi) = -\int g_1(\bar{K}, \bar{\rho}_{12})e^{-V(\rho_{12})\xi}g_2(\bar{K}, \bar{\rho}_{12})\mathrm{d}^2\rho_{12}. \tag{C.7}$$

If $g_1(\bar{K}, \bar{\rho}), g_2(\bar{K}, \bar{\rho}) \sim V(\rho)$ at large ρ, then

$$f_2(K, \xi) \sim \frac{1}{\xi^2}\rho_0(\xi), \qquad \text{(C.8)}$$

where $\rho_0(\xi) = R\ln(v_0\xi)$ if $V(\rho) = v_0\exp(-\rho/R)$, $v_0\xi > 1$, i.e. $f_2(K, \xi)$ falls with increasing energy despite the absence of the shrinkage of the peak.

Similarly, when considering a contribution of the three-reggeon cut (Fig. C.5) if the reggeon interaction is accounted for, we obtain for the three-reggeon Green function an expression of the form

Fig. C.5

$$G(\bar{\rho}_1, \bar{\rho}'_1, \bar{\rho}_2, \bar{\rho}'_2, \bar{\rho}_3, \bar{\rho}'_3; \xi) = \delta(\bar{\rho}_1 - \bar{\rho}'_1)\delta(\bar{\rho}_2 - \bar{\rho}'_2)\delta(\bar{\rho}_3 - \bar{\rho}'_3)$$
$$\times \exp\{[V(\rho_{12}) + V(\rho_{13}) + V(\rho_{23})]\xi\}, \qquad \text{(C.9)}$$

and accordingly the contribution of the three-reggeon cut

$$f_3(K, \xi) = \int g_1(\bar{K}, \bar{\rho}_{12}, \bar{\rho}_{13})\exp\{-[V_{12} + V_{13} + V_{23}]\xi\}$$
$$\times g_2(\bar{K}, \bar{\rho}_{12}, \bar{\rho}_{13})\,\mathrm{d}^2\rho_{12}\mathrm{d}^2\rho_{13}. \qquad \text{(C.10)}$$

If $g \sim V^2$ at all large $\rho_{12}, \rho_{13}, \rho_{23}$, then $f_3 \sim \frac{1}{\xi^4}\rho_0^2(\xi)$, that is, it falls even more rapidly.

We see that non-enhanced diagrams are inessential at large ξ and $\alpha' = 0$. The fall in the contribution of the non-enhanced cuts with increasing energy is due to the fact that at small energies the fluctuations occurring at large ρ_{12} (within the tail area) have no time to interact but an increasingly large region of ρ_{12} becomes correlated as ξ increases.

C.3 Estimation of enhanced cuts at $\alpha' = 0$

Let us now discuss the amplitude for the decay of one reggeon to two. It would be natural to present it in the form of the diagrams shown in Fig. C.6.

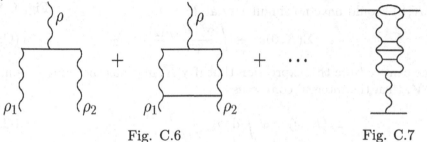

Fig. C.6   Fig. C.7

It is obvious that

$$\Gamma(\bar{\rho}, \bar{\rho}_1, \bar{\rho}_2; \xi) = \gamma(\bar{\rho}, \bar{\rho}_1, \bar{\rho}_2)\delta(\xi) - \int_0^\xi V(\rho_{12})\Gamma(\bar{\rho}, \bar{\rho}_1, \bar{\rho}_2; \xi')\mathrm{d}\xi', \quad (\text{C.11})$$

or in the ω space

$$\Gamma(\bar{\rho}, \bar{\rho}_1, \bar{\rho}_2; \omega) = \frac{\omega\gamma(\bar{\rho}, \bar{\rho}_1, \bar{\rho}_2)}{\omega + V(\rho_{12})}, \quad (\text{C.12})$$

i.e. it tends to zero at $\omega = 0$. It is clear from this that the contribution of the half-enhanced cuts in Fig. C.7 is also not large. In the ω space it has the form

$$-\int g_1(\bar{K}, \bar{\rho}_{12})\frac{1}{\omega}\Gamma(\bar{K}, \bar{\rho}_{12})\mathrm{d}^2\rho_{12}\frac{1}{\omega}g_2(\bar{K})$$

$$= -\int \mathrm{d}^2\rho_{12}\frac{g_1(\bar{K}, \bar{\rho}_{12})\gamma(\bar{K}, \bar{\rho}_{12})}{\omega + V(\rho_{12})}\frac{1}{\omega}g_2(K). \quad (\text{C.13})$$

The coefficient for $1/\omega$ converges at $\omega = 0$ and yields a renormalization of the constant, coupling the pomeron to a particle. After renormalization, the contribution of a half-enhanced cut has the form

$$\int \frac{g_1(\bar{K}, \bar{\rho}_{12})\gamma(\bar{K}, \bar{\rho}_{12})\mathrm{d}^2\rho_{12}}{V(\rho_{12})[\omega + V(\rho_{12})]}g_2(K) \sim \ln^2\frac{1}{\omega}, \quad (\text{C.14})$$

if $\gamma \sim g \sim V$, that is, it falls with increasing energy as $(1/\xi)\ln\xi$.

We now turn to discussing corrections to the pomeron Green function. The first correction to the self-energy Σ (Fig. C.8) has the form

$$\Sigma(K, \omega) = -\int \frac{\gamma(\bar{K}_1, \bar{\rho}_{12})\gamma(\bar{K}, \bar{\rho}_{12})}{\omega + V(\rho_{12})}\mathrm{d}^2\rho_{12}. \quad (\text{C.15})$$

If based on our assumption, $\alpha' = 0$, i.e. the pole position is independent of K, we should renormalize $\Sigma(K, \omega)$ so that when taking into account the interaction, α' remains equal to 0, i.e. the renormalized self-energy $\Sigma_c(K, \omega) = \Sigma(K, \omega) - \Sigma(K, 0)$, as distinct from the conventional method of renormalization when we subtract $\Sigma(0, 0)$. For this to be done $\Sigma(K, 0)$ should have no singularity at $K = 0$,

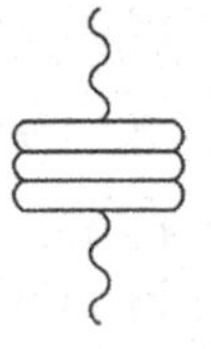

Fig. C.8

$$\Sigma(K, 0) = -\int \frac{\gamma^2(\bar{K}, \bar{\rho}_{12})}{V(\rho_{12})}\mathrm{d}^2\rho_{12}. \quad (\text{C.16})$$

One can see from this expression that if $\gamma(\overline{K}, \rho_{12})$ has the same exponents as V, then the integral converges:

$$\Sigma_c(K, \omega) = \omega\int \mathrm{d}^2\rho_{12}\frac{\gamma^2(\bar{K}, \bar{\rho}_{12})}{V(\rho_{12})[\omega + V(\rho_{12})]}. \quad (\text{C.17})$$

If $\gamma \sim V \sim \exp(-\rho/R)$, then $\gamma = \beta V$ and, at $K = 0$,

$$\Sigma_c(0,\omega) = \beta^2 \omega \ln^2 \frac{1}{\omega}, \tag{C.18}$$

that is

$$G(K,\omega) \sim \frac{1}{\omega} + \frac{\beta^2 \ln^2(1/\omega)}{\omega}. \tag{C.19}$$

Hence, the corrections to G are significant although they are at the $\ln(1/\omega)$ level. This means that there are higher approximations to Σ_c. Prior to taking them into account let us discuss the possible structure of $\gamma(\bar{K}, \bar{\rho}_{12})$ and its relation with $V(\rho_{12})$ in more detail.

C.4 Structure of the transition amplitude of one pomeron to two

As noted above, taking account of the interaction $V(\rho_{12})$ between two pomerons caused the probability of a fluctuation within which, for instance, two pomerons remain independent during the 'time' to decrease with increase in ξ. In this case the pomerons convert to their normal state, i.e. one pomeron.

The vertex $\gamma(\bar{\rho}, \bar{\rho}_1, \bar{\rho}_2)$ describes the same process explicitly, therefore the two quantities $\gamma(\rho, \rho_1, \rho_2)$ and $V(\rho_{12})$ should be intimately connected. The existence of a relationship between them is particularly clear from the viewpoint of the t-channel.

From the point of view of the t-channel the pomeron can be described by a set of Feynman diagrams containing in the intermediate state, for instance, two, four, $\ldots \pi$-mesons. The contribution of the cut is a special selection of the diagrams in which mesons 1, 2 do not interact with mesons 3, 4. Taking account of the pomeron interaction corresponds to the inclusion of these interactions. As a result, the state of the four mesons becomes correlated and is nothing but the contribution of the four-meson state to the pomeron. This suggests that the transition amplitude of one pomeron to two does not correspond to the blocks A and B in Fig. C.9, but is determined by the residue in the pomeron pole of the four-particle interaction amplitude. The difference here is the same as that between the point amplitude of the transition of two particles to one in field theory and the amplitude of the transition of two particles to the bound state determined by the interaction potential. The situation is aggravated by the fact that whereas a neutron and a proton, located

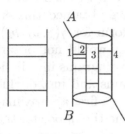

Fig. C.9

at particular points ρ_1 and ρ_2, can either remain in the continuous spectrum or convert to a deuteron, in our case two pomerons necessarily convert to one.

In the light of the above it seems quite natural that the quantity $\gamma(0, \rho_{12}) = \int d^2\rho \, \gamma(\bar{\rho}, \bar{\rho}_1, \bar{\rho}_2)$ determining the total conversion probability of two reggeons to one per unit interval ξ depends on ρ_{12} just like $V(\rho_{12})$, in any case at large ρ_{12}, i.e.

$$\gamma(0, \rho_{12}) = \beta V(\rho_{12}). \qquad \text{(C.20)}$$

We now consider the structure of $\gamma(\bar{\rho}, \bar{\rho}_1, \bar{\rho}_2)$ in more detail using elementary examples so as to have an idea about the possible dependence of $\gamma(\bar{K}, \bar{\rho}_{12})$ upon K and hence $G(K, \omega)$. Let us write $\gamma(\bar{\rho}, \bar{\rho}_1, \bar{\rho}_2)$, for instance, in the form

$$\gamma(\bar{\rho}, \bar{\rho}_1, \bar{\rho}_2) = \gamma_0(\bar{\rho}, \bar{\rho}_1, \bar{\rho}_2) \exp\left[-\frac{1}{R}|\bar{\rho} - \bar{\rho}_1| - \frac{1}{R}|\bar{\rho} - \bar{\rho}_2| \right], \qquad \text{(C.21)}$$

where γ_0 contains no exponential factors. At large ρ_{12},

$$\gamma(\bar{\rho}, \bar{\rho}_1, \bar{\rho}_2) = \gamma_0(\rho_{12}) e^{-\rho_{12}/R} \exp\left[-\frac{2(\bar{\rho} - \bar{\rho}_c)^2}{R\rho_{12}} \sin^2 \varphi \right], \qquad \text{(C.22)}$$

where $\bar{\rho}_c = \frac{1}{2}(\bar{\rho}_1 + \bar{\rho}_2)$, φ is the angle between $\bar{\rho}_{12}$ and $\bar{\rho} - \bar{\rho}_c$. From this expression it will be obvious that over the angular range $\varphi \sim \sqrt{R/\rho_{12}}$, $\gamma(\bar{\rho}, \bar{\rho}_1, \bar{\rho}_2)$ contains no exponential smallness in $|\bar{\rho} - \bar{\rho}_c|$ up to $|\bar{\rho} - \bar{\rho}_c| \sim \rho_{12}$. This means that $\gamma(K, \rho_{12})$ will significantly change at large ρ_{12} in the region of small $K \sim 1/\rho_{12}$ and decrease at $K\rho_{12} \gg 1$. This in turn indicates that $\Sigma_c(K, \omega)$ in (C.17) will fall rapidly with increasing K at $KR\ln(1/\omega) > 1$ because essentially the ρ_{12} in (C.17) are of the order of $R\ln(1/\omega)$ assuming (C.20). As a result, if the pomeron Green function $G(\omega, K)$ is written as $(1/\omega)Z(\omega, K)$ then $Z(\omega, K)$ will vary essentially at $K \sim 1/(R\ln(1/\omega)) \sim 1/(R\ln\xi)$, and the elastic scattering cross section at $K^2 \sim 1/(R^2 \ln^2 \xi)$. In other words, despite the fact that we put $\alpha' = 0$, we obtain a shrinkage of the diffraction peak, but much less significant, $K^2 \sim 1/(R^2 \ln^2 \xi)$ instead of $K^2 \sim 1/(\alpha'\xi)$. We have obtained this effect under the assumption of a definite exponential form for $\gamma(\rho, \rho_1, \rho_2)$. Should we choose γ in the Gaussian form, this effect would not occur. However, a more detailed analysis of the higher order diagrams in γ indicates that it would nevertheless occur in these diagrams and just of this order. Therefore, there is no reason to believe that even a first approximation does not have this feature. Moreover, if we assume the opposite, the problems related to the unitarity condition would occur due to the sign changing higher approximations. We shall assume below, unless otherwise specified, that

$$\gamma(\bar{K}, \bar{\rho}_{12}) = V(\rho_{12})\beta(K\rho_{12}, \bar{K}\bar{\rho}_{12}), \quad \beta(0,0) \equiv \beta. \tag{C.23}$$

C.5 The Green function and the vertex part at $\alpha' = 0$

We have shown above that the elementary ladder diagrams in Figs. C.6 and C.8 give a significant contribution to the vertex part and the pomeron Green function.

Let us now consider higher orders. We start with the diagram in Fig. C.10. On renormalizing the inner self-energy, the contribution from this diagram may be written in the form

$$
\begin{aligned}
&\Sigma(\omega, \bar{\rho} - \bar{\rho}') \\
&= \int \prod_i \mathrm{d}^2\rho_i \, \gamma(\bar{\rho}, \bar{\rho}_1, \bar{\rho}_2) \frac{1}{\omega + V_{12}} \gamma(\bar{\rho}_1, \bar{\rho}_3, \bar{\rho}_4) \frac{\omega + V_{23} + V_{24}}{V_{34}(\omega + V_{34} + V_{23} + V_{24})} \\
&\qquad \times \gamma(\bar{\rho}_3, \bar{\rho}_4, \bar{\rho}_5) \frac{1}{\omega + V_{25}} \gamma(\bar{\rho}_2, \bar{\rho}_5, \bar{\rho}').
\end{aligned}
\tag{C.24}
$$

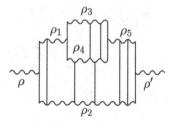

Fig. C.10

To find the singular contribution to $\Sigma_c(\omega, \bar{\rho} - \bar{\rho}') = \omega\sigma(\rho_0, \bar{\rho} - \bar{\rho}')$, $V(\rho_0) = \omega$, we can differentiate $\Sigma(\omega, \bar{\rho} - \bar{\rho}')$ with respect to ω and then seek for the integration region yielding only the logarithmic dependence on ω. Such a contribution in $\sigma(\rho_0(\omega), \bar{\rho} - \bar{\rho}')$ occurs in the region where all potentials cancel and only integration with respect to $\mathrm{d}^2\rho_i$, over the region of the order of ρ_0^2, remains. If potentials remain in the expression for σ, then the corresponding regions give contributions either to renormalizations or to the terms of the order of ω. Differentiating ω in the numerator of (C.24) gives

$$
\begin{aligned}
\sigma_2(\rho_0, \bar{\rho} - \bar{\rho}_1) = -\int \prod_i \mathrm{d}^2\rho_i \, \gamma(\bar{\rho}, \bar{\rho}_1, \bar{\rho}_2) \frac{1}{\omega + V_{12}} \gamma(\bar{\rho}_1, \bar{\rho}_3, \bar{\rho}_4) \\
\times \frac{1}{V_{34}(\omega + V_{34})} \gamma(\bar{\rho}_3, \bar{\rho}_4, \bar{\rho}_5) \frac{1}{\omega + V_{25}} \gamma(\bar{\rho}_2, \bar{\rho}_5, \bar{\rho}') \\
(V_{34} \gg V_{23}, V_{24}).
\end{aligned}
\tag{C.25}
$$

If we differentiate $1/(\omega + V_{12})$ or $1/(\omega + V_{25})$, then at $\omega \to 0$ we obtain an expression different from the first approximation diagram in Fig. C.8 only in the replacement of one of the interaction potentials with a non-local potential

$$V(\bar{\rho}_1\bar{\rho}'_1,\rho_2) = \int \mathrm{d}^2\rho_2\mathrm{d}^2\rho_4\gamma(\bar{\rho}_1,\bar{\rho}_3,\bar{\rho}_4)\frac{V_{23}+V_{24}}{V_{34}(V_{23}+V_{24}+V_{34})}\gamma(\bar{\rho}_3,\bar{\rho}_4,\bar{\rho}'_1).$$

(C.26)

We shall see later what would change in our analysis if we considered non-local potentials from the very beginning. Now we assume that this change is insignificant, and then (C.26) merely renormalizes the interaction potential and thus can be neglected.

Differentiating the denominator $\omega + V_{34} + V_{23} + V_{24}$ would give an expression containing instead of $1/V_{34}^2$ in (C.25) ($\omega \to 0$) the factor

$$\frac{1}{V_{34}^2}\frac{V_{34}(V_{23}+V_{23})}{(V_{34}+V_{23}+V_{24})^2},$$

which incorporates additional smallness everywhere except in the region $V_{34} \sim V_{23} + V_{24}$ which is small compared with the integration region in (C.25). Thus, the main contribution to $\sigma_2(\rho_0(\omega), \bar{\rho}-\bar{\rho}')$ is given by (C.25). This contribution is easy to calculate:

$$\sigma_2(\rho_0,\bar{\rho}-\bar{\rho}')$$

$$= -\int \prod_i \mathrm{d}^2\rho_i \frac{\gamma(\bar{\rho},\bar{\rho}_1,\bar{\rho}_2)}{V_{12}}\frac{\gamma(\bar{\rho}_1,\bar{\rho}_3,\bar{\rho}_4)}{V_{34}}\frac{\gamma(\bar{\rho}_3,\bar{\rho}_4,\bar{\rho}_5)}{V_{34}}\frac{\gamma(\bar{\rho}_5,\bar{\rho}_2,\bar{\rho}')}{V_{25}}$$

$$\simeq -\int \mathrm{d}^2\rho_1\mathrm{d}^2\rho_2\mathrm{d}^2\rho_5 \frac{\gamma(\bar{\rho},\bar{\rho}_1\bar{\rho}_2)}{V_{12}}\sigma_1(\rho_{12},\bar{\rho}_1-\bar{\rho}_5)\frac{\gamma(\bar{\rho}_5\bar{\rho}_2\bar{\rho}')}{V_{25}}$$

$$(\rho_{12} < \rho_0, \quad \rho_{25} < \rho_0, \quad \rho_{34} < \rho_{12},\rho_{25}),$$

(C.27)

$$\sigma(\rho_0, K = 0) = -\beta^2\int_0^{\rho_0} \mathrm{d}^2\rho_{12}\sigma_1(\rho_{12}, K=0) \simeq -\frac{1}{2}\beta^4\pi^2\rho_0^4(\omega)$$

$$(\rho_{12} < \rho_0).$$

(C.28)

To understand what occurs in higher approximations, let us consider the diagram in Fig. C.11.

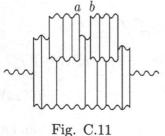

Fig. C.11

An interesting feature of the theory under consideration, distinguishing it from the conventional logarithmic field theories, is that the diagrams of the type presented in Fig. C.11 are inessential to the main logarithmic approximation. This will be obvious from what follows. The diagram in Fig. C.11 contains two factors of the form

$$\frac{\omega + V_{23} + V_{24}}{V_{34}(V_{34}+V_{23}+V_{24})}$$

as against one such factor in (C.24). This factor is of the order of ω/V_{34}^2 over the main region. In the case of (C.24) $(\omega+V_{23}+V_{24})$ in the numerator ensured the wanted smallness of $\Sigma_c \sim \omega\sigma$. As to the diagram in Fig. C.11, two such factors will give a contribution of the order of ω^2.

Should we neglect the reggeon interaction between points a and b, the additional smallness would be compensated by $1/\omega$ of the contribution from the intermediate state between points a and b. If we take into account the interaction $1/\omega \rightarrow 1/(\omega + V)$ we have the smallness $\omega/(\omega + V) \sim \omega/V$ in logarithmic approximation. The situation changes if we consider, for instance, the diagrams in Fig. C.12 corresponding to simultaneously occurring fluctuations; hence there is no interaction between them. Diagrams renormalizing inner self-energies are given in Figs. C.13(a), C.13(b). The diagram in Fig. C.13(c) renormalizes the diagram similar to Fig. C.11, but with self-energy inserts in different lines.

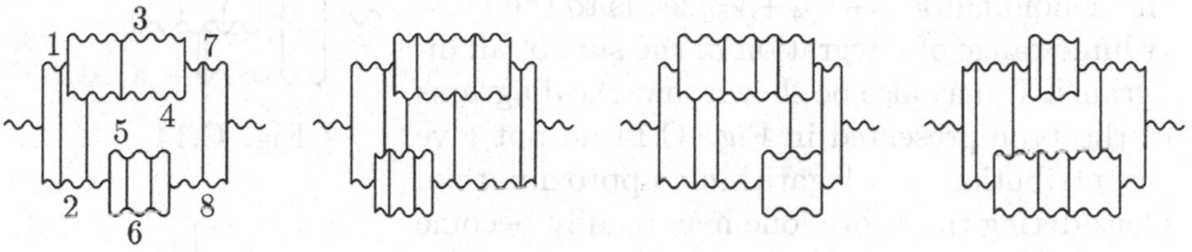

Fig. C.12

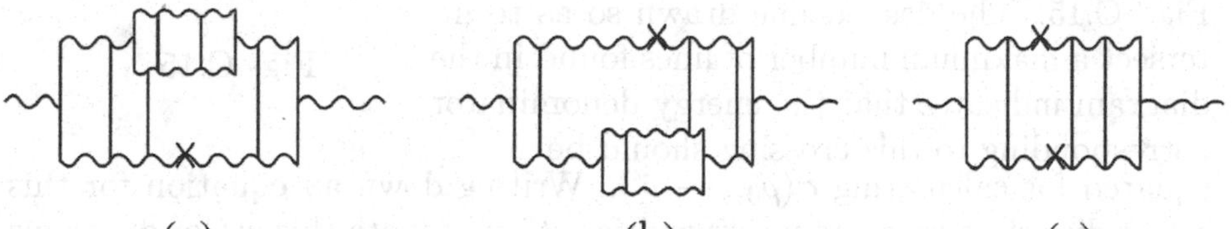

(a) (b) (c)

Fig. C.13

Let us write out the energy denominators from the diagrams in Fig. C.12, assuming from the very beginning for simplicity that

$$V_{12}, V_{78} \ll V_{34}, V_{56},$$

$$\frac{1}{\omega + V_{12}} \cdot \frac{1}{\omega + V_{34}} \cdot \frac{1}{\omega + V_{34} + V_{56}} \cdot \frac{1}{\omega + V_{34}} \cdot \frac{1}{\omega + V_{78}} + \text{permutations}$$

$$= \frac{1}{\omega + V_{12}} \cdot \frac{1}{(\omega + V_{34})^2} \cdot \frac{1}{(\omega + V_{56})^2} \cdot \frac{(2\omega + V_{34} + V_{56})^2}{\omega + V_{34} + V_{56}} \cdot \frac{1}{\omega + V_{78}}.$$

Similar energy denominators in renormalizing diagrams give

$$\frac{1}{\omega + V_{12}} \cdot \frac{1}{V_{56}} \cdot \frac{1}{(\omega + V_{34})^2} \cdot \frac{1}{\omega + V_{78}} + \frac{1}{\omega + V_{12}} \cdot \frac{1}{V_{34}} \cdot \frac{1}{(\omega + V_{56})^2} \cdot \frac{1}{\omega + V_{78}}.$$

As a result, after renormalizing, instead of energy denominators we obtain the following expression:

$$-\frac{1}{\omega + V_{12}} \cdot \frac{1}{V_{34}(\omega + V_{34})} \cdot \frac{1}{V_{56}(\omega + V_{56})} \cdot \frac{\omega(V_{34} + V_{56})}{\omega + V_{34} + V_{56}} \cdot \frac{1}{\omega + V_{78}}.$$

$$\simeq -\omega \frac{1}{V_{12}} \cdot \frac{1}{V_{34}^2} \cdot \frac{1}{V_{56}^2} \cdot \frac{1}{V_{78}} \quad (V_{56}, V_{34} \gg V_{12}, V_{78} \gg \omega).$$

We would obtain the same result through calculating the derivative with respect to ω instead of performing renormalizations explicitly. In so doing it is readily seen that only differentiating the denominator $\omega + V_{34} + V_{56}$ leads to the logarithmic range of integration in the sum of all diagrams. It may also be shown that the diagrams of the type presented in Fig. C.14 do not give a contribution to a logarithmic approximation. Considering the above, one may readily become aware of the total set of diagrams giving a contribution to a logarithmic approximation. This set comprises all diagrams of the type shown in Fig. C.15. The dashed line drawn so as to intersect a maximum number of lines found in the diagram indicates that the energy denominator corresponding to this crossing should be

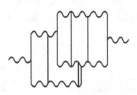

Fig. C.14

Fig. C.15

squared for calculating $\sigma(\rho_0, \bar{\rho} - \bar{\rho}')$. Writing down an equation for this set of diagrams presents no difficulties. If we denote this set of diagrams with n lines in the maximum crossing by $f_n(\rho_0, \bar{\rho} - \bar{\rho}')$, then

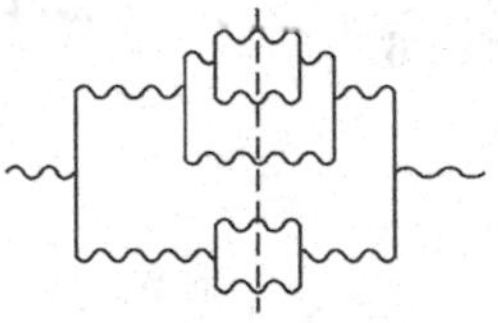

$$f_n(\rho_0, \bar{\rho} - \bar{\rho}') = \sum_{\substack{n_1, n_2 = 1 \\ n_2 + n_2 = n}}^{n} \int d^2\rho_1 d^2\rho_2 d^2\rho_1' d^2\rho_2' \frac{\gamma(\bar{\rho}, \bar{\rho}_1, \bar{\rho}_2)}{V_{12}} f_{n_1}(\rho_{12}, \bar{\rho}_1 - \bar{\rho}_1')$$

$$\times f_{n_2}(\rho_{12}, \bar{\rho}_2 - \bar{\rho}_2') \frac{\gamma(\bar{\rho}_1, \bar{\rho}_2, \bar{\rho}_1')}{V_{1'2'}},$$

$$f_1 = \delta(\bar{\rho}_1 - \bar{\rho}_1'), \quad \rho_{12} < \rho_0, \quad \text{(C.29)}$$

where ρ_{12} is substituted on the right as an argument into $f_{n_1}(\rho_{12}, \bar{\rho}_i - \bar{\rho}_i')$ instead of ρ_0 so as to comply with the condition of the type $\rho_{34}, \rho_{56} < \rho_{12}$ determining the logarithmic range of integration in the internal integration.

If the pomeron Green function is written as

$$G = Z/\omega, \tag{C.30}$$

then

$$Z^{-1} = \sum_{n=1}^{\infty} f_n,$$

and hence in momentum space

$$Z^{-1}(\rho_0, K) = 1 - \int d^2\rho_{12} d^2\rho'_{12} \left(\beta(\bar{K}, \bar{\rho}'_{12}) \right.$$

$$\times \int \frac{d^2q}{(2\pi)^2} \frac{e^{i\,\bar{q}(\bar{\rho}_{12} - \bar{\rho}'_{12})}}{z(\rho_{12}, \frac{1}{2}\bar{K} + \bar{q})z(\rho_{12}, \frac{1}{2}\bar{K} - \bar{q})} \beta(\bar{K}, \bar{\rho}'_{12}) \, \theta\left(\rho_0 - \rho_{12}\right) \Big). \tag{C.31}$$

We have written (C.29) and (C.31) asymmetrically with respect to $\bar{\rho}_{12}$ and $\bar{\rho}'_{12}$ assuming that since in all integrations $\rho_{34}, \ldots$ is less than ρ_{12}, $|\bar{\rho}_1 - \bar{\rho}'_1| < \rho_{12}$ and hence, the functions $Z^{-1}(\rho_{12}, \bar{\rho}_1 - \bar{\rho}'_1)$, $Z^{-1}(\rho_{12}, \bar{\rho}_2 - \bar{\rho}'_2)$ automatically ensure $\rho'_{12} \sim \rho_{12}$. At $K = 0$, $\beta(0, \rho_{12})$ equals, from the assumption, a constant and

$$Z^{-1}(\rho_0, 0) = 1 - \pi\beta^2 \int_0^{\rho_0} d\rho_{12}^2 \, Z^{-2}(\rho_{12}, 0), \tag{C.32}$$

which gives

$$Z(\rho_0, 0) = 1 + \pi\beta^2 \rho_0^2. \tag{C.33}$$

Thus, the total cross sections increase as $\rho_0^2 \sim \ln^2 \xi$. At $K \neq 0$, $Z(\rho_0, K)$ determined by (C.31) exhibits the following significant property. Since $\beta(\bar{K}, \bar{\rho}_{12})$, as discussed above, decreases at $K\rho_{12} > 1$ (shrinkage of the cone) in (C.31), $\theta(\rho_0 - \rho_{12})$ may be omitted at $K\rho_0 \gg 1$ and hence $Z(\rho_0, K)$ is independent of ρ_0. This means that if

$$Z(\rho_0, K) = \pi\beta^2 \rho_0^2 F(K, \rho_0), \qquad \rho_0 \ll R,$$

then

$$F(K, \rho_0) = c/K^2\rho_0^2 \quad \text{at} \quad K^2\rho_0^2 \gg 1, \quad \text{but} \quad K^2R^2 \ll 1.$$

Consequently,

$$Z(\rho_0, K) = c/K^2 \quad \text{for} \quad 1/\rho_0 \ll K \ll 1/R, \tag{C.34}$$

i.e. the elastic scattering amplitudes cease to be dependent on energy at $K\rho_0 \gg 1$. The elastic scattering cross section is

$$\sigma_{\text{el}} \sim \int Z^2(K, \rho_0) \, d^2K \sim \rho_0^2,$$

i.e. the solution we obtained corresponds to scattering on a black disk of radius $\rho_0 \ln(v_0\xi)$, $v_0\xi > 1$.

We shall return later to discussing the properties of the solution obtained and we now consider the behaviour of the vertex part. Following the analysis made above, the calculation of Γ presents no special problems. For calculating this quantity it suffices to note that when computing Σ_c for the elementary diagram only were we required to renormalize the 'mass'; for other diagrams it was sufficient to renormalize the mass in the internal lines, that is to perform the same renormalization which is required when calculating Γ.

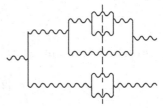

Fig. C.16

This means that the total set of diagrams for Γ is the same as for Σ (Fig. C.16). In this case,

$$\omega\sigma(\rho_0, K) = \int \mathrm{d}^2\rho_{12}\beta(\bar{K}_1, \bar{\rho}_{12})\Gamma(\bar{K}, \bar{\rho}_{12}). \tag{C.35}$$

Comparison between (C.35) and (C.31) gives

$$\Gamma(\bar{K}, \bar{\rho}_{12}) = \omega \int \frac{\mathrm{d}^2q}{(2\pi)^2} \frac{\beta(\bar{K}, \bar{q})\theta(\rho_0 - \rho_{12})}{Z(\rho_{12}, \frac{1}{2}\bar{K} + \bar{q})Z(\rho_{12}, \frac{1}{2}\bar{K} - \bar{q})}. \tag{C.36}$$

At $K = 0$,

$$\beta(\bar{K}, \bar{q}) = (2\pi)^2\delta(\bar{q})\beta,$$

$$\Gamma(0, \rho) = \frac{\omega\beta}{Z^2(\rho, 0)}\theta(\rho_0 - \rho). \tag{C.37}$$

Prior to considering the physical consequences of the solution obtained, let us discuss the question as to whether the character of the solution will change if we go beyond the scope of application of the logarithmic solution. Formally the logarithmic approximation is valid at $\beta^2 R^2 \ll 1$, $\beta^2\rho_0^2 \sim 1$. Essentially, the problem is the following. We have found the main contribution to G and Γ due to the logarithmically large integration regions ignoring the regions ρ_0/R times less in each approximation. As a result of taking into account these regions, the Green function turned out to be rising, $G \sim \rho_0^2/\omega$, and the vertex part $\Gamma \sim \omega/\rho_0^4$ falling. Will the relative contributions of large and small ranges change if we take into account the new behaviour of G and Γ? Such situations are possible. For example, in theories where the Green function in logarithmic approximation acquires a pole, the region close to the pole, despite its smallness, begins to give an essential contribution. We shall show that no such situation occurs in our case.

Let us ignore the dependence of $Z(\rho_0, K)$ on K, that is, we shall disregard the pomeron motion arising from fluctuations in $\bar{\rho}$ space for it only reduces the difference between the Green functions and the initial ones,

$$G(\omega, \bar{\rho} - \bar{\rho}') = \delta(\bar{\rho} - \bar{\rho}')\frac{Z(\omega)}{\omega}. \tag{C.38}$$

We have started by calculating the Green function for two pomerons $G_{12}(\omega, \rho_{12})$ (Section C.2). In so doing, we obtained for $G_{12}(\omega, \rho_{12})$ the following expression:

$$G_{12}(\omega, \rho_{12}) = \frac{1}{\omega + V(\rho_{12})}.$$

If we substitute into the equation for $G_{12}(\omega, \rho_{12})$ expression (C.38) instead of $G_0 = 1/\omega$, it is easily shown that due only to the logarithmic dependence of Z on ω, $G_{12}(\omega, \rho_{12})$ will have the form

$$G_{12}(\omega, \rho_{12}) = \frac{1}{\omega/Z^2(\omega) + V(\rho_{12})}.$$

Similarly the three-particle Green function will take the form

$$G_{123}(\omega, \rho_{ik}) = \frac{1}{\omega/Z^3 + V_{12} + V_{13} + V_{23}}.$$

If we substitute these expressions instead of the old ones into the diagrams for $\sigma(\rho_0, K)$, obviously nothing will change, except the equation for the boundary of the integration region $V(\rho_0) = \omega$. Now it will be determined by the condition $V(\rho_0) = \omega/Z^2(\omega)$. Besides, we have omitted the diagrams of the type shown in Fig. C.14. This diagram was of the order of $\beta^4 \rho_0^3$ whereas the diagram in Fig. C.10 which was taken into account was of the order of $\beta^4 \rho_0^4$. If we calculate the contribution of the diagram in Fig. C.14 to Σ with the new Green functions, a renormalization needs to be performed which is equivalent to finding the derivative $\partial\Sigma/\partial\omega$. Then calculating $\partial\Sigma/\partial\omega$ we shall have to differentiate the denominators corresponding to various intermediate states. Differentiating the denominators for two-particle intermediate states will give, as in the case of the diagram in Fig. C.10, the renormalization of the interaction potential. Differentiation of the denominator corresponding to the three-particle intermediate state will yield the expression containing the factor $1/Z^3(\omega)$. As a result, this diagram, besides the narrower integration range, will contain an additional small factor β^2/Z which is small at low $\omega \sim \beta^2/1 + \pi^2\beta^2\rho_0^2 \sim 1/\rho_0^2$ and independent of β. The arguments adduced here show that the solution obtained is self-consistent and the only condition for its application is $\rho_0/R \gg 1$, i.e. $\omega/V_0 \ll 1$.

So far it has been assumed that $\beta(\bar{K}, \bar{\rho})|_{\bar{K}=0} = \beta = \text{const}$, i.e. the integral

$$\int \gamma(\bar{\rho}, \bar{\rho}_1, \bar{\rho}_2) \frac{\mathrm{d}^2\rho}{V(\rho_{12})}$$

is independent of ρ_{12}. Let us discuss what happens if we abandon this hypothesis. If we assume that $\beta = \beta(\rho_{12})$ rises with increasing ρ_{12}, $\beta(\rho_{12}) \sim \rho_{12}^{\delta}$, $\delta > 0$, it will be found that $Z(\rho_0, K = 0) \sim \rho_0^{2(1+\delta)}$, i.e. the cross section rises faster than ρ_0^2, but the interaction radius remains of the order of ρ_0. This means that the elastic cross section is larger than the total one. Assuming that $\beta(\rho_0)$ falls with the increase of ρ, $\beta(\rho_0) \sim \rho_0^{-\delta}$,

$$Z(\rho_0, K = 0) \sim \rho_0^{2(1-\delta)} \quad \text{at} \quad 0 < \delta < 1.$$

It will be shown below, when calculating the inclusive cross section in the three-pomeron limit, that this cross section is positively determined only at $\delta < \frac{2}{3}$. Consequently, in the theory with $\alpha' = 0$ the cross section always rises slowly,

$$\sigma_t \sim (\ln \xi)^\nu, \quad \frac{2}{3} < \nu < 2.$$

C.6 Properties of high energy processes in the theory with $\alpha' = 0$

Let us formulate the principal results required for further discussion. The pomeron Green function

$$G(\omega, K) = Z(\rho_0, K)/\omega,$$

where ρ_0 is determined by the condition $V(\rho_0) = \omega$. The vertex part

$$\Gamma(\omega, \bar{K}, \bar{\rho}_{12}) = \omega \int \frac{\mathrm{d}^2 q}{(2\pi)^2} \frac{\beta(\bar{K}, \bar{q})\Theta(\rho_0 - \rho_{12})}{Z(\bar{\rho}_{12}, \frac{1}{2}\bar{K} + \bar{q})Z(\bar{\rho}_{12}, \frac{1}{2}\bar{K} - \bar{q})}, \tag{C.39}$$

$$\beta(\bar{K}, \bar{q}) = \int \mathrm{d}^2\rho_{12} \frac{\gamma(\bar{K}, \bar{\rho}_{12})}{V(\rho_{12})} e^{i\bar{q}\bar{\rho}_{12}}, \tag{C.40}$$

where $V(\rho_{12})$ is the pomeron interaction potential, and $\gamma(K, \rho_{12})$ is the bare three-pomeron vertex.

Assuming that

$$\beta(0, q) = (2\pi)^2 \delta(\bar{q})\beta, \tag{C.41}$$

$$\Gamma(\omega, 0, \rho_{12}) = \frac{\omega\beta\Theta(\rho_0 - \rho_{12})}{Z^2(\rho_{12}, 0)}. \tag{C.42}$$

C.6.1 Two-particle processes, total cross sections

We consider now the main diagrams making contributions to the two-particle process amplitudes and hence to the total cross sections.

Calculating the diagrams in Figs. C.17(b,c,d) presents no special problems because it is coincident with the calculation of $\Sigma(\omega, K)$. The renormalization of the pole position is replaced by that of the particle coupling constant. After renormalizing, we obtain, for instance, for the contribution of the diagram in Fig. C.17(b),

$$\frac{1}{\omega} \int \mathrm{d}^2 \rho_{12} \frac{g_1(\bar{K}, \bar{\rho}_{12})}{V(\rho_{12})} \Gamma(\omega, \bar{K}, \bar{\rho}_{12}) Z(\rho_0 K) g_2(K)$$

$$\equiv g_1(\rho_0, K) Z(\rho_0, K) g_2(K). \quad \text{(C.43)}$$

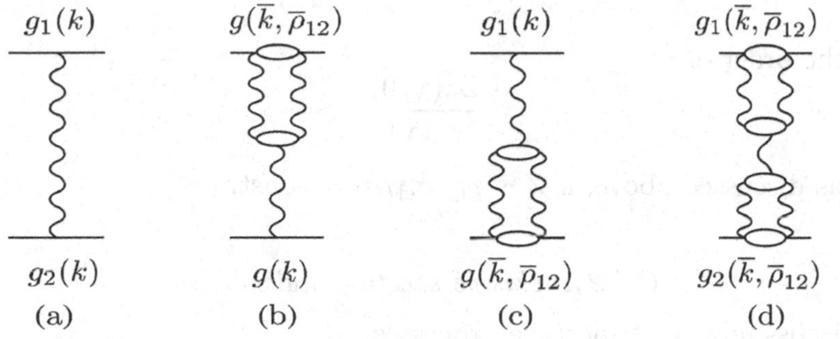

Fig. C.17

The contribution from all diagrams in Fig. C.17 is

$$f(\omega, K) = \frac{1}{\omega} Z(\rho_0, K)[g_1(K) + \omega g_1(\rho_0, K)][g_2(K) + \omega g_2(\rho_0, K)]$$

$$\simeq \frac{1}{\omega} Z(\rho_0, K) g_1 g_2 + Z(\rho_0, K)[g_1(K) g_2(\rho_0, K) + g_2(K) g_1(\rho_0, K)]. \quad \text{(C.44)}$$

The first 'pole' term results in a slowly changing amplitude

$$f_p(\xi, K) = g_1 g_2 \left(i - \frac{1}{2}\pi\frac{\partial}{\partial \xi} \right) \int \frac{\mathrm{d}\omega}{2\pi i} \frac{e^{\omega\xi}}{\omega} Z(\rho_0, K)$$

$$= g_1 g_2 \left(i - \frac{1}{2}\pi\frac{\partial}{\partial \xi} \right) \tilde{Z}(\chi, K),$$

$$\chi = \ln(v_0 \xi), \quad v_0 \xi > 1, \quad \tilde{Z}(\chi, K) \simeq Z(\rho_0(\xi), K), \quad \text{(C.45)}$$

where $\rho_0(\xi)$ is determined by the condition $V(\rho_0)\xi = 1$. The correction term $\tilde{f}(\xi, K)$ falls as a power of ξ. The extent to which it decreases depends on the behaviour of the 'coupling constant' for two pomerons

$g_1(\bar{K}, \bar{\rho}_{12})$ at large ρ_{12}. If $g_1(\bar{K}, \bar{\rho}_{12})$ at large ρ_{12} coincides with $V(\rho_{12})$, then the correction falls as $1/\xi$. If $g_1(\bar{K}, \bar{\rho}_{12}) \sim e^{-\rho_{12}/R_1}$ and $V(\rho_{12}) \sim e^{-\rho_{12}/R}$ with $R_1 < R$, $\tilde{f}(\xi, 0) \sim 1/\xi$ while for $R_1 > R$, $\tilde{f}(\xi, 0) \sim \xi^{-R/R_1}$,

$$\tilde{f}(\xi, K)$$
$$= \left(i - \frac{1}{2}\pi\frac{\partial}{\partial\xi} \right) \frac{1}{\xi}\frac{\partial}{\partial\chi} Z(\rho_0, K)[g_1(K)g_2(\rho_0, K) + g_1(\rho_0, K)g_2(K)],$$
$$\rho_0 = \rho_0(\chi), \quad \chi = \ln v_0\xi. \tag{C.46}$$

The contribution of non-enhanced diagrams is less than (C.46). It is of the order of $1/\xi^2 g_1(K, \rho_0)g_2(K, \rho_0)$. The elastic scattering cross section

$$\sigma_{\rm el} = \int \tilde{Z}^2(\chi, K)\frac{{\rm d}^2 K}{(4\pi)^2}$$

is of the order of

$$\frac{Z^2(\chi, 0)}{\rho_0^2(\chi)},$$

and, as discussed above, if $Z \sim \rho_0^2$, $\sigma_{\rm el}/\sigma_t \sim$ const.

C.6.2 Inclusive spectra, multiplicity

We discuss now multi-particle processes.

An elementary characteristic of multi-particle processes is the inclusive spectrum in the central region determined by the Kancheli–Muller diagram (Fig. C.18). In the theory under consideration,

$$f(\xi, \eta) = \tilde{Z}(\chi_\eta)\tilde{Z}(\chi_{\xi-\eta})g_1\varphi(K_1)g_2. \tag{C.47}$$

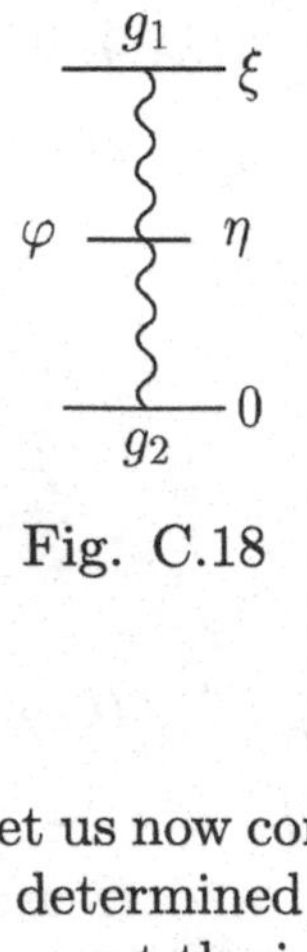

Fig. C.18

This spectrum has a maximum at $\eta = \xi/2$ rising slowly with increasing energy. The average multiplicity is

$$\bar{n} = \frac{1}{\tilde{Z}(\chi_\xi)} \int_0^\xi \tilde{Z}(\chi_\eta)\tilde{Z}(\chi_{\xi-\eta}){\rm d}\eta \sim \xi Z(\chi_\xi). \tag{C.48}$$

At $Z(\rho_0, 0) \sim \rho_0^2$,

$$\bar{n} \sim \xi \ln^2 \xi. \tag{C.49}$$

C.6.3 Correlation, multiplicity distribution

Let us now consider the correlation function in a two-particle spectrum. It is determined by the diagrams in Fig. C.19. There is no need to take into account the interactions between points η_1 and η_2 in the second diagram, because the appropriate diagrams cancel when calculating the inclusive spectrum. This statement was proved for three-reggeon interactions in [2].

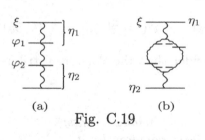

(a)

(b)

Fig. C.19

Generally speaking, for four-reggeon interactions this statement was considered to be wrong. It will be shown elsewhere that this statement is right irrespective of the type of interaction.

The correlation corresponding to the diagram in Fig. C.19(a) has a simple form

$$f_1(\xi, \eta_1, \eta_2) = \varphi(K_1)\varphi(K_2)$$
$$\times \left[\frac{Z(\eta_1)Z(\eta_2)Z(\eta_3)}{Z(\xi)} - \frac{Z(\eta_1)Z(\xi - \eta_1)Z(\eta_2)Z(\xi - \eta_2)}{Z^2(\xi)} \right],$$
$$Z(\eta) \equiv Z(\chi_\eta), \quad \eta_3 = \xi - \eta_1 - \eta_2. \tag{C.50}$$

It is easily seen that this correlation is always negative and if the second diagram were not essential, the multiplicity distribution would be narrower than the Poisson distribution $\overline{n^2} - \overline{n}^2 < \overline{n}$. In fact, the second diagram in Fig. C.19 is more important and leads to the opposite result. The correlation corresponding to the diagram in Fig. C.19 can be written in the form

$$f_2(\xi, \eta_1, \eta_2) = \frac{Z(\eta_1)Z(\eta_2)}{Z(\xi)} \int \frac{d^2 q}{(2\pi)^2} \beta(\eta_1, q)\beta(\eta_2, q)Z^2(\eta_3, q)$$
$$- \frac{1}{Z^2(\xi)} Z(\eta_1)Z(\xi - \eta_1)Z(\eta_2)Z(\xi - \eta_2),$$
$$\beta(\eta, q) = \beta \int d^2 \rho \, e^{2\bar{q}\,\bar{\rho}}\theta(\rho_0 - \rho), \quad V(\rho_0)\eta = 1. \tag{C.51}$$

We have omitted $\varphi(K_1)\varphi(K_2)$. The region of integration with respect to q is determined by the largest of the quantities η_1, η_2, η_3. At $\eta_1, \eta_2 < \eta_3$,

$$f_2(\xi, \eta_1, \eta_2) = \frac{\pi^2 \beta^2}{Z(\xi)} Z(\eta_1)\rho_0^2(\eta_1)Z(\eta_2)\rho_0^2(\eta_2) \int \frac{d^2 q}{(2\pi)^2} Z^2(\eta_3, q)$$
$$- Z(\eta_1)Z(\xi - \eta_1)Z(\eta_2)Z(\xi - \eta_2)\frac{1}{Z^2(\xi)}, \tag{C.52}$$

and the order of magnitude in this region is

$$f_2(\xi, \eta_1, \eta_2) = Z(\eta_1)Z(\eta_2)[cZ(\eta_1)Z(\eta_2) - 1], \tag{C.53}$$

i.e. there is a large positive correlation. In the region $\eta_1 > \eta_2, \eta_3$,

$$f_2(\xi, \eta_1, \eta_2) = cZ^2(\eta_2)Z^2(\eta_3) - Z(\eta_2 + \eta_3)Z(\eta_1). \tag{C.54}$$

Again, the correlation is positive at large Z. We can estimate $\overline{n(n-1)}$, using expression (C.53):

$$\overline{n(n-1)} \sim c \int d\eta_1 d\eta_2 Z^2(\eta_1) Z^2(\eta_2) = c' \overline{n}^2(\xi) Z^2(\xi), \qquad (C.55)$$

i.e. the multiplicity distribution is much wider than the Poisson distribution.

C.6.4 Probability of fluctuations in individual events: the inclusive spectrum in the three-pomeron limit

A characteristic property distinguishing fluctuations in the particle production processes at high energies from those occurring in statistical systems even close to the phase transition points is diffraction and quasi-diffraction processes which take place with a considerable probability and comprise fluctuations in the distribution of particles covering the entire rapidity range, i.e., the fluctuations of the order of the system volume.

In order to understand the relation between these fluctuations and normal fluctuations, let us consider the inclusive spectrum of the particles produced in the so-called three-reggeon limit where a particle scattered with an energy transfer which is small compared with the total energy transfer is observed. If an incident particle has rapidity ξ, then the smallness of the energy transfer indicates that no particles can be produced over the large rapidity range $\xi - \eta$ and those produced have rapidities below η. If $\xi - \eta$ is large, we are concerned with a large fluctuation whose size is $\xi - \eta$. The cross section of such a process is described by a diagram of the type shown in Fig. C.20(a). It coincides with that in Fig. C.20(b) for the vertex part with one significant difference, namely, here the largest value of η at which the interaction occurs is fixed. In terms of field theory this quantity is the amplitude of the process in which not only the initial and final states, but also the 'time' of the first interaction, is fixed.

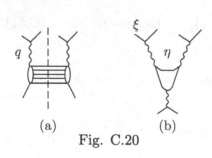

(a)   (b)

Fig. C.20

Let us try to calculate this quantity in our theory. We write $f(\xi, \eta, q)$ in the form

$$f(\xi, \eta, q) = \int e^{\omega_1(\xi-\eta)+\omega_2(\xi-\eta)+\omega\eta} f(\omega_1, \omega_2, \omega, q) \frac{d\omega_1 d\omega_2 d\omega}{(2\pi i)^3}. \qquad (C.56)$$

Upon integrating $f(\xi, \eta, q)$ with respect to η, we obtain the usual expression in the form of an integral of $f(\omega_1, \omega_2, \omega_1 + \omega_2, q)$ over ω_1 and ω_2. The elementary diagrams to a logarithmic approximation (Fig. C.21) give for $f(\omega_1, \omega_2, \omega, q)$

$$f_1(\omega_1, \omega_2, \omega, q) = g^2(q) \frac{Z(\omega_1)}{\omega_1} \frac{Z(\omega_2)}{\omega_2} \int e^{i\bar{q}\bar{\rho}_{12}} \frac{\omega}{Z^2(\rho_{12})} \beta(\rho_{12}) d^2\rho_{12} \frac{Z(\omega)}{\omega}.$$
$$\text{(C.57)}$$

The factor $1/Z^2(\rho_{12})$ in (C.57) reflects the fact that of all the diagrams (Fig. C.21(a)) with the exact pomeron Green functions, the pomeron multiplication occurs only in a particular cutting in Fig. C.21(b). Should we integrate with respect to η, then (C.57) would cover the main contribution to f. However, if η is fixed, we must add the contribution of the diagrams in Fig. C.21(c) in which multiplication occurs precisely in the rapidity range of the order of η. We now consider a contribution from the simplest diagram of those presented in Figs. C.21(c), C.22. Prior to integrating with respect to ρ_{12} the contribution of this diagram is

$$\frac{Z(\omega_1)}{\omega_1} \frac{Z(\omega_2)}{\omega_2} \int \frac{\gamma(\bar{\rho}_1, \bar{\rho}_3, \bar{\rho}_4)}{\omega_1 + V_{34}} \frac{V_{23} + V_{24}}{\omega + V_{34} + V_{23} + V_{24}} \frac{\gamma(\bar{\rho}_3, \bar{\rho}_4, \bar{\rho}_5)}{\omega + V_{25}}$$
$$\times \gamma(\rho_{25}, K = 0) \, d^2\rho_3 d^2\rho_4 d^2\rho_5 \frac{Z(\omega)}{\omega}. \qquad \text{(C.58)}$$

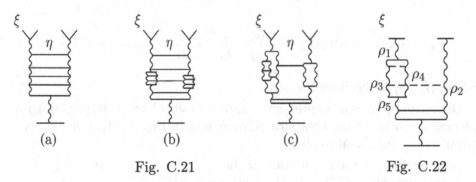

Fig. C.21   Fig. C.22

The integral over $\bar{\rho}_4$, $\bar{\rho}_5$ in (C.58) at $\omega_1, \omega_2 \to 0$ is non-singular and represents a renormalization of the interaction potential. Subtracting this renormalizing contribution from (C.58) gives the expression

$$\frac{Z(\omega_1)}{\omega_1} \frac{Z(\omega_2)}{\omega_2} \omega_1 \int \frac{\gamma(\bar{\rho}_1, \bar{\rho}_3, \bar{\rho}_4)\gamma(\bar{\rho}_3, \bar{\rho}_4, \bar{\rho}_5)}{V_{34}(\omega_1 + V_{34})} d^2\rho_3 d^2\rho_4 d^2\rho_5$$
$$\times \frac{1}{\omega + V_{25}} \gamma(\rho_{25}, 0) \frac{Z(\omega)}{\omega},$$
$$V_{34} \ll V_{23} + V_{24}, \qquad \text{(C.59)}$$

where we used the fact that the integration region $\rho_{34} > \rho_{23}$ or $\rho_{34} > \rho_{24}$ gives the main contribution. On renormalizing, a logarithmically smaller contribution but proportional to ω (not to ω_1) remains along with (C.59). This term does not differ from (C.57) qualitatively and is less than the latter. If we now sum all possible inserts into lines 3–4, then $\gamma(\bar{\rho}_3, \bar{\rho}_4, \bar{\rho}_5)$ is replaced by the exact vertex part and we obtain an additional contribution to $f(\omega_1, \omega_2, \omega, q)$, of the form

$$
f_2(\omega_1, \omega_2, \omega, q) = g^2(q) \frac{Z(\omega_1)}{\omega_1} \frac{Z(\omega_2)}{\omega_2} \omega_1
$$
$$
\times \int e^{i\,\bar{q}\bar{\rho}_{12}} \sigma(\bar{\rho}_1 - \bar{\rho}_1', \omega_1, \rho_{12}) \frac{\gamma(\rho_{21'})}{\omega + V(\rho_{21'})} d^2\rho_{12} d^2\rho_{1'} \frac{Z(\omega)}{\omega} g, \quad \text{(C.60)}
$$

$$
\sigma(\bar{\rho}_1 - \bar{\rho}_1', \omega_1, \rho_{12}) = \int \frac{d^2 q'}{(2\pi)^2} e^{i\,\bar{q}'(\bar{\rho}_1 - \bar{\rho}_1')} \int_{\rho_{12}}^{\rho_1^0} d^2\rho_{34}\, \beta(\bar{q}', \bar{\rho}_{34}) \Gamma(\bar{q}', \bar{\rho}_{34}),
$$
$$
V(\rho_1^0) = \omega_1. \quad \text{(C.61)}
$$

In the simplest case $\gamma(\rho_{21'})/V(\rho_{21'}) = \beta(\rho_{21'}) = \text{const}$ and $\omega \ll \omega_1$; the expression for $f_2(\omega_1, \omega_2, \omega, q)$ is simplified and we have

$$
f_2(\omega_1, \omega_2, \omega, q) = g^2(q) Z(\omega_1) \frac{Z(\omega_2)}{\omega_2} \int_0^{\rho_0^0} \frac{d^2\rho}{Z^2(\rho)} \rho^2 \frac{\mathcal{J}_1(q\rho)}{q\rho} \pi\beta^3 \frac{Z(\omega)}{\omega} g,
$$
$$
\text{(C.62)}
$$

where $\mathcal{J}_1(q\rho)$ is the Bessel function.

Obviously, a similar expression should be added to (C.60)–(C.62) replacing ω_1 by ω_2. More complicated diagrams in Fig. C.21(c) give a contribution proportional to $\omega_1\omega_2$.

Let us return to the calculation of the inclusive cross section $f(\xi, \eta, q)$; in compliance with (C.56) note that either (C.57) or (C.60)–(C.62) determine $f(\xi, \eta, q)$ according to the relation between quantities $\xi - \eta$ and η.

If $\eta \ll \xi - \eta$ (in the production of small masses), then $\omega_1, \omega_2 \ll \omega$ and we may ignore $f_2(\omega_1, \omega_2, \omega, q)$ compared with $f_1(\omega_1, \omega_2, \omega, q)$. In this case the inclusive cross section has the form

$$
f_1(\xi, \eta, q) = g^2(q) \tilde{Z}^2(\xi - \eta, q) \psi_1(\eta, q) g, \quad \text{(C.63)}
$$
$$
\psi_1(\eta, q) = \int \frac{d\omega}{2\pi i} e^{\omega\eta} \int_0^{\rho_0} e^{i\,\bar{q}\bar{\rho}} \frac{\beta(\rho)}{Z^2(\rho)} d^2\rho\, Z(\rho_0),
$$
$$
V(\rho_0) = \omega. \quad \text{(C.64)}
$$

Considering that in the integral over ω every term except $\exp(\omega\eta)$ depends solely on $\ln \omega$,

$$\psi_1(\eta, q) = \frac{1}{\eta} \frac{\partial}{\partial \ln \eta} Z(\rho_0) \int_0^{\rho_0} d^2\rho \, e^{i\bar{q}\bar{\rho}} \frac{\beta(\rho)}{Z^2(\rho)},$$

$$\eta V(\rho_0) = 1, \tag{C.65}$$

$$\psi_1(\eta, q) = \frac{c}{\eta} \left\{ Z'(\rho_0) \int_0^{\rho_0} \rho d\rho J_0(q\rho) \frac{\beta(\rho)}{Z^2(\rho)} + J_0(q\rho_0) \frac{\beta(\rho_0)\rho_0}{Z(\rho_0)} \right\},$$

$$Z' = \frac{\partial Z}{\partial \rho_0}, \qquad c = 2\pi \frac{\partial \rho_0}{\partial \ln Z}, \tag{C.66}$$

that is, the cross section falls with η and increases slowly with $\xi - \eta$. In the opposite limiting case $\eta \gg \xi - \eta$ (production of large masses) we may ignore $f_1(\omega_1, \omega_2, \omega, q)$ and the inclusive cross section has the following form (for instance, at $\beta = $ const):

$$f_2(\xi, \eta, q) = 2g^2(q)\tilde{Z}(\eta)\tilde{Z}(\xi - \eta)\psi_2(\xi - \eta, q)g, \tag{C.67}$$

$$\psi_2(\xi - \eta, q) = \frac{c\beta^3}{\xi - \eta} \frac{\partial}{\partial \rho_1^0} \int_0^{\rho_1^0} \frac{\rho^2 d^2\rho}{Z^2(\rho)} \frac{J_1(q\rho)}{q\rho},$$

$$V(\rho_1^0)(\xi - \eta) = 1, \tag{C.68}$$

i.e. the inclusive cross section decreases with increasing $\xi - \eta$. Thus, we are facing an interesting situation. If a fluctuation is not large, $\xi - \eta < \eta$, its probability falls with increase in its size $\xi - \eta$, i.e. the events free from this fluctuation are most probable. If the fluctuation is sufficiently large, its probability rises with the increase in its size, that is the fluctuation tends to take up the whole rapidity range and convert a quasi-diffraction process into a diffraction process.

At relatively small produced masses the inclusive cross section falls with increasing mass and is slightly dependent on the total energy. However, it ceases to be dependent on mass and begins to fall as a function of the total energy with increasing mass. Hence, in the approximation $\alpha' = 0$, no three-pomeron limit exists.

In conclusion of this section it should be noted that generally speaking, expressions (C.63), (C.66) and (C.67), (C.68) oscillate with changing q. Therefore, the danger of the inclusive cross section becoming negative arises. Let us now see under what conditions this occurs for (C.66). Note that according to (C.32) (at $\beta(\rho) = \beta = $ const)

$$Z'(\rho_0) = 2\pi\beta^2\rho_0,$$

$$\psi_1(\eta, q) = \frac{c}{\eta} 2\pi\beta^3 \rho_0 \left[\int_0^{\rho_0} \rho d\rho \frac{J_0(q\eta)}{Z^2(\rho)} + \frac{J_0(q\eta)}{2\pi\beta^2 Z(\rho_0)} \right]. \tag{C.69}$$

Indeed, if β is so small that even at $\pi\beta^2\rho_0^2 \ll 1$ $\rho_0 = R\ln(v_0\eta) \gg R$, we may use even for this range a logarithmic approximation, i.e. formula (C.69). Considering that over this range $Z \simeq 1$, it follows from (C.69) that

$$\psi_1(\eta, q) = \frac{c\beta\rho_0}{\eta} J_0(q\rho_0)$$

is apparently sign-changing. This shows that within the range of application of the theory the first term in (C.69) should always be larger than the second one, i.e. $Z(\rho_0) \gg 1$ at $\rho_0/R \gg 1$. This is possible only provided $\beta R \gtrsim 1$. Hence, the dimensionless three-pomeron vertex cannot be small:

$$\gamma(K = 0, \rho_{12}) \gtrsim \frac{1}{R} V(\rho_{12}). \tag{C.70}$$

If this condition is fulfilled, then (C.69) is the case only at $\beta\rho_0 \gg 1$ and

$$\psi_1(\eta, q) \simeq \frac{2\pi\beta^3\rho_0 c}{\eta} \int_0^{\rho_0} \rho\mathrm{d}\rho \, \frac{J_0(q\rho)}{Z^2(\rho)}$$

does not change its sign for a wide class of $Z(\rho)$.

It is interesting to note that the relationship is not determined by the bare three-pomeron vertex or the form of the potential $V(\rho)$. Both of these quantities have dropped out from the expression for $\psi_1(\eta, q)$. The behaviour of $\psi_1(\eta, q)$ as a function of q is determined by that of $Z(\rho_{12})$ at $\rho_{12} \sim R$.

We now turn to the question as to whether $\beta(\rho)$, which is not a small quantity, can fall with increasing ρ. If $\beta(\rho)$ decreases with increasing ρ rather slowly ($\beta \sim \rho^{-\delta}$, $0 < \delta < 2$), then, in order of magnitude it follows from (C.31) that

$$\frac{\partial Z(\rho)}{\partial\rho} = 2\pi\beta^2(\rho)\rho, \qquad Z(\rho_0) \simeq 2\pi \int_0^{\rho_0} \beta^2(\rho)\,\rho\mathrm{d}\rho. \tag{C.71}$$

If (C.66) and (C.71) are taken into account, the requirement of positive $\psi_1(\eta, q)$ gives

$$2\pi\beta(\rho_0) \int_0^{\rho_0} \frac{J_0(q\rho)}{Z^2(\rho)} \beta(\rho)\rho\mathrm{d}\rho > \frac{J_0(q\rho_0)}{2\pi \int_0^{\rho_0} \beta^2(\rho)\rho\mathrm{d}\rho},$$

$$\rho_0^{-3\delta+2} > c\frac{J_0(q\rho_0)}{\int_0^{\rho_0} \frac{J_0(q\rho)}{Z^2(\rho)}\beta(\rho)\rho\mathrm{d}\rho}. \tag{C.72}$$

From this it follows that $\delta < 2/3$, $Z(\rho_0) > \rho_0^{2/3}$.

C.6.5 Multi-reggeon processes

Another process in which fluctuations show up in the distribution of the particles produced is the multi-reggeon process (Fig. C.23(a)). The cross section of this process is determined by the diagram in Fig. C.23(b) where the pomeron interaction amplitude appears. In our theory $\lambda(\omega_i, k_i)$ contains ω_i as do all pomeron amplitudes, and thus the probability of such a fluctuation falls with increasing energy.

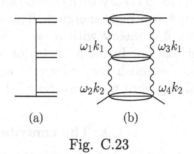

(a)   (b)

Fig. C.23

C.7 The case $\alpha' \neq 0$

So far we have discussed a theory with $\alpha' = 0$ in which the pomeron Green function $G(\omega, R)$ in (C.38) has a singularity at $\omega = 0$ ($j = 1$) with any K. It is well known [3] that, if there are no specially selected compensating cuts, such a singularity is inconsistent with the t-channel unitarity condition. In our theory this ($\omega = 0$) singularity showed up not due to the introduction of special cuts but because we renormalized the pole position identically with respect to K. In fact, we have written $G^{-1}(\omega, K)$ in the form

$$G^{-1}(\omega, K) = \omega + \Delta(K) + \omega\sigma(\omega, K), \qquad (C.73)$$

and then set $\Delta(K) = 0$ referring to the smallness of the shrinkage of the diffraction peak. Obviously for the diagrams that we really summed, $\Delta(K)$ cannot identically be equal to zero merely because these diagrams meet the two-particle t-channel unitarity condition. They have a positive imaginary part at $t > 4\mu^2$ and hence $-\Delta(K)$ has a positive imaginary part over this range. However, $\Delta(K)$ can be numerically small in the essential region $K^2 \sim 1/R^2$. A minimum possible $\Delta(K)$ can be estimated as was done in [4] assuming that the theory is not singular with respect to the mass of a π meson and estimating a two-meson contribution to $\Delta(K)$. In this case for $K^2 > \mu^2$,

$$\Delta_{\min}(K) \simeq \frac{\sigma_{\pi\pi}}{32\pi^3} K^2 \ln \frac{m^2}{K^2}, \qquad (C.74)$$

where $\sigma_{\pi\pi}$ is the cross section of the $\pi\pi$ interaction, m^2 is the characteristic mass in strong interactions, m is of the order of the nucleon mass.

Hence, if $\sigma_{\pi\pi} \sim 1/\mu^2$, then the effective $\alpha'_{min} \simeq 4 \times 10^{-3}/\mu^2$. The experimental α' exceeds this value.

Thus, a theory with $\alpha' = 0$ can be valid only for the intermediate region of not too high energies. Therefore, we should obtain it as a limiting case of a theory with $\alpha' \neq 0$, but small, and thus establish its field of application. In the next section we shall make this clear and at the same time establish the order of magnitude of various interaction constants which govern the character of the interaction at asymptotic energies.

C.8 The contribution of cuts at small α'

We begin, as before, with the two-pomeron Green function. It can be written as

$$G(\omega, \bar{K}, \bar{\rho}_{12}, \bar{\rho}'_{12}) = \int \frac{\mathrm{d}^2 q}{(2\pi)^2} \frac{\psi_{\bar{q}}(\bar{\rho}_{12})\psi_{\bar{q}}^*(\bar{\rho}'_{12})}{\omega + \frac{1}{2}\alpha' K^2 + 2\alpha' q^2}, \qquad (C.75)$$

where K is the total momentum of two pomerons, and $\psi_{\bar{q}}(\bar{\rho}_{12})$ is the wave function of relative motion.

The contribution of an elementary diagram to the self-energy Σ has the form

$$\begin{aligned}
\Sigma_1(\omega, K) &= \int \gamma(\bar{K}, \bar{\rho}_{12}) G(\omega, \bar{K}, \bar{\rho}_{12}, \bar{\rho}'_{12}) \gamma(\bar{K}, \bar{\rho}'_{12}) \mathrm{d}^2\rho_{12} \mathrm{d}^2\rho'_{12} \\
&= \int \frac{\mathrm{d}^2 q}{(2\pi)^2} \frac{\Gamma^2(\bar{K}, \bar{q})}{\omega + \frac{1}{2}\alpha' K^2 + 2\alpha' q^2}, \qquad (C.76)
\end{aligned}$$

$$\Gamma(\bar{K}, \bar{q}) = \int \mathrm{d}^2\rho\, \gamma(\bar{K}, \bar{\rho})\, \psi_{\bar{q}}(\bar{\rho}). \qquad (C.77)$$

It is easy to determine the order of magnitude of $\Gamma(\bar{K}, \bar{q})$ and $\Sigma_1(\omega, K)$ over the asymptotic region ω and $K \to 0$. To this end, it suffices to write a quasi-classical expression for $\psi_{\bar{q}}(\bar{\rho}_{12})$ at $q = 0$:

$$\psi_{\bar{q}}(\bar{\rho}) \sim \frac{1}{\sqrt{\rho}} \exp\left[-\int\limits_{\rho_{12}}^{\infty} \sqrt{\frac{V(\rho')}{2\alpha'}} \mathrm{d}\rho' \right]. \qquad (C.78)$$

This expression is valid at small α' in the region $\rho^2/2\alpha' > 1$.

From (C.78) it is obvious that $\psi_0(\rho)$ is exponentially small in this quasi-classical region and thus in (C.77) only the region of sufficiently large ρ, $\rho > \rho_0$, where $V\rho_0^2/2\alpha' = 1$, is essential.

Considering that for $\bar{K}, \bar{q} \to 0$, $\rho > \rho_0$,

$$\gamma(\bar{K}, \bar{\rho})|_{K=0} \sim V(\rho), \quad \Gamma(\bar{K}, \bar{q}) \sim \int \mathrm{d}^2\rho V(\rho)\psi_0(\rho) \sim \alpha', \qquad (C.79)$$

i.e. after taking into account the interaction, the effective vertex with respect to the three-pomeron processes at $\omega, K, q \to 0$ contains a smallness of the order of α' because the quasi-classical region gives a zero contribution. We may check the validity of (C.79) even without using the quasi-classical expression for $\psi_0(\rho)$, if it is remembered that in conventional quantum mechanics the scattering length determined by the integral $2m \int V\psi d^2\rho$ does not approach infinity at $m \to \infty$ but is governed by the interaction radius. This means that $\int V\psi_0 d^2\rho \sim 1/m \sim \alpha'$. If ω does not tend to 0, but is larger than or of the order of $4\alpha'/R^2$, then the quasi-classical region yields a non-zero contribution, because the effective q^2 is larger than or of the order of $1/R^2$. In the quasi-classical region it is easier to calculate the two-pomeron Green function not in the ω but in the ξ space

$$G(\xi, \bar{K}, \bar{\rho}_{12}, \vec{\rho}'_{12}) = \frac{e^{-(\alpha'/2)K^2\xi}}{2\pi} \left\| \frac{\partial^2 \tilde{S}(\bar{\rho}_{12}, \vec{\rho}'_{12}, \xi)}{\partial^2 \bar{\rho}_{12} \partial^2 \vec{\rho}'_{12}} \right\| e^{\tilde{S}(\bar{\rho}_{12}, \vec{\rho}'_{12}, \xi)},$$

$$\tilde{S}(\bar{\rho}_{12}, \vec{\rho}'_{12}, \xi) = -i\, S(\bar{\rho}_{12}, \vec{\rho}'_{12}, -i\xi), \qquad (C.80)$$

where $S(\bar{\rho}_{12}, \vec{\rho}'_{12}, t)$, is the classical action function for a particle whose mass is $m = 1/4\alpha'$ in the repulsion field $V(\rho)$. It is easily shown through solving classical equations of motion that at large ξ and $\rho_2 > \rho_1$

$$\tilde{S}(\bar{\rho}_1, \vec{\rho}'_2, \xi) = -\frac{|\bar{\rho}_1 - \vec{\rho}'_2|^2}{4\alpha'\xi} - \xi \int_{\rho_1}^{\rho_2} \frac{V(\rho)\mathrm{d}\rho}{\sqrt{|\bar{\rho}_1 - \vec{\rho}'_2|^2 - \frac{1}{\rho^2}[\rho_1 \times \rho_2]^2}}. \qquad (C.81)$$

From (C.81) it is clear that $\tilde{S}(\bar{\rho}_2, \bar{\rho}_1, \xi)$ is different from zero only within the region $V(\rho_1)\xi \lesssim 1$ (ρ_1 being the lesser of ρ_1 and ρ_2), i.e. in the region where the potential is less than or of the order of $1/\xi$. This means that

$$\Sigma(\xi, K) = \sum \gamma(\bar{K}, \bar{\rho}_{12}) G(\xi, \bar{\rho}_2, \vec{\rho}'_1) \gamma(\bar{K}, \vec{\rho}'_1) \mathrm{d}^2\rho_1 \mathrm{d}^2\rho_2 \lesssim \frac{1}{\xi^2}, \qquad (C.82)$$

i.e. the same as with $\alpha' = 0$.

Thus, essential over the preasymptotic (quasi-classical) region are distances running over the range from ρ_0, where $V(\rho_0) \sim 1/\xi$, to $\tilde{\rho}$, where $V(\tilde{\rho}) \sim 4\alpha'/\tilde{\rho}^2$, and over the asymptotic range $\rho \sim \tilde{\rho}$. In respect of the exact pomeron Green function the following can be said:

$$G^{-1}(\omega, R) = \omega + \alpha'K^2 + \omega\sigma(\omega, K, \alpha'/R^2), \qquad (C.83)$$

with $\omega\sigma(\omega, K, \alpha'/R^2)$ calculated for the region $\omega \gg \alpha'/R^2$. In the region $\omega \sim \alpha'/R^2$ this value is of the order of α'/R^2. The calculations of $\omega\sigma$ in the region $\omega \ll \alpha'/R^2$ are treated in the asymptotic theory. This may be both a strong coupling theory involving a slowly rising cross section which has been discussed recently in [5, 6] and a weak coupling theory [7]. The present considerations define only the order of magnitude of unrenormalized constants in asymptotic theories and show that these constants have a smallness α'/R^2. The coupling constants have a similar smallness (a certain power of α'/R^2) for particles with two or more pomerons.

C.9 Conclusion

The solution to the problem of the pomeron interaction in the preasymptotic region considered in this paper explains a number of qualitatively significant properties of high energy experiments such as an approximate factorization of the elastic and inelastic cross sections, a rise in the cross section and interaction radius, a relatively wide (as compared to the Poisson distribution) multiplicity distribution, smallness of the cross section in the three-pomeron limit. From the viewpoint of describing the pomeron using sets of Feynman diagrams, taking into account the pomeron interaction leads to the fact that the pomeron is basically determined by diagrams containing a fairly large number of connected ladder diagrams. It is precisely this fact that leads to a wider multiplicity distribution. In general properties this solution is rather similar to the asymptotic solution for the case of strong coupling [5, 6] and thus its conversion with increasing energy to such an asymptotic solution looks quite natural. Conversion from this solution to the one corresponding to weak coupling and an asymptotically constant total cross section involves a substantial change of the behaviour over the energy region.

Appendix

We have described in this paper the pomeron interaction using the local potential $V(\rho)$. At the same time we have seen that the potential renormalization develops and the effective interaction becomes non-local through the three-pomeron interaction. Therefore, it is natural to put the question whether the main statements will change if we include the non-local interaction right from the start.

The main statement of this study is that the interaction vertices for particles with two and more pomerons and the three-pomeron vertex become small (of the order ω) after taking the pomeron interaction into account. In particular, the three-pomeron vertex $\gamma(\bar{K},\bar{\rho})$ changes to $\omega\gamma(\bar{K},\bar{\rho})/(\omega + V(\rho))$. Let us see whether this statement changes, if the pomeron interaction is described by the non-local potential $V(\bar{K},\bar{\rho}_{12},\bar{\rho}'_{12})$. In this case the equation for the vertex part takes the form

$$\Gamma(\bar{K},\bar{\rho},\omega) = \gamma(\bar{K},\bar{\rho}) - \frac{1}{\omega}\int V(\bar{K},\bar{\rho},\bar{\rho}')\Gamma(\bar{K},\bar{\rho},\omega)\mathrm{d}^2\rho'.$$

Hence, at $\omega \to \infty$

$$\Gamma(\bar{K}, \bar{\rho}, \omega) \to \omega \beta(\bar{K}, \bar{\rho}),$$

where $\beta(\bar{K}, \bar{\rho})$ is determined by the equation

$$\gamma(\bar{K}, \bar{\rho}) = \int V(\bar{K}, \bar{\rho}, \bar{\rho}') \beta(\bar{K}, \bar{\rho}') \mathrm{d}^2 \rho,$$

which is independent of ω. This means that $\Gamma(\bar{K}, \bar{\rho}, \omega)$ has the same smallness. If instead of the condition $\gamma(0, \rho) = c V(\rho)$ we require that

$$\gamma(0, \rho) = c \int V(0, \bar{\rho}, \bar{\rho}') \mathrm{d}^2 \rho',$$

then

$$\beta(0, \rho) = c,$$

and we have exactly the same result as for the local potential.

References

[1] V.N. Gribov, *ZhETF* **53** (1967) 654.

[2] V.A. Abramovsky, V.N. Gribov, and O.V. Kancheli, *Proceedings of the Sixteenth International Conference on High-Energy Physics*, FNAL, US, 1972.

[3] V.N. Gribov, *Nucl. Phys.* **22** (1961) 249.

[4] A.A. Anselm and V.N. Gribov, *Phys. Lett.* **B 40** (1972) 487.

[5] A.A. Migdal, A.M. Polyakov, and K.A. Ter-Martirosian, *Phys. Lett.* **B 49** (1974) 239.

[6] H.D.I. Abarbanel and J.B. Bronzan, *Phys. Lett.* **B 48** (1974) 345; *Phys. Rev.* **D 9** (1974) 2397.

[7] V.N. Gribov and A.A. Migdal, *Yad. Fiz.* **8** (1968) 1002.

This paper was previously published in the following.

Proceedings of the Tenth LNPI Winter School on Nuclear and Elementary Particle Physics, Leningrad, 1975, pp. 5–47 (in Russian).

Nucl. Phys. **B 106** (1976) 189 .

V.N. Gribov. *Gauge Theories and Quark Confinement*, Phasis, Moscow, 2002, pp. 97–128.

Index